The First Men on the Moon

The Story of Apollo 11

David M. Harland

The First Men on the Moon

The Story of Apollo 11

David M. Harland
Space Historian
Kelvinbridge
Glasgow
UK

SPRINGER–PRAXIS BOOKS IN SPACE EXPLORATION
SUBJECT *ADVISORY EDITOR*: John Mason B.Sc., M.Sc., Ph.D.

ISBN 10: 0-387-34176-5 Springer Berlin Heidelberg New York

Springer is a part of Springer Science + Business Media (*springeronline.com*)

Library of Congress Control Number: 2006929205

Cover design: Jim Wilkie
Copy editing: Alex Whyte
Typesetting: BookEns Ltd, Royston, Herts., UK

Printed in Germany on acid-free paper

"I think man has always gone where he has been able to go, and I think that when man stops going where he can go he will have lost a lot. Man has always been an explorer"

Michael Collins, Apollo 11 astronaut.

Other books by David M. Harland

The Mir Space Station – Precursor to Space Colonisation
The Space Shuttle – Roles, Missions and Accomplishments
Exploring the Moon – The Apollo Expeditions
Jupiter Odyssey – The Story of NASA's Galileo Mission
The Earth in Context – A Guide to the Solar System
Mission to Saturn – Cassini and the Huygens Probe
The Big Bang – A View from the 21st Century
The Story of the Space Shuttle
How NASA Learned to Fly in Space – The Gemini Missions
The Story of Space Station Mir
Water and the Search for Life on Mars

Creating the International Space Station
with John E. Catchpole

Apollo EECOM – Journey of a Lifetime
with Sy Liebergot

NASA's Voyager Missions – Exploring the Outer Solar System and Beyond
with Ben Evans

Lunar Exploration – Human Pioneers and Robotic Surveyors
with Paolo Ulivi

Space Systems Failures
with Ralph D. Lorenz

Apollo – The Definitive Sourcebook
with Richard W. Orloff

"I believe that this nation should commit itself to achieving the goal, before this decade is out, of landing a man on the Moon, and returning him, safely, to the Earth. No single space project in this period will be more impressive to mankind, or more important for the long-range exploration of space; and none will be so difficult or expensive to accomplish."

John F. Kennedy, 35th President of the United States of America, in a speech to a joint session of Congress on the theme of *Urgent National Needs*, 25 May 1961

Table of contents

List of illustrations

(*) These illustrations run over several pages.

COLOUR SECTION
(between pages 208 and 209)

List of tables

Author's preface

On 17 December 1903 on the beach at Kitty Hawk, North Carolina, Orville Wright achieved the first flight in a 'heavier than air' machine. On 20 May 1927 Charles Augustus Lindbergh took off in *Spirit of St Louis* at the start of the first successful solo flight across the Atlantic, tracing the 'great circle' route from New York to Paris. On 12 April 1961 Yuri Alekseyevich Gagarin became the first man to orbit Earth. In response, the following month President John F. Kennedy challenged his own nation to land a man on the Moon before the decade was out – and on 16 July 1969 Apollo 11 set off to do so.

By demonstrating that it was feasible to land on the Moon, it cleared the way for the later missions that undertook more ambitious lunar surface activities. Yet Apollo was an anachronism – an element of 21st century exploration provoked by the geopolitical tensions of the 1960s. When Sir Arthur C. Clarke was asked what event in the 20th century he would never have predicted, he said: "That we would have gone to the Moon and then stopped." Nevertheless, at the time of Apollo 11 the Moon was viewed merely as the first step. At a press conference just beforehand, Thomas O. Paine, NASA's Administrator, said: "While the Moon has been the focus of our efforts, the true goal is far more than being first to land men on the Moon, as though it were a celestial Mount Everest to be climbed. The real goal is to develop and demonstrate the capability for interplanetary travel." This task remains to be fulfilled. In 2004 President George W. Bush directed NASA to resume human lunar exploration as a stepping stone to Mars. Perhaps by the time of the 50th anniversary of Apollo 11, mankind will once again be able to enjoy the excitement of a lunar landing.

As the mission of Apollo 11 is a story of exciting times, I have drawn on the mission transcript to recreate the drama. Quotations have been edited for clarity, for brevity, and to eliminate the intermingling that is characteristic of spontaneous conversation, but I have endeavoured to preserve the sense of the moment.

David M Harland
July 2006

Acknowledgements

I would like to thank: Frank O'Brien, W. David Woods, Robert Andrepont, Ken MacTaggart, Hamish Lindsay, Gene Kranz, Gerry Griffin, Dave Scott, Mark Gray, Mick Hyde, Rich Orloff, Mike Gentry, Ed Hengeveld, Eric Jones, Stanley Lebar, Kipp Teague, Marc Rayman and, last but not least, Clive Horwood of Praxis.

1

The Apollo 11 crew

THE HAND OF FATE

It was Saturday, 21 December 1968, and some 2 hours 27 minutes into the mission when astronaut Mike Collins made the call, "Apollo 8, you're Go for TLI." This cryptic one-liner relayed the momentous decision that Frank Borman, Jim Lovell and Bill Anders were cleared to attempt the translunar injection (TLI) manoeuvre that would make them the first humans to head out to the vicinity of the Moon. If all went to plan, in three days the spacecraft would enter lunar orbit to conduct a reconnaissance for the missions that would follow, one of which would hopefully accomplish the challenge made by President John F. Kennedy on 25 May 1961 "to achieving the goal, before this decade is out, of landing a man on the Moon, and returning him, safely, to the Earth".

By a cruel irony, Collins had been assigned to Apollo 8. In early 1968 he had been the astronauts' handball champion, but his game had deteriorated. "My legs felt peculiar, as if they did not belong 100 per cent to me. I had heard prize-fighters talk about their legs going, and I thought, well, instant old age." On seeking medical advice, he was told that a disk had worked completely loose from its vertebra, fallen down into the spinal tunnel, and was impinging on his spinal cord. He suspected this derived from ejecting in 1956 from an F-86 which caught fire. Not only would it be necessary to undergo surgery, but if he was to retain flight status in jet aircraft then the adjacent vertebrae would have to be fused together to enable his weakened neck to withstand another ejection. The surgery in July was completely successful. In the meantime, he had been dropped from the Apollo 8 crew and replaced by his backup, Lovell. While recovering, Collins read in a newspaper that NASA had decided to send Apollo 8 out to orbit the Moon at Christmas, instead of simply orbiting Earth. On returning to work in October, he was assigned as one of the CapComs for the mission.

The commander of the backup crew was Neil Armstrong. On 23 December, as Apollo 8 was nearing the Moon, Deke Slayton, Director of Flight Crew Operations at the Manned Spacecraft Center in Houston, Texas, took Armstrong aside in the Mission Operations Control Room and enquired whether he wished to command Apollo 11; the answer was enthusiastically in the affirmative. To the next question – What did he think of flying with Buzz Aldrin, who was also on the Apollo 8 backup crew? – Armstrong had no objection. Finally, Armstrong was asked if he would retain Fred Haise, who was the third member of the backup crew, or would he prefer Mike Collins? When the assignments had been made Aldrin had been the lunar module pilot (LMP), but when Lovell replaced Collins as the command module pilot (CMP) and Fred Haise was added, the fact that Haise was a lunar module specialist had resulted in Aldrin being reassigned to back up the CMP. If Haise were to fly on Apollo 11, Aldrin would remain as CMP, but if Collins were to join the crew then he would do so as CMP and Aldrin would revert to LMP. Armstrong first had a word with Collins, who was very enthusiastic, then he told Slayton, who submitted his recommendation for the Apollo 11 prime crew of Armstrong, Collins and Aldrin. This crew was endorsed by Robert R. Gilruth, Director of the Manned Spacecraft Center.

When Apollo 8 entered lunar orbit as planned, the crew were awed by the spectacle of Earth rising over the lunar horizon. Prior to heading home, they marked Christmas Eve by reciting the opening verses of the *Book of Genesis*.[1] On their return, Lovell and Anders were teamed with Haise to back up Apollo 11; but as Anders intended to leave NASA soon after this assignment, Ken Mattingly was assigned to shadow him, with a view to joining Lovell and Haise as the prime crew of a subsequent mission.

On Monday, 6 January 1969, Slayton called Armstrong, Collins and Aldrin to his office, confirmed their flight assignments, and told them to assume that their mission would involve a lunar landing. Thomas O. Paine, NASA's Administrator, made the announcement three days later in Washington, and the following day, in Houston, the crew gave their first press conference.

At that time, Armstrong was by no means confident that Apollo 11 would make the first lunar landing. The lunar module (LM), whose development had been so protracted, had yet to be tested in its manned configuration, Mission Control had yet to show that it could operate the LM in parallel with the command and service modules (CSM), and Apollo 9 and Apollo 10 would have to demonstrate many procedures and systems before Armstrong's own mission could be finally specified. Collins estimated that Apollo 10 had a 10 per cent chance of attempting the historic landing, while Apollos 11 and 12 had, respectively, a 50 and a 40 per cent chance.

NEIL ARMSTRONG

Stephen Koenig Armstrong and Viola Louise Engel were married on 8 October 1929, and their son Neil Alden Armstrong was born on 5 August 1930 on his maternal

[1] Madalyn Murray O'Hair, a militant atheist, described by *Life* magazine in 1964 as "the most hated woman in America", sued the federal government over Apollo 8's reading from *Genesis*, arguing that this violated the separation of state and church. This was rejected by the Supreme Court.

On 10 January 1969 Buzz Aldrin, Neil Armstrong and Mike Collins pose in front of a mockup of the LM at the Manned Spacecraft Center following their first press conference as the crew of Apollo 11.

grandmother's farm, some 6 miles from the small town of Wapakoneta, Ohio. The Armstrong family hailed from the border country of Scotland, and the Engels from Germany. As an auditor, Stephen Armstrong was constantly travelling the state (it took about a year to audit the books for a county) setting up temporary home in a succession of small towns. June was born in 1932, and Dean 19 months later. Neil was a non-conformist, spending his time playing the piano and voraciously reading books. He developed an early passion for flying, and by 9 years of age he was building his own model aircraft. "They had become, I suppose, almost an obsession with me," he later reflected. He read everything he could lay his hands on about aviation, filling notebooks with miscellany.

When Neil was 14, the family settled in Wapakoneta (although born nearby, he had not actually lived there). The money from out-of-school jobs, initially stocking shelves at 40 cents per hour in a hardware store, and later working at a pharmacy, helped to pay for flying lessons at $9 each. He gained his student pilot's licence on his sixteenth birthday, but had not yet felt the need for a driver's licence. It was apparent that he would need a technical education if he was to become a professional pilot, but the family did not have the resources to send him through college. Although he was not specifically interested in military aviation, the Navy offered scholarships for university in return for time in service afterwards. Neil applied, and in 1947 was accepted. On the advice of a high school teacher, he went to Purdue University in Indiana because it had a strong aeronautical engineering school. After he had been there 18 months, the Navy – as it was entitled to do – interrupted his studies and sent him to Pensacola in Florida for flight training. He opted for single-seat rather than multi-engine aircraft because he "didn't want to be responsible for anyone else" by having a crew. The Korean War broke out on 25 June 1950 and he gained his 'wings' soon thereafter. In view of the situation, his return to college was deferred and he was sent to the West Coast for additional training. In mid-1951 he was sent to the USS *Essex* to fly F9F Panthers with Fighter Squadron 51, one of the early 'all jet' carrier squadrons. Although he had been trained for air combat, most of his missions were low-level strikes against bridges, trains and armour. On 3 September 1951 he flew so low that he struck a cable and damaged one wing, but was able to nurse his stricken aircraft back over friendly lines before ejecting. In all, he flew 78 combat missions.

In early 1952 he returned to the USA. Rather than attend a military academy in order to receive a commission, he resigned from the Navy and resumed his studies at Purdue, where he met fellow student Janet Elizabeth Shearon. He was then 22 and she was 18; her father was a physician in Welmette, Illinois, and Janet was the youngest of three sisters. On graduating in 1955 with a degree in aeronautical engineering, he was recruited as a research pilot at the High-Speed Flight Station operated by the National Advisory Committee for Aeronautics at Edwards Air Force Base in the high desert of the San Gabriel mountains of California. On his drive west, Neil detoured to Wisconsin, where Janet was working, to ask her to marry him; she agreed to think it over. They were married on 28 January 1956, and their first home was a small cabin with neither electricity nor running water, off base among the Joshua trees and rattlesnakes of the Juniper Hills. This was "the most fascinating time of my life," Armstrong later reflected. "I had the opportunity to fly almost every kind of high-performance airplane, and at the same time to do research in aerodynamics." The X-15 was a sleek black rocket-powered aircraft which, in a zooming climb following release from a B-52, was able to rise above the bulk of the atmosphere. Armstrong first flew the X-15 in 1960, and in all he tested the aircraft seven times. His highest altitude was 207,000 feet, but this did not set a record. However, "above 200,000 feet, you have essentially the same view you'd have from a spacecraft when you are above the atmosphere. You can't help thinking, by George, this is the real thing. Fantastic!" Armstrong helped in the development of the advanced flight control system for the vehicle. Like many at Edwards Air Force Base, he felt that the route into space would be by ever faster aircraft. When a NASA recruiter arrived at Edwards seeking Project Mercury 'astronauts' to ride in a 'capsule' that would parachute into the ocean, Armstrong was not interested. "We reckoned we were more involved in space flight research than the Mercury people, but after John Glenn orbited Earth three times in a little less than 5 hours on 22 February 1962, we began to look at things a bit differently." In April 1962 NASA sought its second intake of astronauts. The first group had all been military test pilots. Although test pilot experience was still a requirement, civilians were now allowed to apply. Candidates had to have a college degree in an engineering subject, be no taller than 6 feet, and not exceed 35 years of age at the time of selection. Armstrong was blond, blue eyed, 165 pounds, 5 feet 11 inches tall, and had a few years to spare. He submitted his application. Of all the civilian applicants, he had by far the greatest experience. On 17 September he was announced as one of nine new astronauts. By the end of the year, the Armstrongs had relocated to El Lago, a housing development near the Manned Spacecraft Center at Clear Lake, which, being neither a lake nor clear, was an alluvial mud flat on Galveston bay about 30 miles from Houston.

Although about the same age as his group, Armstrong looked much younger. He did not match the popular image of an astronaut as a hard-drinking, adrenaline-primed partier. In fact, he was notable for *not* jogging or doing pushups (which the others did eagerly in pursuit of physical fitness) and his social life was spent with his family.

Each astronaut 'tracked' some aspect of the space program to ensure that the astronauts' points of view were represented, and to report back in order to enable the astronaut office to be aware of everything that was going on. While a civilian research test pilot at Edwards, Armstrong had been involved in the *development* of new flight simulators, whereas military test pilots merely used them. It was logical, therefore, that he should be assigned to monitor the development of trainers and simulators.

Deke Slayton opted to fly the military pilots of the second group ahead of the civilians. After jointly backing up Gemini 5 Armstrong and Elliot See were given separate assignments, with Armstrong commanding Gemini 8 and See commanding Gemini 9. On 16 March 1966 Armstrong and Dave Scott were launched into orbit and, after a perfect rendezvous with an Agena target vehicle, they achieved the first docking between vehicles in space. Unfortunately, several minutes later, and now in

Neil Armstrong with the X-15-1 research aircraft on 1 January 1960.

On 9 April 1959 NASA announced the recruitment of its first group of astronauts: (left to right, seated) Leroy Gordon Cooper Jr, Virgil Ivan 'Gus' Grissom, Malcolm Scott Carpenter, Walter Marty Schirra Jr, John Herschel Glenn Jr, Alan Bartlett Shepard Jr and Donald Kent 'Deke' Slayton.

On 17 September 1962 the second group was announced: (left to right, standing) Edward Higgins White II, James Alton McDivitt, John Watts Young, Elliot McKay See Jr, Charles 'Pete' Conrad Jr, Frank Frederick Borman II, Neil Alden Armstrong, Thomas Patten Stafford and James Arthur Lovell Jr.

On 17 October 1963 NASA announced its third group of astronauts: (left to right, standing) Michael Collins, Ronnie Walter Cunningham, Donn Fulton Eisele, Theodore Cordy Freeman, Richard Francis Gordon Jr, Russell Louis 'Rusty' Schweickart, David Randolph Scott and Clifton Curtis Williams; (seated) Edwin Eugene 'Buzz' Aldrin Jr, William Alison Anders, Charles Arthur Bassett II, Alan LeVern Bean, Eugene Andrew Cernan and Roger Bruce Chaffee.

darkness, the docked combination became unstable. Thinking that the fault must be associated with the Agena they undocked, only to find themselves in an accelerating spin owing to the fact that one of their thrusters was continuously firing. By the time the rate of spin had reached one rotation per second, 'tunnel vision' had set in and a black-out was imminent, but Armstrong was able to regain control by shutting off the primary attitude control system and switching to the thrusters designed for use during atmospheric re-entry, which in turn necessitated an emergency return, which was carried out successfully.

At the time of Apollo 11, the Armstrong family comprised Neil and Jan, and sons Ricky, aged 12, and Mark, 6.

BUZZ ALDRIN

Edwin Eugene Aldrin was born in Worchester, Massachusetts, in 1896, not long after his parents, brother, and two sisters had immigrated to the USA from Sweden. After World War One he became a friend of Orville Wright. Later, while serving in the Philippines, he married Marion Gaddys Moon, the daughter of an Army chaplain. On his return to the USA in 1928 Aldrin left the Army to become a stockbroker. Three months prior to the financial crash of August 1929 he sold his stocks, bought a large house in Montclair, New Jersey, and joined Standard Oil to expand the market for petroleum by promoting commercial aviation. In 1938 he left Standard Oil to become an aviation consultant, and in World War Two joined the Army as a colonel in the Air Force.

Edwin Eugene Aldrin Jr was born on 20 January 1930 – a new brother for 3-year-old Madeline and 1-year-old Fay Ann. As Fay Ann pronounced 'brother' as 'buzzer', he gained the nickname 'Buzz'. He had his first ride in an aeroplane at 2 years of age, when his father flew to Florida, but was sick for most of the journey. At school his priority was sports, at which he was extremely competitive, with his father cheering him on – as long as he excelled, his father was content. On leaving high school in 1947 Buzz accepted his father's case for attending a military school, but dismissed his father's recommendation of the Naval Academy at Annapolis, Maryland, opting instead for the Military Academy at West Point, New York. Instead of going to summer camp as he usually did, he attended a 6-week school in order to prepare for the entrance examinations, in which he scored sufficiently well to be accepted. The first-year curriculum gave more or less equal time to scholastics and athletics. One-third of the course work was in mathematics, at which he excelled, with the result that he was rated first in both scholastics and athletics. At his graduation in 1951, at the age of 21, he was rated third in his class of 435 students.

In his final year at West Point, Buzz and his father agreed that he should join the Air Force, but while his father favoured multi-engine school because it would inevitably lead to command of a crew, Buzz wished to be a fighter pilot. After 6 months of basic flight training, 3 months of fighter pilot training, and 3 months at Nellis Air Force Base, Nevada, learning to fly the F-86 fighter-interceptor, he was posted to the 51st Fighter Wing, arriving in Seoul, South Korea, on 26 December 1951. Although the war was less intense by the time he was ready for his first operational mission in February 1952, on 14 May he shot down a MiG during a patrol over North Korea (his gun camera film of the pilot ejecting was featured in *Life* magazine a week later) and on 7 June shot down a second. By the ceasefire on 1 July 1952 he had clocked up a total of 66 missions. He returned to Montclair in December. Prior to his Korean deployment he had accompanied his parents to a cocktail party where one of his father's acquaintances, Mrs Evelyn Archer, invited him to dinner to meet her daughter, Joan, who had just gained her degree from Columbia and was hoping to make a career as a television actress. Michael Archer, her father, was an oil executive. Although Buzz and Joan had not corresponded while he was in Korea, he phoned her on his return and asked her to accompany him to a New Year's Eve party, which she did. They met twice more before he returned to Nellis as a gunnery instructor (he had gained two 'combat kills', after all), and they kept in touch. Some time later, Buzz invited Joan for a week's sightseeing in Las Vegas, which, although nearby for him, represented a major trip for her. As her mother had been killed in an air crash while Buzz was in Korea, Joan asked her father to accompany her. On the penultimate day Buzz proposed marriage, to which Joan agreed with her father's consent. When Buzz's parents were informed, they were delighted. Buzz and Joan were married on 29 December 1954, and two days later they left for Maxwell Field, Alabama, where Buzz was to spend 4 months in squadron officer school. He was then assigned as aide to the Dean of the Air Force Academy in Colorado, and as a flight instructor six months later. In August 1956 he went to Bitburg in West Germany to fly the F-100 with the 36th Fighter Wing. In June 1959 they returned to the USA to enable Buzz to gain a postgraduate degree at the Massachusetts Institute of Technology to advance his military career. One option was a masters degree as a preliminary to attending Experimental Test Pilot School. If he took a doctorate, he would, on graduating, have exceeded the age limit for Experimental Test Pilot School. He opted therefore for a doctorate in astronautics – a new subject that was clearly going to become important to the Air Force. In May 1961, when John F. Kennedy initiated the 'Moon race', Aldrin was 30 years old and well into his doctorate. In December 1962, with a thesis entitled *Line of Sight Guidance Techniques for Manned Orbital Rendezvous* in draft, he was sent to the Air Force's Space Systems Division in Los Angeles. When NASA invited applications for its third intake of astronauts in June 1963, he noted that the requirement for test pilot experience had been relaxed; now 1,000 hours of jet time was sufficient. He applied, and on 17 October was announced as one of 14 new astronauts. The family set up home in Nassau Bay, one of many new housing developments near the Manned Spacecraft Center.

In view of his background, Aldrin's assigned specialism was mission planning, working with the Trajectories and Orbits group led by Howard W. 'Bill' Tindall, which studied every contingency involving the computer that would process either radar tracking or sextant sightings to compute a sequence of manoeuvres designed to make a rendezvous in space – the primary objective of the Gemini program was to demonstrate rendezvous techniques for Apollo. He tutored Wally Schirra and Tom Stafford for Gemini 6, which was to attempt the first rendezvous. As Aldrin noted, "It was essential for the pilot to understand what the computer was doing, and to make sure it made no errors that went unnoticed – i.e. the pilot must know how to guide the computer to the correct conclusion." When Aldrin was assigned as backup pilot for Gemini 10, the frustration was that the system of 'rotation' introduced by Slayton – although not rigidly followed, by which, after serving in a backup capacity, a crew would skip two missions and fly the next – would in this case lead nowhere since the program was to finish with Gemini 12. Nevertheless, Aldrin was delighted to get a crew assignment because, having served in a backup capacity for Gemini he would rank ahead of the total 'rookies' when it came to selecting the early Apollo crews. Fate intervened, however. On 28 February 1966 Elliot See and his partner for Gemini 9, Charles Bassett, died in an air crash. In reshuffling the crews, Slayton advanced Lovell and Aldrin from backing up Gemini 10 to backing up Gemini 9, which put them in line to fly Gemini 12. When the radar on that mission failed, Aldrin completed the rendezvous by computing the manoeuvres manually, and later, during a record three spacewalks, he demonstrated a mastery of the art of working in weightlessness that paved the way for such activities to be included on Apollo missions. Although Aldrin had not been as involved in the development of the LM as some of his peers, his expertise made him well suited to accompany Armstrong on the first lunar landing attempt.

At the time of Apollo 11, the Aldrin family comprised Buzz and Joan, sons Michael, aged 13, and Andrew, 11, and daughter Janice, 11.

MICHAEL COLLINS

General James L. Collins was a career officer who served in the Philippines, in the 1916 Mexican campaign, and in France in World War One. He married Virginia Stewart, whose family had British roots; his own family came from Ireland. Michael Collins was born on 31 October 1930 while his father was Army attaché to Rome, joining siblings James L. Collins Jr, who was 13 years older, and sisters Agnes and Virginia, 10 and 6 years older respectively. The family returned to the USA in 1932. As a child, Michael read a lot, was athletic, and had fun, but in contrast to most of his contemporaries did not develop any great passion for airplanes. His father had graduated from West Point Military Academy, as had his brother, but Michael was inclined towards medicine. His mother suggested a career in the State Department. Although his father put no pressure on him to attend West Point, Louisiana congressman Edward Hébert, a family friend, urged him to follow in the family tradition, which, on leaving high school in 1948, Michael decided to do – more for the free education than for any desire to join the military. After graduating in 1952 he joined the Air Force, gained his 'wings' in the summer of 1953, and was sent to Nellis Air Force Base, Nevada, for advanced fighter training, followed by training for ground attack using nuclear bombs. In December 1954 he was posted to an F-86 fighter squadron at a NATO base in France. In 1956 he met 21-year-old Patricia

Finnegan, a civilian worker in the Air Force who had arrived the previous year and was the eldest of the eight children of Joseph and Julia Finnegan of Boston, Massachusetts. Michael and Patricia were soon engaged, but did not marry until 28 April 1957. On returning to the USA a few months later, Collins was assigned as an instructor, and as he considered a test pilot to be more an engineer than a seat-of-the-pants fighter pilot, in August 1960 he enrolled at the Experimental Test Pilot School at Edwards Air Force Base. When NASA sought a second intake of astronauts in April 1962 he applied, but was rejected. When the agency made another call in June 1963 he applied again, and on 17 October was announced as one of 14 new astronauts. The family moved to Nassau Bay, buying a house not far from that of the Aldrins.

As his specialism Collins was assigned to track the development of space suits and miscellaneous equipment for extravehicular activity. On 18 July 1966, John Young and Collins were launched for the Gemini 10 mission, during which, over a three-day period, they rendezvoused with an Agena target vehicle which was then used to rendezvous with the Agena left by Gemini 8. Collins made two spacewalks, one standing in the hatch and the other involving floating across to the old Agena in order to retrieve an experiment which, if Gemini 8 had not been cut short, Dave Scott would have retrieved.

On being assigned to Apollo 11, Collins was asked whether he was frustrated by having to remain in lunar orbit while his colleagues attempted the landing. "I'd either be a liar or a fool if I said that I think I have the best of the three seats on the mission. On the other hand, all three seats are necessary. I would very much like to see the lunar surface – who wouldn't!? – but I am an integral part of the operation, and am happy to be going in any capacity. I am going 99.9 per cent of the way, and I don't feel frustrated at all."

At the time of Apollo 11, the Collins family comprised Mike and Pat, son Michael, aged 6, and daughters Kathleen, 10, and Ann, 7.

AMIABLE STRANGERS

The crew of Apollo 11 did not become close friends, as some crews did during training, but this was not a prerequisite for mission success – it was required only that each man should know his job, trust his colleagues to do likewise, and work together as part of a team. Collins later described the trio as "amiable strangers". In a sense, they were no more than military men assigned to a mission. Of Armstrong, Collins observed, "Among the dozen test pilots who flew the X-15 rocket ship, Neil was considered one of the weaker stick-and-rudder men, but the very best when it came to understanding the machine's design and how it operated." He was "notable for making decisions slowly, but making them well". Collins considered him "far and away the most experienced test pilot among the astronauts", and the best choice to command the first attempt to land on the Moon.

2

Preliminaries

LUNAR SURFACE EXPERIMENTS

During a meeting in the summer of 1964 at Woods Hole, Massachusetts, the Space Science Board of the National Academy of Sciences listed basic questions relating to the Moon that ought to be studied either by spacecraft placed into lunar orbit or by instruments emplaced on the lunar surface.

On 19 November 1964, after tests conducted on an aircraft providing one-sixth gravity established that astronauts would be able to offload scientific instruments from the descent stage of the LM onto the lunar surface, the Manned Spacecraft Center began to study how instruments might be powered. It was decided that the best source would be a radioisotope thermal generator (RTG) in which heat was converted by thermocouples into electricity. The Grumman Aircraft Engineering Corporation of Bethpage, New York, which was developing the LM, was asked to give some thought to how an RTG might be packaged and carried. Grumman was also asked to develop a prototype for a container in which to return to Earth samples of lunar material. This would require to be carried on the exterior of the vehicle, accessed while on the surface, loaded, hermetically sealed, transferred into the ascent stage, and later passed through the tunnel into the command module and stowed for the flight home.

In January 1965 NASA undertook a time-and-motion investigation in order to assess how best to use the limited time that would be available to the first Apollo crew to land on the Moon. In May, a preliminary list of surface experiments was drawn up, and George E. Mueller, Director of the Office of Manned Space Flight, initiated a two-phase procurement process: the definition phase was to be done in parallel by a number of companies, one of which would be selected to develop the hardware for flight. In June the Manned Spacecraft Center set up the Experiments Program Office within its Engineering Development Directorate to manage all experiments for manned spacecraft, and Robert O. Piland, formerly deputy manager of the Apollo Spacecraft Program Office, was selected to head it. On 7 June Mueller approved the procurement of the Lunar Surface Experiments Package (LSEP) and assigned responsibility for its development to the Experiments Program Office. It was to be an RTG-powered suite of instruments that had to be able to be deployed by two men in 1 hour, and was to transmit data to Earth for 1 year. Overall, it was envisaged as a passive seismometer to monitor moonquakes; an active seismometer that would detonate calibrated explosive charges in order to seismically probe the shallow subsurface; a gravimeter to measure tidal effects that might shed light on the deep interior; an instrument to measure the heat flowing from the interior; radiation and meteoroid detectors; and an instrument to analyse the composition of any lunar atmosphere. The instruments would be electrically connected to a central station that would transmit to Earth. Mueller specified that the package should be available for the first landing mission. On 3 August NASA announced that Bendix Systems, TRW Systems and Space–General Corporation had each been given a 6-month contract worth $500,000 to propose designs. On 14 October NASA contracted the General Electric Company to supply the RTG under the supervision of the Atomic Energy Commission. An instrument to investigate any lunar magnetic field was added to the suite on 15 December. By early 1966 the instrument suite had been renamed the Apollo Lunar Surface Experiments Package (ALSEP). On 16 March NASA Administrator James E. Webb decided that, in view of the company's experience in developing experiments for automated lunar spacecraft, Bendix of Ann Arbor, Michigan, would receive the contract to design, manufacture, test and supply four ALSEPs (three flight units and one in reserve), the first of which was to be delivered no later than 1 July 1967.

Homer E. Newell, Associate Administrator for Space Science and Applications, wrote to Mueller on 6 July 1966, "the highest scientific priority for the Apollo mission is the return to Earth of lunar surface material", with the position of each sample being carefully documented prior to sampling. Newell recommended that on the first moonwalk the astronauts start by collecting an assortment of readily accessible samples (a 'grab bag' in the vernacular of field geology), deploy the ALSEP, and end with a 'traverse' to collect a number of 'documented samples', utilising a range of tools, including core tubes.

By the autumn of 1966 the magnetometer was having serious developmental problems, and the central data-processor was in a critical state. At the end of the year, NASA headquarters suggested that an instrument on the second ALSEP be brought forward as a replacement for the magnetometer, but as the scientists said that the magnetometer would be required to properly interpret the data from the other instruments, it was decided to develop a simpler magnetometer as a stand-by. It was also necessary to consider the 'fuel cask' of plutonium-238 for the RTG. The cask gave structural support and thermal insulation to the fuel capsule: in the case of the SNAP-27 unit for the ALSEP this comprised 8.4 pounds of plutonium. On the Moon, an astronaut would require to remove the 500°C fuel capsule from the cask on the exterior of the LM and insert it into the thermocouple assembly. When simulations revealed flaws in this procedure, the design had to be modified, and after several launch failures unrelated to the Apollo program the cask had also to be 'hardened' to ensure that it would not spill its contents. The Manned Spacecraft Center established the Science and Applications Directorate in December, which took over the activities of the Experiments Program Office and, as Newell had long urged, put science on a par with engineering and operations. Wilmot N. Hess, formerly of the Goddard Space Flight Center, was appointed as Director of the Science and Applications Directorate, with Piland as his deputy.

On 4 January 1967 Christopher C. Kraft, Director of Flight Operations at the Manned Spacecraft Center, said that if a lunar landing was to involve two surface excursions, the first outing should facilitate lunar environment familiarisation, an inspection of the vehicle, photographic documentation and contingency sampling. The ALSEP should be deployed on the second outing, and be followed by a more systematic geological survey. Conversely, if only one excursion was planned, that mission should *not* be provided with an ALSEP since its deployment would use a disproportionate amount of the time. This rationale applied particularly to the first landing, when the mass saved by deleting the instruments would undoubtedly be able to be put to good use. It was also decided that the astronauts should be provided with a rough time line but be allowed to make real-time decisions; the surface operations must not be micro-managed by Mission Control, at least not on the first mission, when there would be so many unknowns and the people on the spot would be best positioned to make decisions. On 16 March NASA announced that 110 scientists, including 27 working in laboratories outside the USA, had been selected to receive lunar samples. In June, Apollo Program Director Samuel C. Phillips formed an *ad hoc* team to review the status of the magnetometer. It was concluded that while the technical problems were certain to be resolved, the instrument was unlikely to be ready for the first landing, which at that time was thought might occur in the latter part of 1968. Unfortunately, neither would the simpler magnetometer be ready for that date, so work on this was terminated. Leonard Reiffel, on Phillips's science staff, recommended on 20 June that in view of the uncertainties concerning an astronaut's ability to work in one-sixth gravity, "an uncrowded time line" would be "more contributory to the advance of science than attempting to do so much that we do none of it well".

By mid-September 1967, on the basis of the LM spending 22.5 hours on the lunar surface, the planners recommended that two excursions should be defined, but the second, to follow a sleep period, should not be listed as a primary objective. The decision on whether to conduct the second excursion – on which the ALSEP would be deployed – should be made on the basis of the astronauts' performance during the first outing. However, one year later, on 6 September 1968, with the LM significantly overweight and the development of the RTG behind schedule, Robert R. Gilruth, Director of the Manned Spacecraft Center, recommended that the first landing should make a single excursion of 2.5 hours; the ALSEP should not be carried (as it could not function without the RTG); the high-gain antenna for the television should not be carried (instead, the 210-foot-diameter antenna at Goldstone in California could receive a transmission from a smaller antenna on the LM); and the geological activities be restricted to the 'minimum lunar sample'. As Gilruth put it, "I'm sure all will agree that if we successfully land on the Moon, transmit television directly from the surface, and return with lunar samples and detailed photographic coverage, our achievement will have been tremendous by both scientific and technological standards." However, Hess argued for a compromise in which, in view of the development problems of the ALSEP, a smaller package should be assigned to this mission using instruments that would be easier to deploy, with the duration of the outing being open ended. On 9 October the Manned Space Flight Management Council, chaired by Mueller, agreed to the development of three lightweight experiments for the first landing mission – a solar-powered passive seismometer, an unpowered laser reflector, and a solar wind composition experiment that would be deployed and later retrieved for return to Earth. It was decided to carry the erectable antenna for the television transmission in case the time of the moonwalk did not coincide with a line-of-sight to Goldstone. The mass saved by not carrying the ALSEP would allow more fuel to be carried, and thereby increase the time available for the hovering phase of the descent. In effect, the first landing was to be an 'operational pathfinder' for its successors. On 5 November Bendix was told to make the three-instrument Early Apollo Surface Experiments Package (EASEP), which was to be shipped by mid-May 1969. On 6 December Phillips said that if the special tools under development for the geological investigation were ready, and if the astronauts had sufficient time to train in their use, they would be carried. One such item was a camera designed by Thomas Gold, an astronomer at Cornell University. In the early 1960s he had argued, on the basis of radar reflections, that the lunar surface was a thick blanket of extremely fine dust into which a spacecraft would sink without trace, and he maintained this position even after automated landers settled on firm ground. His camera was designed to take stereoscopic close-up pictures of the lunar dust.

FIRST MAN OUT

At the press conference in Houston on 10 January 1969 that introduced the crew of Apollo 11, a reporter enquired about which of them would be first to set foot on the Moon. Armstrong turned to Deke Slayton, Director of Flight Crew Operations, for guidance. Slayton said the matter had not yet been decided, but would be resolved by the training exercises. This ambiguity provoked much speculation in the media. The Gemini precedent was that a commander remained in the spacecraft while his copilot undertook extravehicular activity. In March, after the success of Apollo 9 increased the likelihood of Apollo 11 being assigned the first lunar landing, Kraft and George M. Low, Manager of the Apollo Spacecraft Program Office in Houston, had an informal discussion and both felt that since the first man to set foot on the Moon should be a Lindbergh-like figure, Armstrong would be preferable to Aldrin. On hearing a rumour that Armstrong had been chosen to egress first because (despite his being a former naval aviator) he was "a civilian", Aldrin discussed the issue with Armstrong, who said simply that since it was not their decision to make they must wait and see. Several days later, Aldrin went to Low and urged that a decision be made in order to facilitate training. This was a reasonable request, because one of Aldrin's assignments in planning the mission was to refine procedural issues. Low and Kraft then met with Gilruth and Slayton, and they formally decided that the first man to exit the LM would be Armstrong, if only for the fact that the hatch was hinged to open towards the man on the right, meaning that the man on the left, the commander, must exit first. When Slayton called the astronauts into his office, he cited the hinge on the hatch as the reason for Armstrong being first out and last in.[1] On Monday, 14 April, Low announced to the press that if all went well, Armstrong would be the first man to set foot on the lunar surface.

PORTABLE LIFE-SUPPORT SYSTEM

On 15 October 1962 Hamilton Standard of Windsor Locks, Connecticut, initiated development of the Portable Life-Support System (PLSS) for use by an astronaut on the lunar surface. It had to be able to accommodate the metabolic heat liberated by a man doing the equivalent of shovelling sand and, for short periods, sawing wood without overheating or fogging the visor. An attempt to use the oxygen circulation system of the space suit proved to be inadequate, and in September 1964 it was decided to develop an undergarment incorporating a network of fine tubes through which cool water could be pumped. In 1965, with the PLSS growing in size and complexity, consideration was given to cancelling it in favour of just providing the astronauts with 50-foot umbilicals that would snake out of the hatch, even though this would have restricted lunar surface activity to the immediate vicinity of the LM. Fortunately, the pace of development promptly improved. The backpack was 26 inches high, 18 inches wide and 10 inches deep, and contained: (a) a primary oxygen system to regulate the suit at 3.7 pounds per square inch; (b) a ventilator to circulate oxygen, both for breathing and to cool, dehumidify, and cleanse the suit of carbon dioxide and other contaminants; (c) a loop to circulate 4 pounds of water per minute through the liquid-coolant garment; (d) a sublimator to shed waste heat to vacuum; and (e) a communications system to provide primary and backup voice relay via the LM. Each internal system was covered by a thermal insulator of fire-resistant beta cloth, and the entire pack was covered with aluminised kapton to minimise heat transfer and fibre-glass as protection against incidental damage. It had sufficient water and oxygen for 4 hours of nominal operation, but this would begin at the time of disconnecting from the LM's life-support system, prior to egress, and run on after ingress until switching back to the LM. However, as no one could be certain of the metabolic rate of a man on the lunar surface, and therefore of the rate at which oxygen and coolant would be consumed, it was decided to limit the first moonwalk to half of this time. If a second moonwalk were to be scheduled then the PLSS would be replenished as necessary from the LM's resources.

When Apollo 9 lifted off on 3 March 1969 with LM-3, mission commander Jim McDivitt thought that if they achieved only 50 per cent of their demanding program they would still be able to declare the mission a success. Rusty Schweickart was to test the PLSS by emerging from the forward hatch of the LM, translating along a

[1] Nevertheless, if it had been decided that Aldrin should egress first, it would have been possible for them to switch places prior to donning their bulky backpacks.

Rusty Schweickart tests the PLSS after exiting the forward hatch of LM-3 during Apollo 9.

handrail onto the roof of the vehicle, grasping a shorter rail on the CSM and entering the command module through its side hatch, so rehearsing the external transfer that would be used in the event of a returning lunar crew being unable to employ the tunnel in the docking system. However, when Schweickart suffered 'space sickness' early in the flight his spacewalk was limited to the 'porch' of the LM. Nevertheless, the 38-minute excursion was sufficient to demonstrate the PLSS in the space environment, and no one seriously doubted that an external transfer between vehicles was feasible.

LANDING SITE

At the dawn of the 'space age', despite centuries of telescopic observations very little was known for certain about the Moon. For example, there were competing theories for how the craters were made, and the origin of the smooth dark plains that together cover 30 per cent of the visible surface was disputed. Because the Moon's axial rotation is synchronised with its orbital motion around Earth, we can never view its far side. When the Soviet Union sent a spacecraft beyond the Moon in October 1959 and transmitted photographs of its far side, this was revealed to be virtually devoid of dark plains, thereby posing the mystery of why there should be such a dichotomy. One thing was certain: the Moon represented a new frontier to be explored.

Initial reconnaissance

When NASA initiated the Ranger project in December 1959, this was intended to serve as the flagship for its reconnaissance of the Moon. The first two missions in August and November 1961 were to test the spacecraft's basic systems in the deep space environment, but the Agena rocket stages failed and stranded their payloads in low 'parking orbit'. Nevertheless, the Jet Propulsion Laboratory (JPL) decided to proceed with the second batch of spacecraft, whose plunging dive to the Moon was to be documented by a television camera and, just prior to hitting the surface, the spacecraft was to release a shock-resistant 'hard landing' capsule that contained a seismometer. Unfortunately, Ranger 3's Agena overperformed and the spacecraft missed the Moon by 20,000 nautical miles. On the next attempt the trajectory was so accurate that Ranger 4 hit the Moon, but by then an electrical fault had already crippled the spacecraft. Ranger 5, which missed the Moon by 420 nautical miles, was also disabled by a power failure. In December 1962, with its best result being an inert spacecraft striking the Moon, the project was at risk of cancellation. After a review of spacecraft assembly procedures, NASA redefined the project's goals: the next batch of vehicles would have only the television package, and their single objective would be to gain close-up pictures of the lunar surface in order to assess whether this was capable of supporting the weight of a spacecraft. The location of the target was constrained by flight dynamics considerations. The initial television view was to match the best telescopic pictures, and the spacecraft was to execute a near-vertical dive in order to reduce 'smearing' in the final phase, which required a target in the western hemisphere. Unfortunately, the television on Ranger 6 was disabled by an electrical arc at launch, but this did not become evident until the system failed to start as the vehicle neared the Moon. The project's luck changed on 31 July 1964, when Ranger 7 dived into the Sea of Clouds. Its final image showed detail only a few feet across – an improvement in resolution by a factor of a *thousand* over the best telescope. The terrain was fairly soft and rolling, with none of the jagged features portrayed by science fiction. A set of shallow ridges suggested that the dark plain of the 'sea' was a lava flow, but this was disputed. The presence of boulders indicated the surface was likely to support a spacecraft. Although an automated craft might well come to grief by setting down on a rock or in a crater, there were evidently many open spaces and an Apollo crew ought to be able to manoeuvre to a safe spot on which to set down. As Apollo's dynamical constraints favoured eastern sites, on 20 February 1965 Ranger 8 took a shallow trajectory that crossed the central highlands *en route* to the Sea of Tranquility, east of the lunar meridian. Although this approach increased the surface coverage, it also created substantial smearing in the final frames. Satisfied that the dark plains would support the weight of an Apollo spacecraft, NASA released the final probe to the scientists, and on 24 March 1965 Ranger 9 was sent to dive into Alphonsus, a 60-nautical-mile-diameter crater having a central peak and a flat floor displaying interesting rilles and 'dark halo' craters that appeared (to some researchers) to be volcanoes. For the first time, the television was fed to the commercial networks, which broadcast it with the banner 'LIVE FROM THE MOON'. JPL had hoped to reinstate a 'hard landing' instrument package and mount a series of follow-on flights, but funding was denied. Originally intended to be the primary means of studying the Moon, the project had been overtaken by the incredible pace of events following President John F. Kennedy's challenge to send astronauts to the Moon.

Orbital investigations

With Ranger, JPL had seized the initiative in the development of space probes for missions in deep space. In May 1960 it took on a much more adventurous project to develop two related spacecraft: one to enter lunar orbit to conduct mapping and the other to land. Unfortunately, the development of the powerful Centaur stage to dispatch these new probes proved protracted. In coming to terms with Kennedy's time scale for Apollo, NASA cancelled JPL's mapper, and instructed the Langley Research Center to build a lightweight orbiter capable of being dispatched by the Atlas–Agena. This new spacecraft was not to be a global mapper, it was simply to chart predetermined areas as potential Apollo landing sites. This unimaginatively named Lunar Orbiter project was initiated in August 1963. Although Ranger had yet to prove itself, it was apparent that developing an orbiter would not just be a matter of fitting a motor to insert Ranger into lunar orbit. While JPL's television camera package was ideal for documenting a 20-minute plunging dive that would result in the destruction of the craft, it was capable of providing the required high surface resolution only in its final few seconds, by which time its field of view was extremely constrained. To survey wide areas with such resolution from an altitude of 35 nautical miles, Lunar Orbiter would expose film that would be developed, scanned and transmitted back to Earth. Furthermore, since the orbiter had to be lightweight, the camera could not be heavily shielded from radiation in space, and very fine-grained 'slow' film was employed, which in turn necessitated long exposures and a mechanism to enable the camera to compensate for the spacecraft's motion. A twin-lens system was used, with the images from a wide lens providing the context for those from a narrower lens. In December 1963, at the same time as it cancelled the follow-on Rangers, NASA awarded the contract for Lunar Orbiter to Boeing; in effect, the budget was transferred. As with Ranger, Lunar Orbiter would fly as a 3-axis stabilised platform, but a spacecraft's configuration is intimately related to its payload and although it was possible to use many off-the-shelf systems, the orbiter was necessarily very different from its predecessor. The budget allowed for five operational spacecraft, plus a spare for engineering trials. It was expected that three successful flights would be sufficient to survey all the candidate sites for the first Apollo landing, which was as far ahead as the agency was thinking at the time. To achieve this, Langley wrote three flight plans, designated 'A', 'B' and 'C'.

Lunar Orbiter was to fly an almost equatorial elliptical orbit with a 35-nautical-mile perilune on the near side timed to enable the spacecraft to expose its pictures at a low Sun angle to highlight the surface relief, and let the perilune point drift to more western longitudes to follow the sunrise terminator. After 10 days the perilune point would have travelled the length of the equatorial zone in which the targets were located, documenting each under ideal illumination. Furthermore, because the zone extended 80 nautical miles to each side of the equator, it was necessary to tilt the trajectory. In fact, it was decided to adopt an inclination of 11 degrees for the first two missions, with the perilune of the first south of the equator and the perilune of the second north of the equator, and then incline the orbit of the third mission as required to fill in gaps and make follow-up studies.

Lunar Orbiter 1 was launched on 10 August 1966, and entered lunar orbit on 14 August. A motion compensator fault smeared the pictures from the narrow lens. Although the flight controllers considered raising the perilune in order to reduce the smearing and map the *entire* Apollo zone with a resolution of about 80 feet, it was decided to remain at low level and document the designated targets using the wide lens. On 29 August the spacecraft photographed the ninth target on its list, processed its film and transmitted the results, thereby completing its primary mission. It transmitted telemetry for a further two months to enable the degradation of its systems to be monitored, and was then de-orbited to clear the radio frequencies for its successor, which began its program on 18 November. In addition to inspecting the remaining 11 candidate targets, Lunar Orbiter 2 was able to snap a number of secondary sites which, while of no immediate interest to the Apollo planners, were of 'scientific' interest and possible candidates for later missions. Lunar Orbiter 2 completed its photography on 26 November. In addition to its own targets, it had taken high-resolution images of the most interesting sites photographed at medium resolution by its predecessor. Site 'A3' (now labelled 2P-6) was confirmed to be promising and (of the new targets) 2P-2 was deemed to be suitable. In addition to photographing the most promising targets from different angles in order to permit stereoscopic analysis of the topography, Lunar Orbiter 3 charted the routes that an Apollo spacecraft might fly to approach these sites. The US Geological Survey (USGS) produced terrain maps for the Apollo planners.

With the primary objective of the project achieved by the first three spacecraft, NASA released the remaining spacecraft to the scientists, who opted to fly them in near-polar orbits at higher altitudes in order to conduct more general mapping, particularly of the far side of the Moon which permanently faces away from Earth. Even after they had finished imaging, this series of spacecraft provided insight into the lunar *interior*. Although Lunar Orbiter 1 was de-orbited prior to the arrival of its successor, it was noted that its orbit was being perturbed, indicating that the Moon's gravitational field was uneven. To study this phenomenon, the subsequent spacecraft were not de-orbited until their attitude-control propellant was almost exhausted, and by virtue of flying vehicles in both equatorial and polar orbits it was possible to chart the field in sufficient detail to infer that the dark plains in the circular basins were the loci of the most intense gravity. The discovery of these 'mascons' (i.e. the excess of mass concentrated in these basins) was fortunate, as otherwise their perturbations of the Apollo mission would have come as a surprise.

Surface investigations

As the mechanical properties of the lunar surface would influence the design of the LM the Apollo planners said, in October 1962, that the development of JPL's 'soft landing' spacecraft should receive a higher priority than the orbital spacecraft. However, the development of the Centaur stage was so protracted that the surface investigations could not start until 1966. The planners for these Surveyor missions were faced with the same dilemma as their Apollo counterparts: where should they send their first mission? Although safety issues obliged them to select one of the dark plains, this was consistent with characterising the surface in the equatorial zone in which the Apollo targets were located. When Surveyor 1 was launched on 30 May 1966, the 'old hands' at JPL might well have wondered whether they were in for a rerun of the teething troubles that had plagued Ranger, but on 2 June the spacecraft landed safely near Flamsteed, in a crater that appeared to have been breached by the Ocean of Storms. As with Ranger, the single instrument was a television camera. Its first picture showed the spacecraft's foot pad resting on the surface, which was barely indented. It then proceeded to take a multitude of individual frames from which a panoramic mosaic was later produced. There was a profusion of small craters and rocks, but the area was generally flat. Although the site seemed to be consistent with a flow of low viscosity lava, this was disputed. The camera continued to send panoramas to document the appearance of the surface under different illumination, and then the solar-powered spacecraft went into hibernation for the long lunar 'night' – and, to everyone's surprise, not only did it awaken with the return of the Sun, it did so each 'morning' for the rest of the year. Having succeeded at the first attempt, the engineers were disappointed when Surveyor 2 tumbled during a course correction on its way to the Moon, and was lost. On 20 April 1967 Surveyor 3 set down in a 660-foot-diameter crater in the

Ocean of Storms, bouncing several times prior to coming to a halt. The inner wall was pocked by smaller craters, one of which had excavated large blocks of rock. This vehicle had an arm with which to determine the mechanical properties of the loose surficial material, dig trenches to reveal the subsurface, and roll rocks to determine the extent to which their state of erosion was selective. In contrast to its hardy predecessor, Surveyor 3 survived only one lunar night. Next, contact was lost with Surveyor 4 several minutes before it was scheduled to land. Having sampled two dark plains in the western hemisphere and failed twice to reach a site on the meridian, JPL dispatched Surveyor 5 to sample the Sea of Tranquility. It landed on 11 September 1967, just 14 nautical miles from the 2-P6 site on the short-list for the first Apollo landing. Instead of an arm, it had an instrument to investigate the chemical composition of the surficial material. After taking one reading, the spacecraft 'pulsed' its thrusters in order to 'hop' several feet, to sample a second patch. The results indicated calcium, silicon, oxygen, aluminium and magnesium, which implied basalt, but the high ratios of iron and titanium meant that the lunar basalt was subtly different from its terrestrial counterpart. Surveyor 6 was sent to the Meridian Bay to fill in for its lost forerunners, and landed without incident on 10 November 1967. The results of its chemical analysis indicated an iron-rich basalt. Since the dark plains across the Apollo landing zone had proved to be remarkably similar, NASA released the final spacecraft to the scientists, who decided to send it to Tycho, a bright 'ray' crater in the southern highlands, where it landed on 10 January 1968. By cutting margins, JPL enabled it to employ both the robotic arm and the chemical analyser – which proved fortunate because the analyser became stuck, and if it were not for the arm nudging it free the scientific study would have been undermined. In addition, rather than make the spacecraft hop so as to sample different patches of surface, the arm was used to place the instrument on a patch of excavated soil in order to check that this was the same as the material on the surface, and later to place it on top of a rock. Some researchers interpreted the elemental abundance data to mean that the lunar highlands were an alumina-rich basalt, but Eugene M. Shoemaker, head of the Astrogeology Branch of the USGS, argued that the dominant rock in the Tycho ejecta was anorthositic gabbro, which had interesting implications.[2]

Apollo requirements

The objective of the Lunar Orbiter series was to reconnoitre possible landing sites for Apollo. As they had insufficient film to *search* for sites, they concentrated on sites that appeared suitable on the basis of telescopic observations. The Apollo Site Selection Board reduced an initial list of 30 candidates – all located on the near side of the Moon within 45 degrees of the meridian and 5 degrees of the equator – to a short-list of five by applying the following *operational* factors:

[2] Of which more in chapter 10.

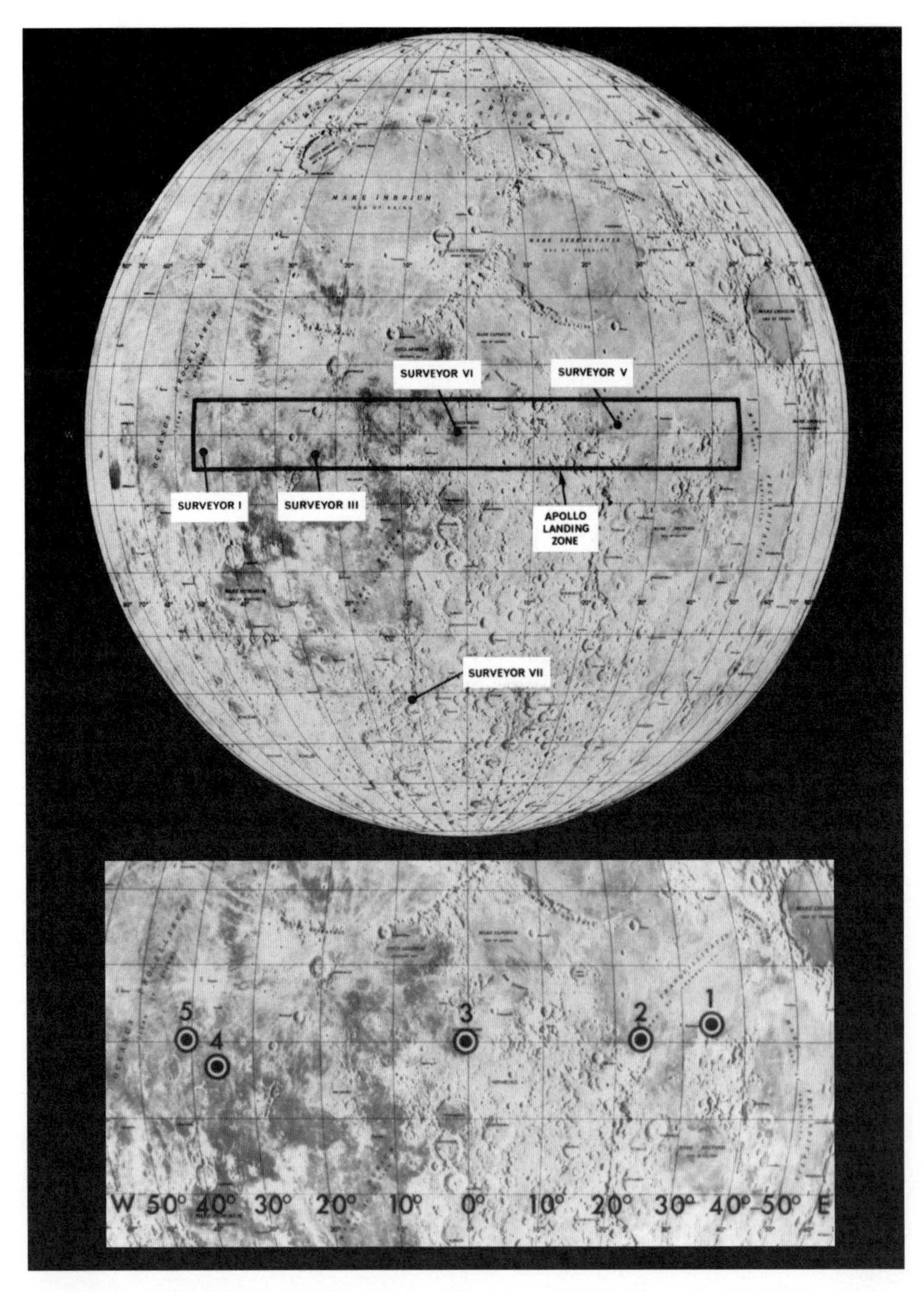

The primary objective of the Surveyor missions was to investigate the dark plains. Five Apollo Landing Sites (ALS) were selected in the equatorial zone for the early Apollo landings (bottom).

- *Smoothness:* Relatively few craters and boulders.
- *Approach:* No large hills, high cliffs, or deep craters that could result in incorrect altitude signals to the lunar module landing radar.
- *Propellant requirements:* Least potential expenditure of spacecraft propellants.
- *Recycling:* Effective launch preparation recycling if the countdown were to be delayed.
- *Free return:* Sites within reach of the spacecraft launched on a free-return translunar trajectory.
- *Slope:* Less than 2-degree slope in the approach path and landing area.

The flight dynamics team insisted that the primary site be located in the eastern hemisphere, in order to allow room further west for one or two suitably lit backup sites in the event of the launch being postponed by several days. As the launch 'windows' for a given site occurred only once per month, it was thought better to go for a secondary site several days late than to wait a month for the primary site to present itself again. The time of landing had to be just after local sunrise, as the Sun was required to be low on the horizon to cast sufficient shadows to reveal surface topography. Because the Sun traverses the lunar sky at a rate of 12 degrees per 24 hours, the backup sites were set 12, 24 or 36 degrees apart in lunar longitude to ensure that the illumination would be right for a delayed mission. On the other hand, the primary site could not be too far east, as this would not allow sufficient time after the final limb crossing to perform the navigational checks prior to initiating the powered descent. All sites had to be within 5 degrees of the lunar equator because a higher latitude would require a less propellant-efficient trajectory, and propellant economy was a priority for the first landing mission. Furthermore, the approaches to the sites had to be flat in order not to complicate the task of the landing radar. These operating constraints restricted the first landing to an eastern dark plain near the equator, which put the primary landing site in either the Sea of Tranquility or the Sea of Fertility, the backup site in the Meridian Bay, and the reserves in the Ocean of Storms. However, the Sea of Fertility was too far east to accommodate the final navigational update, which left the Sea of Tranquility, where there were two sites. The trajectory of Apollo 8 had been timed to inspect the easternmost site, ALS-1, called 2P-2 by the Lunar Orbiter team, at ideal illumination.

The time of the landing was determined by the location and the acceptable range of Sun elevation angles. The range of these angles ran from 6 to 14 degrees, east to west. Under these conditions, the visible shadows of craters would aid the crew in recognising topographical features. As the Sun angle approached the descent angle, the mean value of which was 16 degrees, the viewing conditions would be degraded by a 'washout' phenomenon in which backward reflectance was high enough to eliminate contrast. Sun angles above the flight path were not as desirable, because shadows would not be readily visible unless the Sun was significantly outside the descent plane. Because lunar sunlight incidence changed about 0.5 degree per hour, the Sun elevation angle restriction defined a 16-hour period, recurring every 29.5 days, when landing at a given site could be attempted. The number of Earth-launch opportunities for a given lunar month was equal to the number of candidate landing sites. The time of launch was primarily determined by the allowable variation in launch pad azimuth. A total launch pad azimuth variation of 34 degrees afforded a launch period of 4 hours 30 minutes. Two launch windows occurred each day. One was available for a translunar injection over the Pacific Ocean, and the other over the Atlantic Ocean. The injection opportunity over the Pacific Ocean was preferred because it usually permitted a daytime launch.

FLIGHT DIRECTORS

Clifford E. Charlesworth was appointed as lead flight director for Apollo 11. Cool headed with an easy smile, he had been nicknamed the Mississippi Gambler by the flight controllers on account of the fact that, although he always appeared relaxed, he was focused and confident. As planning firmed up in early 1969, he shared the principal tasks among the available flight directors. Of the eight major phases of the mission, five had either been demonstrated by Apollo 8 or soon were to be by Apollo 10, and the three unrehearsed phases were the powered descent to the lunar surface, the moonwalk, and the lunar liftoff. As, by Apollo 11, Charlesworth would be most familiar with the Saturn V, he took launch on through to the translunar injection manoeuvre, plus the subsequent surface excursion. Eugene F. Kranz had most experience with the LM, including its unmanned test on Apollo 5 and manned test on Apollo 9, and was therefore assigned the lunar landing and transearth injection manoeuvre. As Glynn S. Lunney would have been to the Moon twice, both times focusing on the CSM, he was given responsibility for the lunar liftoff and rendezvous. Gerald D. Griffin and Milton L. Windler were assigned to other miscellaneous tasks. The flight directors met the branch chiefs of the flight control division to create their teams of flight controllers, balancing their individual areas of expertise to each phase of the mission.

REHEARSAL

The original concept for Apollo 10 called for the spacecraft to enter lunar orbit and for LM-4 to undock, enter a slightly different orbit, return and redock as a test of operating in lunar orbit. In December 1968, however, the mission planning and analysis division of Mission Control successfully argued the case for putting the descent propulsion system through a realistic rehearsal in which the perilune would be lower. This would test the ability of the landing radar to lock onto the surface, with the illumination on the low passes exactly as it would be on the landing mission in order to document the primary site and identify landmarks on the approach route. Howard W. 'Bill' Tindall, the assistant division chief, had also suggested that the LM should initiate the powered descent and abort by 'fire-in-the-hole' staging, but this was not pursued. After three outstandingly successful manned missions, consideration was given to assigning Apollo 10 the landing mission. However, because LM-4 was incapable of landing – the software was not ready for either the simulator or the

On 3 June 1969 the Apollo 10 crew discuss their 'dress rehearsal' in lunar orbit with the Apollo 11 crew.

vehicle, and in any case LM-4 was too heavy to carry sufficient propellant to lift off again – Apollo 10 commander Tom Stafford argued against waiting for LM-5. "There are too many 'unknowns' up there," he insisted. "We can't get rid of the risk element for the men who will land on the Moon but we can minimise it; our job is to find out everything we can in order that only a small amount of 'unknown' is left." The plans, procedures, mission rules, manoeuvres, thermal regime and communications would provide a high-fidelity rehearsal of the landing mission. On 24 March 1969 it was announced that Apollo 10 would conduct this dress rehearsal, and if it achieved its primary objectives then Apollo 11 would attempt to land.

One aspect of the Apollo 10 mission was to assess the operation, tracking and communications of two spacecraft in lunar orbit. Apollo 8 had confirmed that the mascons significantly perturbed the orbit of a spacecraft. By having Apollo 10 fly the profile planned for the landing mission, it would be possible to assess how the guidance and navigation system of the LM coped with these gravitational effects while making the low passes of the descent orbit. Apollo 10 lifted off on schedule on 18 May, and on the fifth day the LM separated in a circular parking orbit at an altitude of 60 nautical miles, entered an elliptical orbit with a 50,000-foot perilune, made two low passes, discarded the descent stage, and made a perfect rendezvous. The first low pass rehearsed an approach to ALS-2 (site 'A3', later 2P-6), and while the aim point itself was acceptable, the western end of the 'landing ellipse' was rougher, and Stafford told Armstrong that if he were to find himself coming in 'long', his best option might be to abort.

3

Preparations

SIMULATION

With Apollo 10 having mitigated the risks, Armstrong and Aldrin were able to focus their training on the powered descent and lunar lift off. However, because Apollo 10 had first call on the simulators until early May, Clifford Charlesworth initiated training in April with the Saturn V launch phase. Two months then remained in which to conduct the specialised training because, with a target launch date of 16 July, the most intensive training using the simulators would be completed about 10 days earlier in order to enable the crew and flight control teams to finish other activities. Simulation explored two basic scenarios: 'nominal' and 'contingency'. The nominal part occupied only a few days, and defined the Go/No-Go decision points, the procedures, and the timings for the interactions between the crew and the flight controllers. The first full set of mission rules for Apollo 11 was issued on 16 May, but was preliminary pending methodical testing by simulation. Because the nominal powered descent was to last only 12 minutes, it was possible to perform many runs and debriefings during a single day's training. While Apollo 10 was performing its rehearsal in lunar orbit, Armstrong and Aldrin were routinely landing by flying the nominal profile. Contingency training was designed to test how the crew and flight controllers dealt with departures from the nominal profile involving trajectory and systems problems. The Simulation Supervisor (SimSup) for the powered descent was Dick Koos, an early recruit of the Space Task Group to train control teams. As there were then no graduates with computer degrees, NASA had hired engineers with experience, and his background was the computerisation of ground-to-air missiles for the Army Missile Command at Fort Bliss, Texas. Koos and his five support staff occupied a glassed-in partition at the front of the Mission Operations Control Room, and their role was to develop realistic mission scenarios that would assess the mission strategy, rules and procedures, the knowledge and coordination of the individual flight controllers, the ability of the team as a whole to develop real-time solutions to technical difficulties, and generally to probe the psyches of everyone involved. It was considered that a fully trained team of flight controllers ought to be able to function as a single 'mind'.

The first contingency training was on 10 June. A succession of runs introduced a wide variety of issues for different flight controllers – sometimes in parallel, in an attempt to overload flight director Gene Kranz. On the fourth session the LM crashed onto the Moon. Even travelling at the speed of light, it takes a radio signal 1.3 seconds to cross the space between Earth and the Moon. As a consequence of this transmission delay, the flight control team's abort call had been made too late, leaving insufficient time for the crew to follow through. On the next session, Koos made the guidance system malfunction while the flight controllers were analysing another issue, resulting in another crash – not because the situation had been intrinsically unrecoverable, but because the flight controllers on this occasion had been distracted. Before the end of the day they had suffered another two crashes. As Kranz wrote of this first day, "We were learning the hard way about the 'dead man's box', the seconds-critical relationship of velocity, time and altitude where the spacecraft will always hit the surface before Mission Control can react, and call an abort." Christopher C. Kraft, Director of Flight Operations at the Manned Spacecraft Center, George M. Low, Manager of the Apollo Spacecraft Program Office, and Deke Slayton, Director of Flight Crew Operations, all listened in to the flight director's loop over their office 'squawk boxes'. Afterwards Kraft called Kranz. "Chris," Kranz said, "you have had these types of days. It is just a matter of time and training, we will work it out." In fact, because the 'dead man's box' was dependent on several variables, it proved difficult to determine just when the LM entered it. All that could be done was to explore the parameters and gain a feel for it. It was concluded that in the final phase of the descent it was impracticable for Mission Control to call an abort and hence, after the locus of decision-making had switched to the LM, the flight *controllers* became *spectators*.

Although everyone was eager to explore problems that might arise late in the descent, Koos sometimes presented a flight controller with an earlier issue, with a view to seducing him into making an incorrect selection from several remedial options. Often, after provoking an abort call, Koos would point out that the abort had not been justified and that the flight could have continued. Other times, after the team had either spent too long analysing an issue, or had decided to press on regardless and crashed, Koos would point out precisely where they should have aborted. In one run in late June, just as the LM was manoeuvring to a 'viewing' orientation at an altitude of 7,500 feet, Koos caused a thrusters to continue firing, making the vehicle unstable. Although Armstrong knew he should abort before the tumbling caused the spacecraft to crash, he deliberately waited. Aldrin stared at his commander in amazement, then urged him to abort, but Armstrong continued to wait – he wished to know how long it would take Mission Control to call the abort; it came too late, and they crashed.

Meanwhile, Collins was training solo for CSM operations. Although his task was less demanding in the sense that the trail had been blazed by previous crews, there was nevertheless a great deal to learn. A simulator at the Langley Research Center had full-scale replicas of the CSM and LM slung on wires in a hangar, to enable CMPs to rehearse the retrieval of the LM from the upper stage of the Saturn V following the translunar injection manoeuvre. With missions being launched at 2-monthly intervals, by the time a crew gained priority in the simulators there was

During training, Mike Collins prepares to enter the gondola of a centrifuge.

Mike Collins in the gondola of a centrifuge.

precious little time to become proficient. Fortunately, as a result of his early training to fly as CMP on Apollo 8, Collins was already familiar with the basics. By March he had first call on one of the two simulators at the Cape, although he had to yield to the Apollo 10 crew whenever their simulator was unavailable. The jumble of boxes attached to the spacecraft to provide the requisite external views had prompted John Young, his Apollo 10 counterpart, to describe it as a "train wreck". It took a team of several hundred engineers to maintain the facility, which used a mainframe computer to drive the instrument displays. Time was so precious that the simulators were made available around the clock, with 'lesser' crews coming in at night. Collins simulated not only the nominal flight plan but also a host of contingencies, including rescuing the crew of a LM stranded in a low, unstable lunar orbit. If the LM's radar were to become inoperative, he would track its flashing beacon optically, but, as the Gemini missions had demonstrated, visual tracking against a sunlit surface was difficult. And, of course, Collins had to learn how to perform the post-transearth injection functions, including atmospheric re-entry, on his own – just in case he had to return home alone.

By the end of May, Armstrong, Aldrin and Collins were routinely spending 14 hours per day in the simulators at the Cape during the week – often in 'integrated' sessions with the two simulators hooked up to Mission Control – and flying home to Houston at weekends. As Jan Armstrong recalled of this period, "Neil used to come home with his face drawn white, and I was worried about him." Of course, as commander, Armstrong bore the greatest psychological load. "The worst period was in early June. Their morale was down. They were worried about whether there was enough time for them to learn the things they needed to learn, to do the things they had to do, if this mission was to work."

On 11 June, having digested the lessons of Apollo 10, NASA announced that Apollo 11 would indeed attempt a lunar landing. Prior to departing Washington to visit Paris, France, NASA Administrator Thomas O. Paine had instructed Apollo Program Director Samuel C. Phillips, that if Phillips had any reservations "about the men, about the equipment, about the launch pad facilities" for Apollo 11, then he must "defer the whole thing to August". At noon on 12 June Phillips chaired a flight readiness review, which was conducted by telephone conference. He began by announcing, "I'm fully prepared to delay if something is not ready, or if we're pushing these men too hard; if we're doing that, we will reschedule for August." Lee B. James, the Saturn V Manager at the Marshall Space Flight Center, Rocco A. Petrone, the Launch Director at the Kennedy Space Center, and Gene Kranz, as chief of the flight control division in Houston, agreed that hardware preparations were proceeding to plan. Charles E. Berry, Director of Medical Research and Operations in Houston, was concerned about the crew. In view of the intense pace of training, he would welcome additional time. However, Deke Slayton, who was at the Cape and had spent several days with the astronauts reviewing their training and state of readiness, said Armstrong had told him that while the schedule was tight and they were tired, they would have time to rest and recuperate when the pace slackened in early July. Berry accepted this, George Low concurred, and the scheduled launch date was reaffirmed. Later that day, Phillips called Armstrong, who expressed his

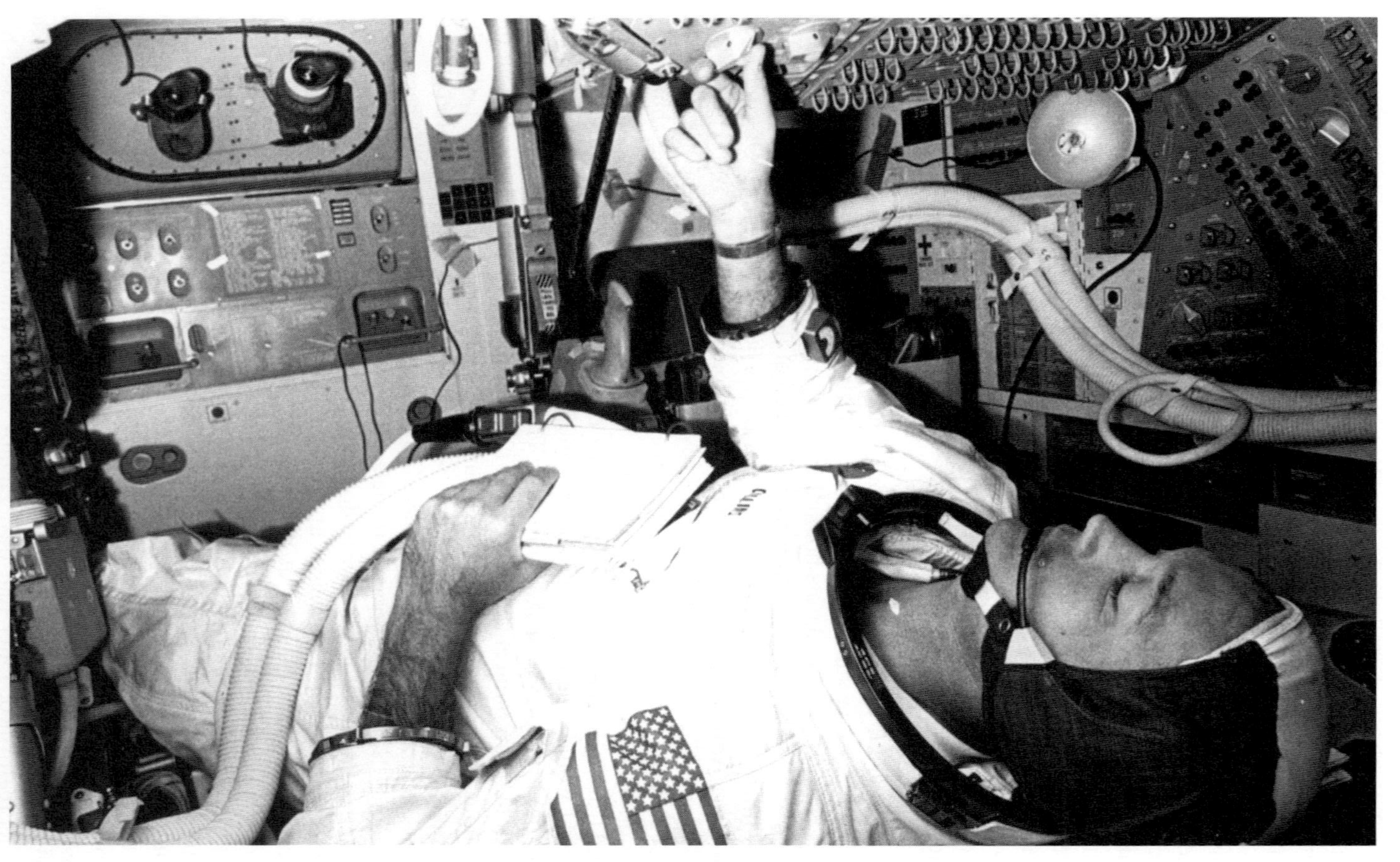

Mike Collins in a CM simulator.

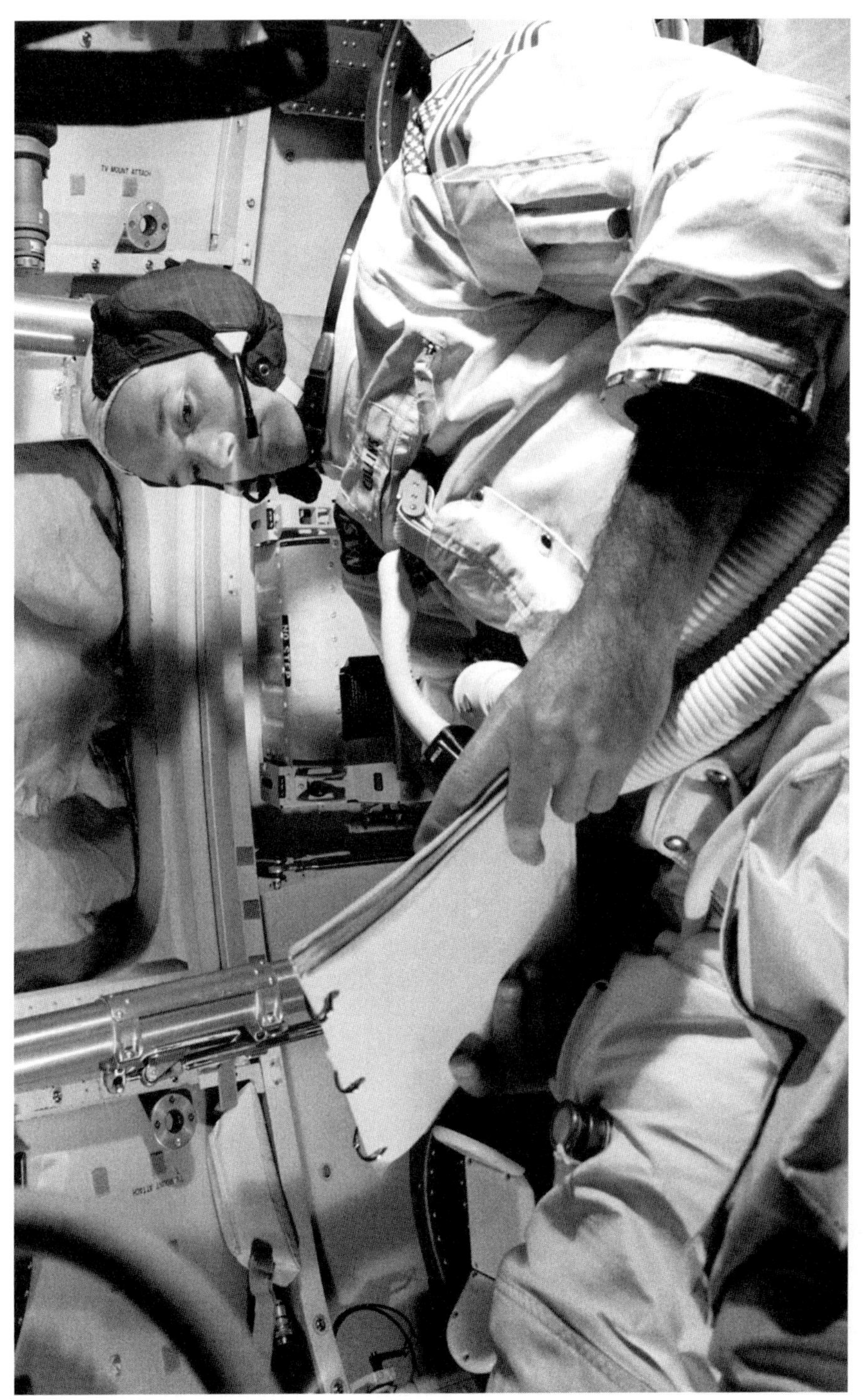

Mike Collins in a CM simulator.

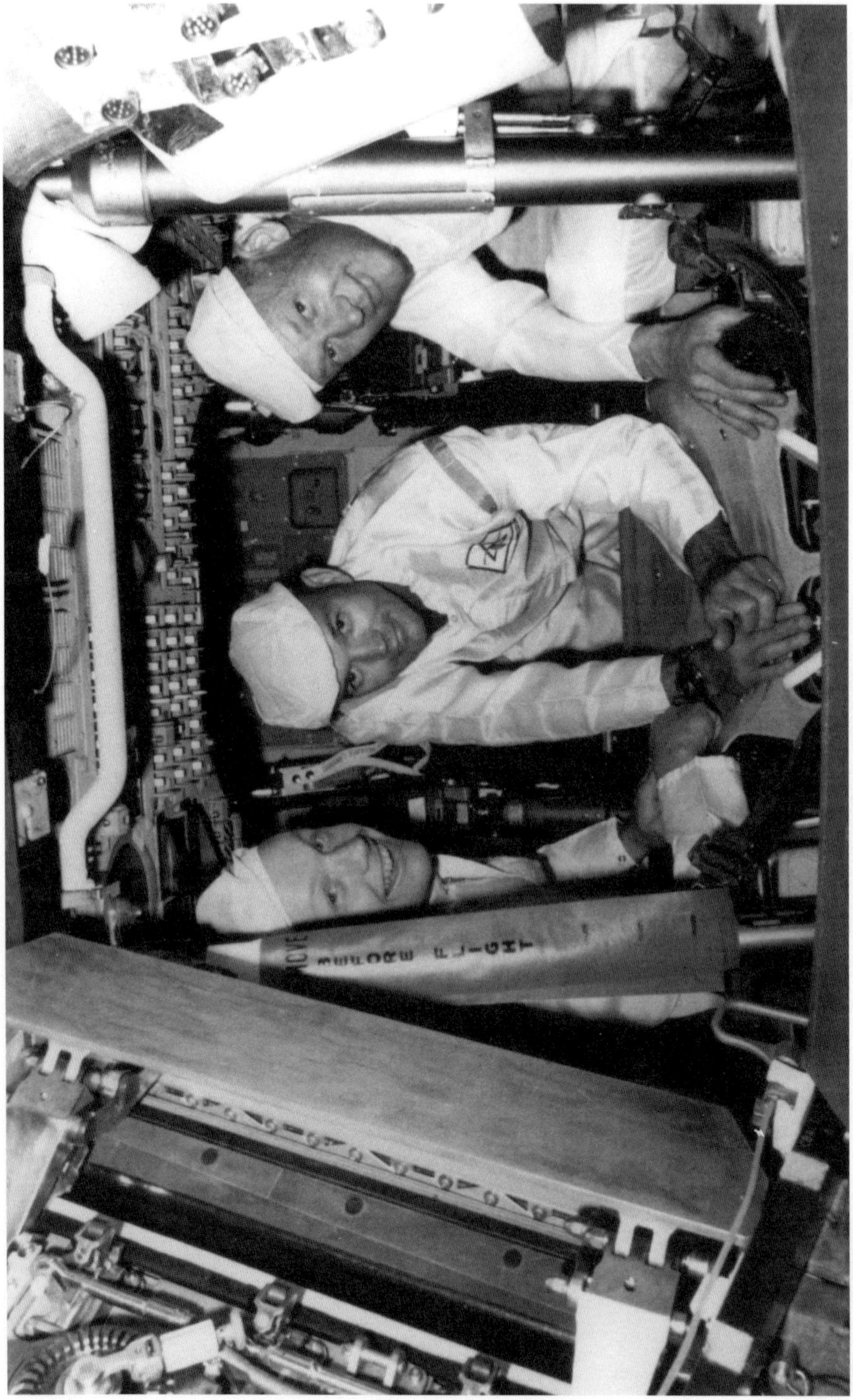

Neil Armstrong, Mike Collins and Buzz Aldrin in CM-107 on 10 June 1969.

Neil Armstrong in a LM simulator on 19 June 1969.

satisfaction. "The turning point", Armstrong's wife observed, "came [on 12 June] with the decision to go. After that, everything seemed to go better. They knew they were going, and this seemed to take the weight off their shoulders."

LUNAR SURFACE ACTIVITY

Because the astronauts who landed on the Moon would be required to act as field geologists, a series of lectures and laboratory exercises were given in 1964 by the US Geological Survey (USGS) in a ramshackle shed of Second World War vintage at Ellington Air Force Base, which served as the airport for the Manned Spacecraft Center. On being introduced to the vocabulary of geology and basic mineralogy, the astronauts were taught how to describe rocks and to characterise a geological setting in terms of the granularity and bearing strength of its surface. Some of the astronauts – mainly those of the first group who were still active,[1] but also some of the second group – argued that there was no requirement for such training because the rocks they returned would be studied by the scientists. But other members of the second group and most of the third group, aware that they were unlikely to be assigned the first landing, looked ahead to the later missions on which science was certain to be a significant factor and reasoned that by taking the subject seriously they would improve their chances of a flight assignment.

The first geology field trip was to the Grand Canyon, incised into the Arizona Plateau by the Colorado River to a depth of some 6,000 feet. Viewing the strata exposed in the canyon wall was undoubtedly awe inspiring, but most of it was sedimentary and (the nomenclature for the lunar features notwithstanding) there were no rivers on the Moon. Later trips included Meteor Crater in Arizona, which seemed more relevant because the Moon *was* pocked by craters. Since there was at that time no consensus as to whether lunar craters were formed by impacts or by volcanism, trips were also made to a wide variety of volcanic features across the American southwest. Jack Schmitt, who joined NASA in 1965 as one of the first group of scientist–astronauts, and had a doctorate in geology from Harvard, was assigned to assist in geological training. He encouraged Armstrong and Aldrin to find time to make field trips. At a volcanic field near Cinder Lake in Arizona, the Astrogeology Branch of the USGS blasted a simulated lunar landscape based on a picture of a potential landing site taken by a Lunar Orbiter. Geologists then made 'traverses' wearing training space suits to evaluate procedures, test the tools that the astronauts were to employ, and determine what could reasonably be done in the time

[1] Of the 'Original Seven' astronauts, Wally Schirra, Gus Grissom and Gordon Cooper were on the active list; Deke Slayton and Al Shepard had been grounded for medical reasons; Scott Carpenter had returned to the Navy; and John Glenn, who had been grounded on the basis that as a national icon he was too valuable to risk on a second mission, had left to pursue a political career.

Buzz Aldrin and Neil Armstrong examine small rocks on a geological field trip to Sierra Blanca in West Texas on 24 February 1969.

Using a mockup of the LM hatch, porch and upper ladder, Buzz Aldrin undergoes one-sixth gravity training in a KC-135 on 10 July 1969.

available to the first moonwalk. A crude LM was constructed as a perch from which to make visual observations. On his Mercury mission, Wally Schirra had been given an off-the-shelf Hasselblad 500C camera manufactured by the Victor Hasselblad Company in Sweden. NASA later asked the company to supply it with a modified version. The mechanism had to be capable of 5,000 'working cycles' in Earth's atmosphere, in pure oxygen, and in a vacuum; accommodate a magazine with a capacity of 160 exposures of 70-millimetre 'thin' film; and incorporate an electric motor to advance the film.[2] This camera was introduced on Gemini, and carried over to Apollo. The geologists conducted tests using a Hasselblad 500EL Data Camera configured for use by a suited astronaut, notably with its view sight deleted. The results were studied to determine how much of what was *known* of the terrain could be *inferred* from just the visual observations and photographs. The trials, conducted early in 1968, were led by Arnold Brokaw, chief of the surface planetary exploration section of the Astrogeology Branch. The conclusions were fed to Houston by Eugene M. Shoemaker, the branch chief, who was seconded to NASA. Shortly prior to the mission, Brokaw visited Armstrong to emphasise the value of photographing rocks, irrespective of whether these were sampled: "It is important to us how a rock got where it is, how and where it lay, how it relates to other things in the area; we can determine a lot about its mineralogy simply from photographs." Aldrin was inspired by geology, because it "opened my eyes to the immensity of time". Collins was not, "I hate geology – maybe that's why they won't let me get out on the Moon." Armstrong, displaying an impishness worthy of Pete Conrad, later admitted that he had been "very tempted to sneak a piece of limestone up" and place it into a rock box as a sample, to see what the scientists would make of it.

The training for lunar surface activities was undertaken in Building 9 of the Manned Spacecraft Center campus, where there was a mockup LM. The astronauts suited up and donned all the extravehicular paraphernalia and, while attended by technicians from the crew systems division and the suppliers of the miscellaneous apparatus, they tested egress and ingress procedures, surface sampling tools, and the deployment of the scientific packages. The scientific community wanted the maximum work from Armstrong and Aldrin while they were on the lunar surface. Each task was timed during training, and integrated into the overall time line. A significant milestone was attained on 18 June 1969 by a full 'walk through' which included deploying the EASEP instruments. However, while the technical fidelity was high, this training was done in full Earth gravity. To familiarise themselves with lunar gravity – which is one-sixth that of Earth – the astronauts flew in a KC-135 aircraft (the military version of the Boeing 707) with its cabin deck cleared and padded. This aircraft would fly a precise arc, zooming, cresting and falling in order to simulate the desired gravitational load. During the climb the suited astronaut was held in position by technicians, and when the desired gravity was reached he had to

[2] Being detachable, the magazine of a Hasselblad is traditionally referred to simply as a 'back'.

Having descended the ladder of a LM mockup in training, Neil Armstrong stands on the foot pad and surveys his situation.

Standing by the MESA, Neil Armstrong rehearses collecting the contingency sample.

Neil Armstrong with a chest-mounted Hasselblad 500EL camera.

Side view of Neil Armstrong with a chest-mounted Hasselblad 500EL camera.

Neil Armstrong having transferred the television camera to its tripod.

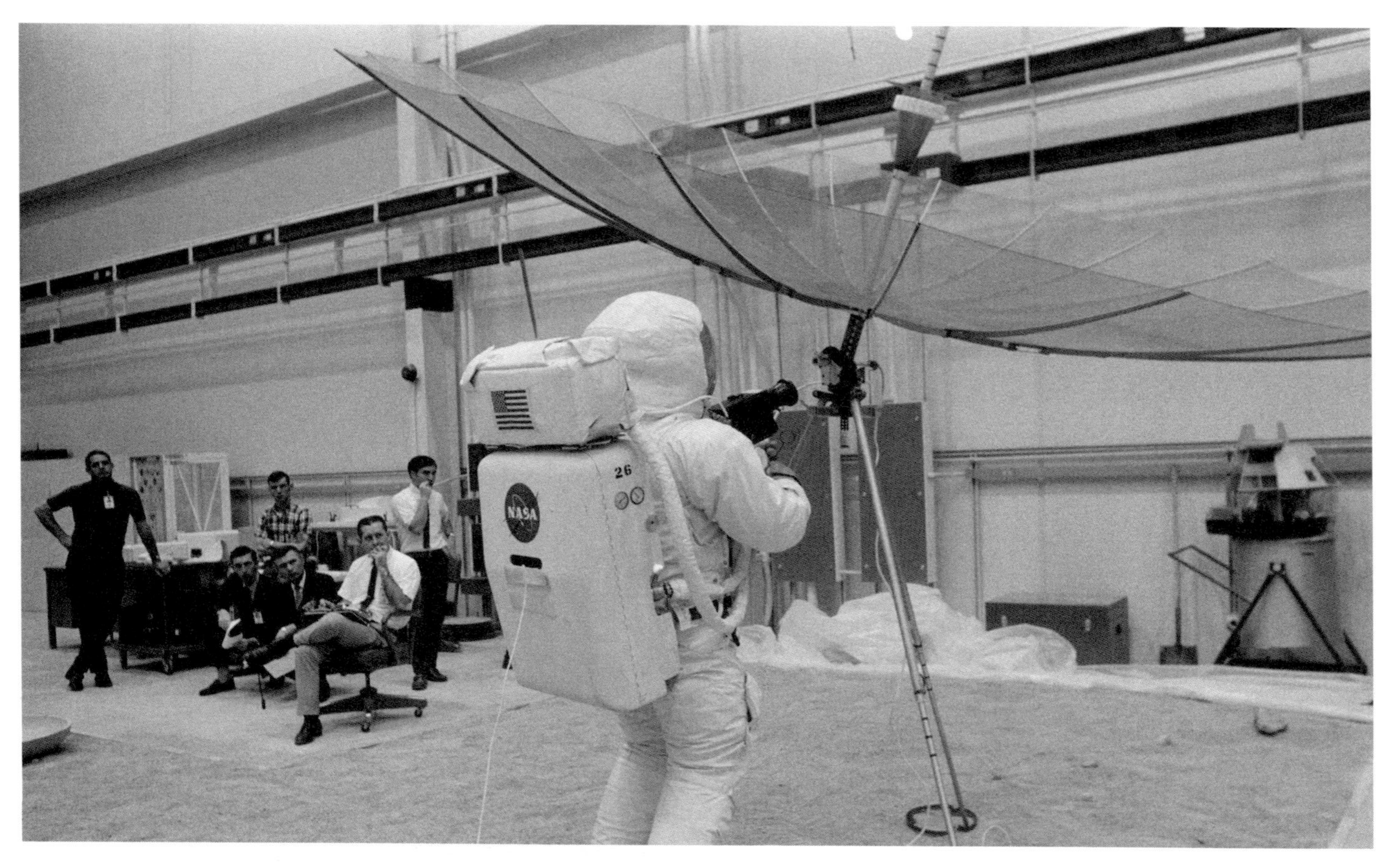

Neil Armstrong rehearses deploying the high-gain antenna for the television.

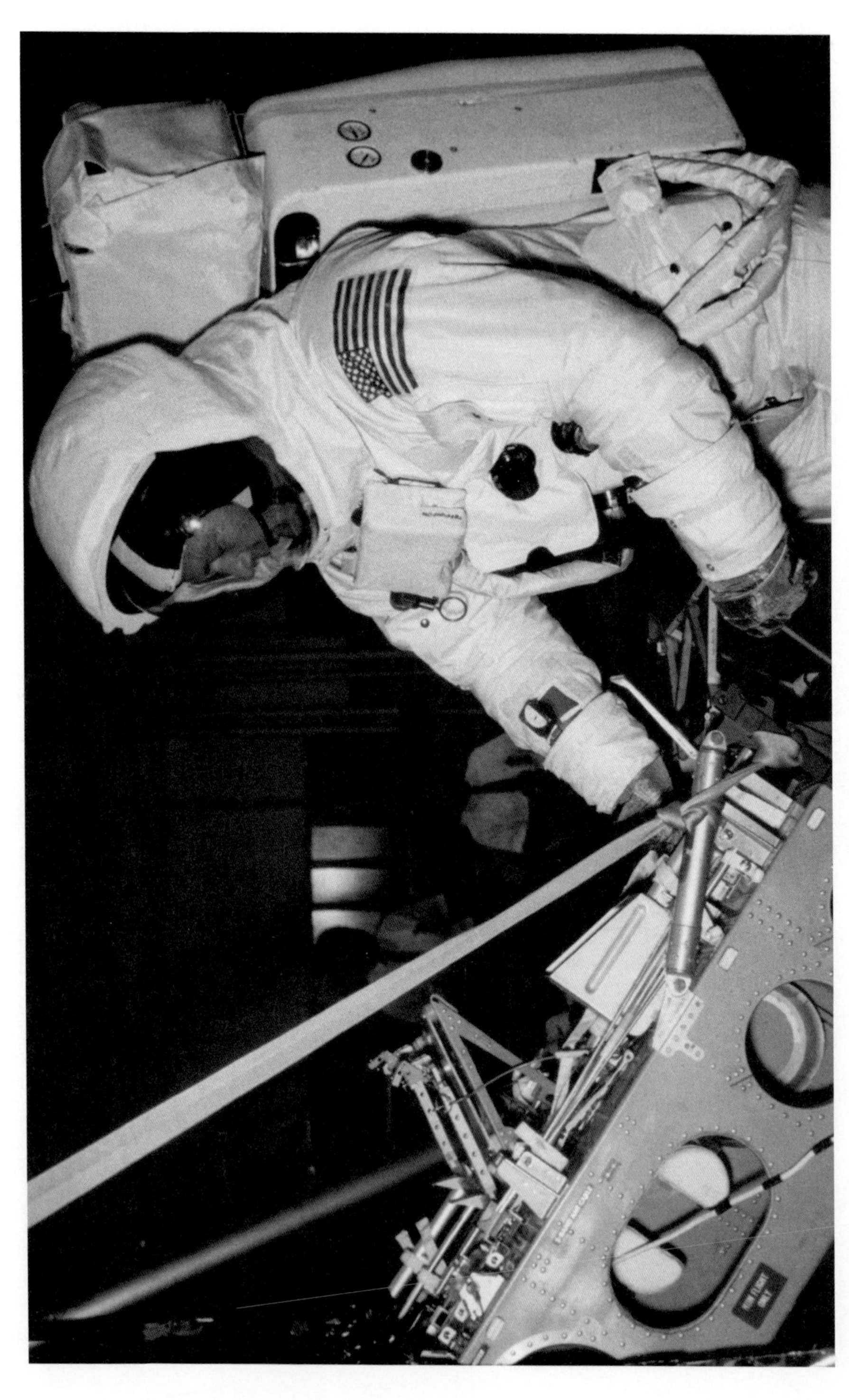

Buzz Aldrin prepares the MESA.

Neil Armstrong rehearses scooping loose material into the 'bulk' sample bag.

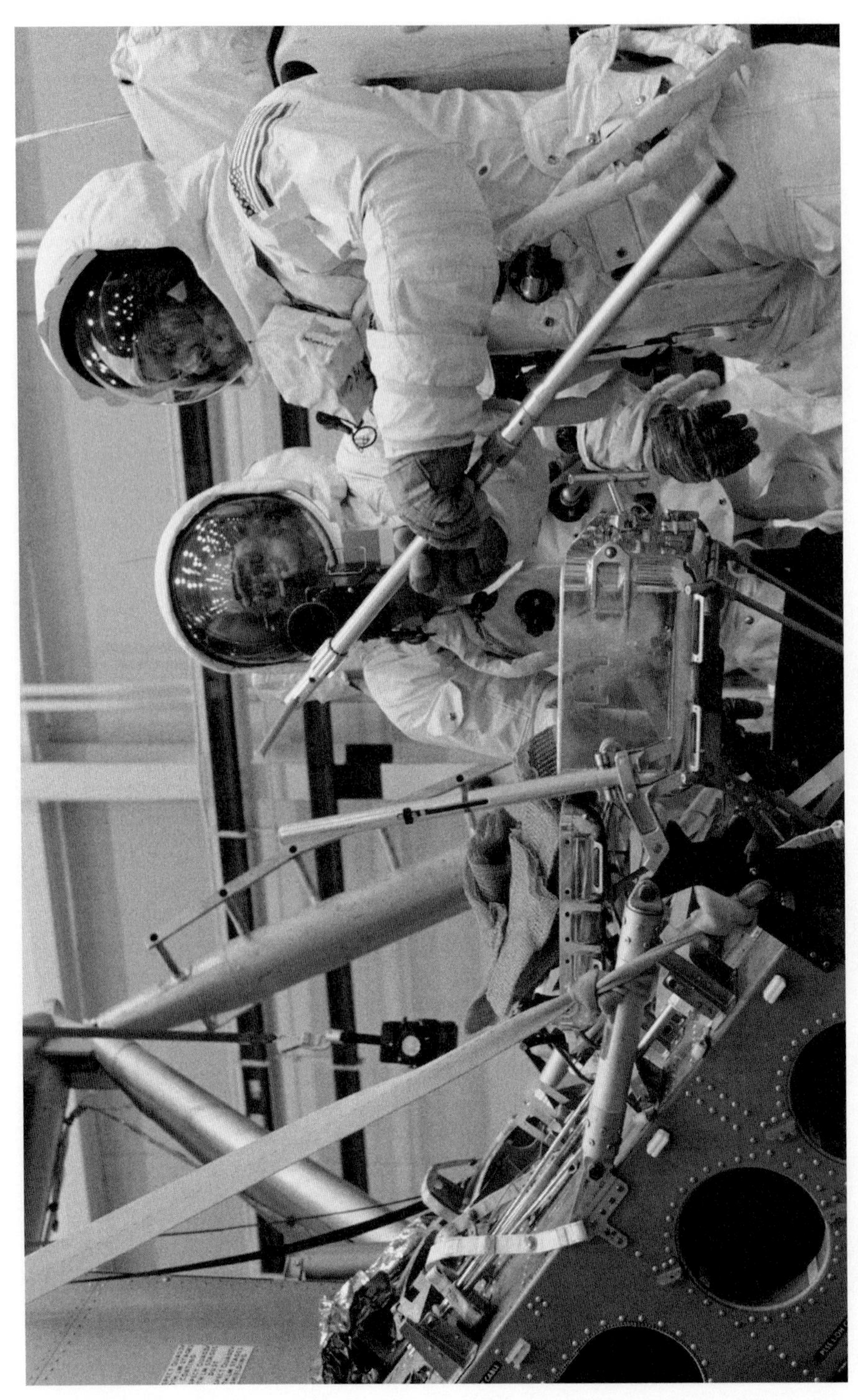

Buzz Aldrin attaches the extension handle to a tube for a 'core' sample.

Having taken a 'core' sample and removed the extension handle, Aldrin caps the tube.

Neil Armstrong and Buzz Aldrin rehearse 'documented' (photographed) sampling using tongs, a gnomon and individual sample bags.

Neil Armstrong and Buzz Aldrin rehearse 'documented' (photographed) sampling using a scoop, a gnomon and individual sample bags.

Neil Armstrong and Buzz Aldrin rehearse 'documented' (photographed) sampling using a scoop, a gnomon and individual sample bags.

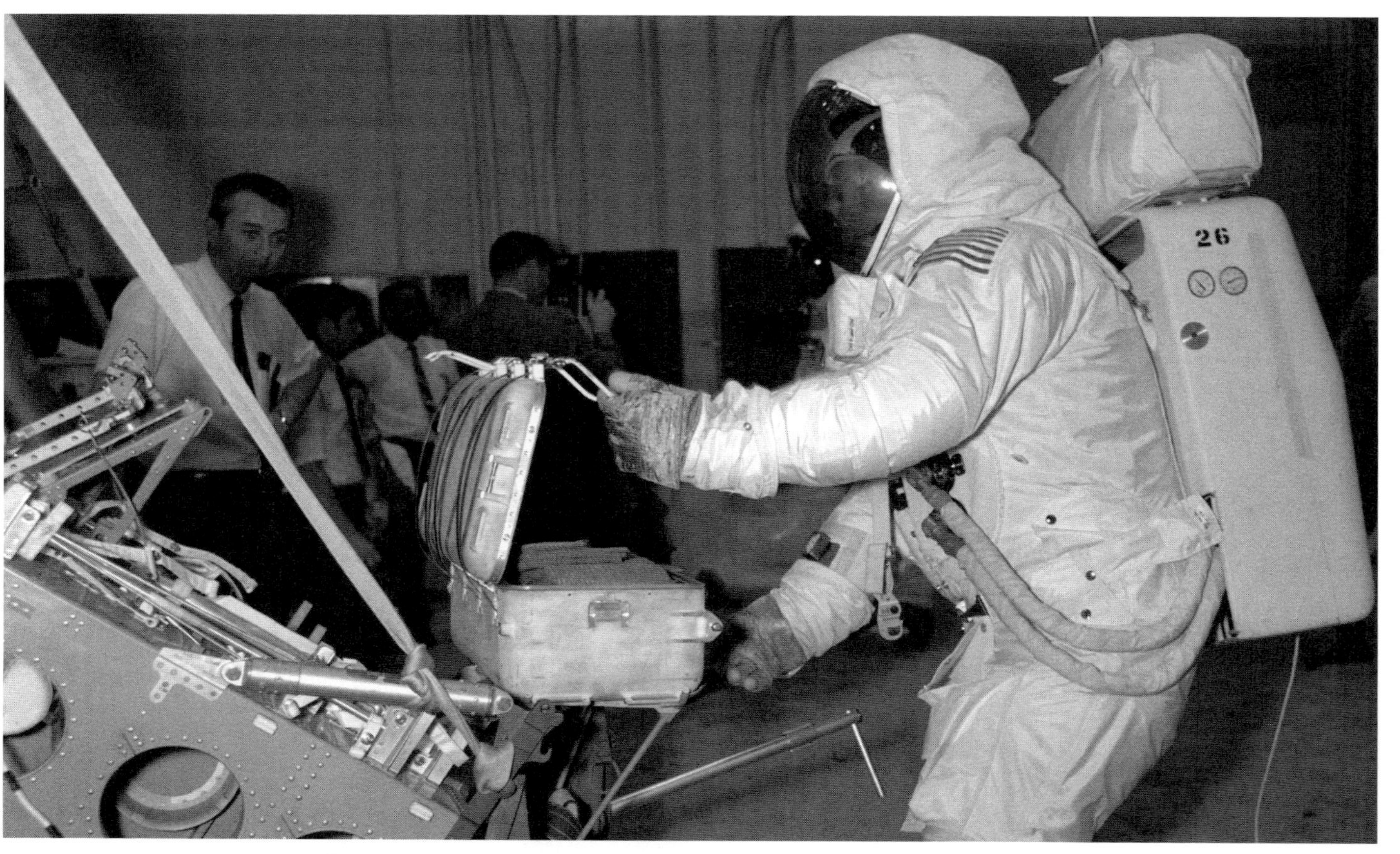

Neil Armstrong prepares to close a 'rock box' mounted on the MESA 'table'.

Buzz Aldrin, having deployed the SWC.

Buzz Aldrin rehearses extracting the EASEP instruments from the SEQ bay.

Buzz Aldrin and Neil Armstrong with the two EASEP instruments ready for deployment.

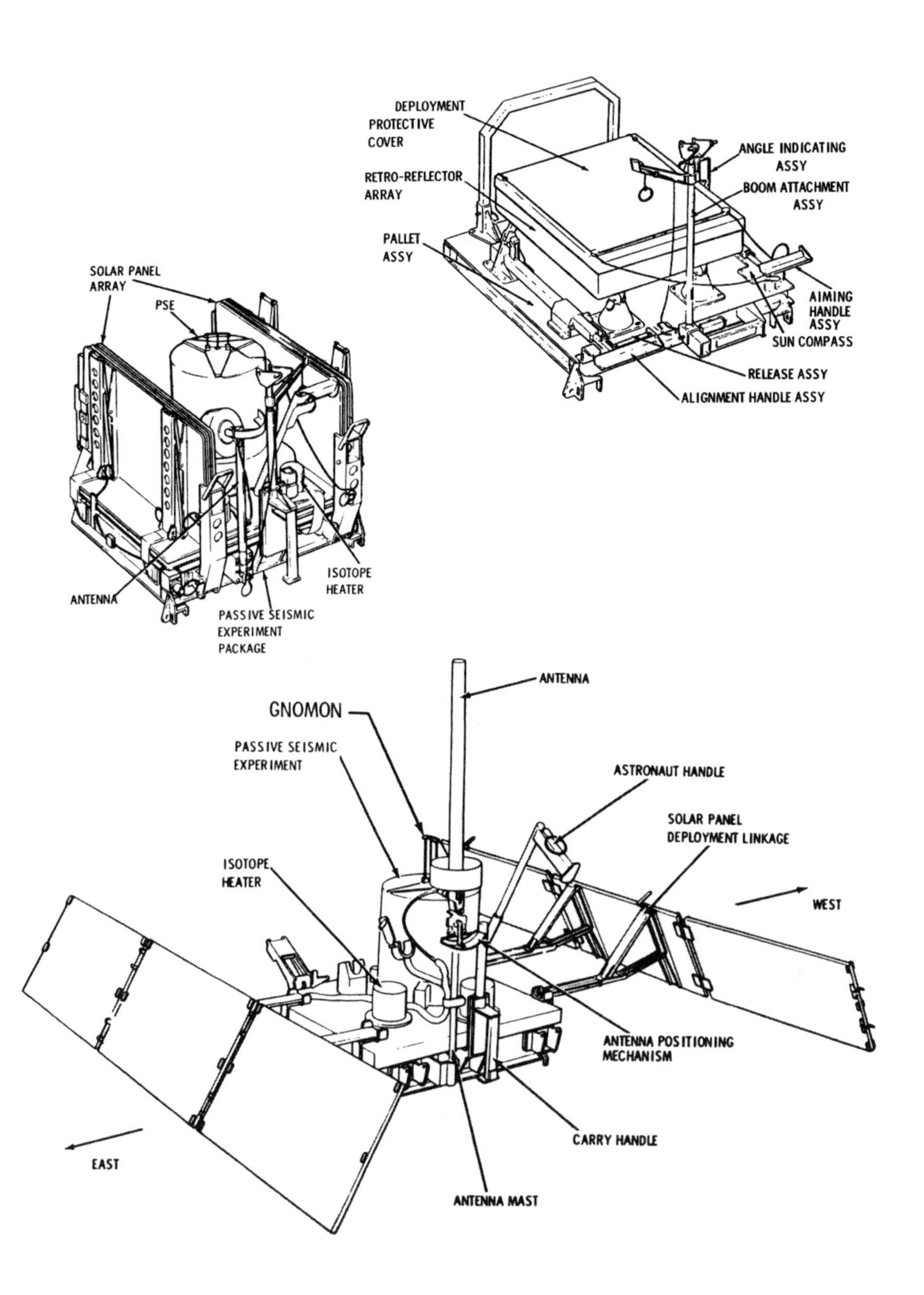

Annotated diagrams of the EASEP instruments.

Neil Armstrong starts to deploy the LRRR.

Neil Armstrong documents the deployed PSE (foreground) and LRRR (behind).

Buzz Aldrin documents the site with a panoramic sequence around the horizon.

Neil Armstrong uses the LEC to transfer a 'rock box' to the LM.

The simulation over, Neil Armstrong prepares to ascend the ladder.

After a long day, Neil Armstrong and Buzz Aldrin are debriefed by senior NASA managers.

set up, conduct the test and then be restrained once more against the load of three gravities as the aircraft pulled out of its dive. The cabin was voluminous, but with technicians, film crew and Air Force supervisors lending assistance it soon became crowded. The aircraft would make several dozen arcs over a period of hours, flying a roller-coaster path through the sky. Inevitably someone would vomit.[3] While this training was valuable, the fact that it simulated lunar gravity for no more than 30 seconds per time meant that it was possible only to test specific tasks, such as using a pair of tongs to lift a rock and pop it into a bag. To rehearse long sequences of tasks, systems using cables and pulleys were built – in some cases with the astronaut operating at an angle against a tilted surface. As Armstrong observed of these 'Peter Pan' rigs: "You had the feeling of being able to jump very high – a very light feeling. You also had the feeling that things were happening slowly, which indeed they were. It was a sort of floating sensation." On the other hand, he was confident, "The lunar setting will become a very easy place to work, I think."

LUNAR LANDING

When John F. Kennedy challenged his nation to land a man on the Moon before the decade was out, Hubert M. 'Jake' Drake at Edwards Air Force Base, who in the 1950s participated in the initial planning for the X-15 rocket plane, concluded that to provide realistic training for flying a lunar module it would be necessary to build a free-flying craft that accurately reproduced the stability and control issues involved in 'flying' in a vacuum and a reduced gravitational field. Drake set up a study group to design such a machine and enrolled Neil Armstrong as one of the team's members. After reviewing 1950s research into vertical takeoff and landing (VTOL) aircraft, it was decided to mount a jet engine in a gimbal to provide vertical thrust. Its throttle would operate in two modes: in 'terrestrial mode' the jet would run conventionally in order to lift off vertically and climb to the altitude needed to simulate the lunar landing, and then be throttled back into 'lunar mode' in order to offset five-sixths of the craft's weight. The rate of descent would be controlled by a pair of throttleable thrusters affixed to the airframe. The attitude control system was based on that developed for the X-15 at the top of its ballistic arc, where aerodynamic control surfaces are useless. It was decided to use 16 thrusters, arranged in pairs, to control roll, pitch and yaw. The project attracted interest precisely because aerodynamics played no part in the craft's operation. To translate, it would have to tilt, and use the angled component of the thrust from the 'descent engines' to impart lateral motion, then tilt back to cancel this motion. By a remarkable coincidence, Bell Aerosystems in Buffalo, New York – which had built the X-1 rocket plane in which Charles E. Yeager had 'broken the sound barrier' on 14 October 1947, and was the only US

[3] It was for this reason that the KC-135 aircraft used for such training was nicknamed the Vomit Comet.

aircraft manufacturer with experience of using jet engines for VTOL – independently submitted to NASA a proposal to develop a vehicle to be used to investigate the issues of making a landing on the Moon. When NASA sent Bell out to Edwards, Drake realised that the company was better placed to develop the vehicle and, as a result, on 18 January 1963 NASA issued Bell with a contract to supply two Lunar Landing Research Vehicle (LLRV) aircraft.

On 15 April 1964 the two LLRVs were shipped to Edwards Air Force Base in crates, because Drake's team wished to do the assembly and install the instruments themselves. Each vehicle stood 10 feet tall on four legs spanning some 13 feet, and weighed 3,700 pounds. The General Electric CF-700-2V turbofan jet delivered a maximum of 4,200 pounds of thrust. The descent engines for 'lunar mode' were non-combustion rocket thrusters using pure hydrogen peroxide propellant, each of which could be throttled between 100 and 500 pounds of thrust in order to control the rate of descent and horizontal translations.[4] The pilot sat on a platform that projected forward between the front legs. In view of the fact that if a vehicle were to get into trouble it would be close to the ground, probably be falling, and certainly be within seconds of crashing, it was fitted with a lightweight ejection seat developed by Weber Aircraft that was not only capable of lifting its user clear of an aircraft on the ground but also from an aircraft that was at low level and falling at 30 feet per second. On 30 October 1964 NASA test pilot Joseph S. Walker, a former X-15 pilot, made three vertical 'hops' in LLRV-1, remaining within 10 feet of the ground for a total duration of 60 seconds to exercise the hydrogen peroxide attitude control thrusters, the steam from which nearly obscured the view of the spectators. Armstrong was no longer at Edwards, but having been assigned the task of overseeing the development of trainers and simulators he closely monitored the test program.

In 1963 NASA began to train astronauts to fly helicopters in the hope that this would enable them to gain a feel for the issues of making a landing on the Moon. However, while a helicopter could duplicate the trajectory of the final phase of a lunar landing, the basic aerodynamic requirements of helicopter flight meant that the controls could not simulate those of a spacecraft. In contrast, the *un*-aerodynamic LLRV did accurately simulate control over the rate of descent, attitude, and lateral movement. On 26 January 1965, Warren J. North, who was in charge of training, ordered that astronauts must have 200 hours of helicopter training prior to trying to fly the LLRV. In October that year NASA drew up the preliminary specifications for a Lunar Landing Training Vehicle (LLTV). Based on the LLRV, this new vehicle was to have an upgraded jet and larger tanks of peroxide for longer endurance in 'lunar mode', a cabin with a similar field of view to that envisaged for the LM, a 3-axis hand controller (instead of the stick and pedals of the LLRV), instruments laid out as in the LM, and as much as possible of the LM's built-in flight control logic in

[4] The engine did not 'burn' its propellant; instead a silver catalyst in the chamber converted the H_2O_2 to superheated steam and oxygen, and the gas passed through the nozzle to produce thrust.

order to enhance its fidelity as a trainer. In August 1966 Armstrong and Joseph S. Algranti, chief of aircraft operations at the Manned Spacecraft Center, worked with Bell to implement these upgrades. To augment helicopter training, a cratered surface based on the highest resolution pictures from the Ranger probes was mocked up, and on climbing to 500 feet the astronauts would cut the throttle and land at various angles and rates of descent and in a variety of lighting conditions to familiarise themselves with visually gauging their height and sink rate over the alien landscape. Meanwhile, it had been decided that once Edwards completed its LLRV tests these vehicles should be sent to Ellington. When LLRV-1 arrived on 12 December 1966, Armstrong was present to watch Algranti perform the formal acceptance trial. LLRV-2 followed in mid-January 1967. In a rationalisation, the two LLRVs were redesignated LLTV A1 and A2, and the three new vehicles were to be B1, B2 and B3. Before being permitted to fly, an astronaut was required to undertake a 3-week helicopter refresher, 1 week of familiarisation with the Lunar Landing Research Facility at Langley,[5] spend 15 hours in a ground simulator and then be cleared by Algranti.

Armstrong made his first flight in LLTV A1 on 27 March 1967, but did not fly again until starting an intensive program of lunar landing rehearsals in early 1968. A typical flight involved using the jet at maximum thrust to lift off vertically and climb to 500 feet altitude, throttling back to balance five-sixths of the weight, and then, as when using the helicopter, flying a profile that would match the trajectory of a LM at that altitude, except that now the rate of descent and lateral manoeuvres were actively controlled employing the 'descent engines'. As Armstrong reflected of his experience:[6]

> "The thing that surprises people on their initial flights in 'lunar mode' is the tendency of the vehicle to float far beyond where you think it is going to go. It takes practice to anticipate the distance required to slow down – you must start to brake much earlier, if you are to stop where you want to stop. Similarly, if you are in a hover, and change your mind, it takes a lot of effort to get moving again. The vehicle is sluggish in its translating ability, so it takes a long time, and big angles, to gain a little speed and translate 50 feet. We hope to have one-and-a-half to two minutes of fuel essentially in hover when we're landing on the

[5] The Lunar Landing Research Facility at the Langley Research Center became operational on 30 June 1965. It was a 260-foot-tall 400-foot-long frame structure with a system of travelling pulleys to suspend a vehicle in such a manner as to balance five-sixths of its weight. It provided a 'flying volume' 180 feet in height and 360 feet in length, with a lateral range of 42 feet. Its main role was to test instruments and software to be used by the LM during the final 150 feet of a lunar descent, but astronauts used it to familiarise themselves with flying in one-sixth gravity prior to advancing to the LLTV.

[6] Based on an account in *First on the Moon: A Voyage with Neil Armstrong, Michael Collins and Edwin E. Aldrin Jr*, by Gene Farmer and Dora Jane Hamblin. Michael Joseph, pp. 216-218, 1970.

> Moon, but you can use that up really fast if you change your mind frequently about where you want to go."

On 6 May 1968 LLTV A1 went out of control during a descent and he had to eject.

> "I lifted the vehicle off the ground and climbed to an altitude of 500 feet in preparation for making the landing profile. I had been airborne for about 5 minutes, and was down to about 200 feet when the trouble began. The first indication was a decreasing ability to control the vehicle. It began to tilt sharply. There was less and less response. The trouble developed rather rapidly, but wasn't an abrupt stop. It was a decay in attitude control. Without attitude control there is no way to remain upright. The vehicle does have two separate systems for doing this, but in this case both systems failed at their common point – the high-pressure helium to pressurise the propellant to the rockets. I was losing both systems simultaneously, and that's where I had to give up and get off. I guess I ejected at 100 feet, plus or minus – we don't have a way of measuring it accurately, even from photographs. How far the ejection throws you depends on your attitude at the time you leave, and also on your upward or downward velocity at the time. If you start from an upright attitude at a hover, it will take you up about 300 feet. The parachute ejector is automatic, although there is a manual override. I had always thought I might be able to match the automatic system, but when I was reaching for the D-ring the automatic system had already fired."

The abandoned vehicle fell straight down.

> "The ejection system threw me somewhat east of the crash, but the wind was from the east and at the time my chute opened I was a bit concerned that I might be drifting down into the fire, but the wind was strong and I actually missed the flames by several hundred feet. After I landed, I got up and walked away. The only damage to me was that I bit my tongue."

As Armstrong had abandoned a stricken Panther jet over Korea, this was his second ejection. Most astronauts would have sought out colleagues and related an enthusiastic account of the event, but Armstrong returned to his office to catch up on paperwork. At the time, observers speculated that there had been an explosion, but they had been misled by the steam issuing from the thrusters as Armstrong was attempting to recover. LLTV A2, which was not yet in operation, was grounded pending an investigation led by Algranti, which concluded that a design flaw had enabled the helium pressurisation of the peroxide system to decay, rendering the thrusters ineffective.

The first B model LLTV was delivered to Ellington in December 1967, but did not become available until mid-1968. A Flight Readiness Review on 26 November declared LLTV B1 ready for astronaut training. On 8 December 1968, on its tenth flight, the vehicle developed an uncontrollable lateral control oscillation, obliging Algranti to eject at an altitude of 200 feet. Kraft and Robert R. Gilruth, Director of the Manned Spacecraft Center, suggested that the LLTV was too dangerous, but the

The Lunar Landing Research Vehicle in 1965 at its maximum altitude during a test flight at Edwards Air Force Base.

The Lunar Landing Research Vehicle in 1967 at Ellington Air Force Base.

Neil Armstrong by the Lunar Landing Research Facility at the Langley Research Center on 12 February 1969.

astronauts, particularly Armstrong, who had most experience with it, insisted it was essential. On 13 June 1969 LLTV B2 was declared ready for astronaut training. As commander of the mission that was to attempt a lunar landing, Armstrong had first call, and he flew it on 14, 15 and 16 June. Since the vehicle carried propellant only for about 6 minutes of flight and it took several minutes to climb and establish the required profile, a descent test lasted at most 4 minutes and often was concluded with only seconds to spare. Although dangerous, the LLTV was the *only* effective training for *flying* the LM in a manual mode.[7]

ISOLATION

On 17 June the Apollo 11 crew had their T–30-day medicals and transferred to the Manned Spacecraft Operations Building, located on the industrial facility 5 miles south of the Vehicle Assembly Building. The third-floor crew quarters, which had a ventilation system designed to maintain a germ-free environment, comprised a living room, dining room, kitchen, briefing room, bathroom, exercise room, equipment room, and a number of small windowless bedrooms. Lewis Hartzell had been hired to cook for the Gemini crews and remained, not for the money, but for the honour of cooking for the astronauts. As a former Marine and a cook on tugboats, Hartzell only did plain cooking, which raised no objections from the astronauts.

A flight readiness review later on 17 June authorised loading the hypergolic propellants into the LM and CSM tanks. This represented a major decision point, because if a mid-July launch should prove impracticable, it would not be safe to retain such corrosive chemicals in the tanks for an additional month – not only would the tanks have to be drained, but certain components would require to be removed and returned to the vendor for refurbishment. Worse, there would be no guarantee that the vehicles could be reassembled in time for the August launch window. The loading operation began on 18 June and, despite delays caused by weather conditions at the Cape, was completed on 23 June.[8]

On 26 June Armstrong, Aldrin and Collins had medical examinations that were not only to confirm their physical state, but also to catalogue the organisms in their systems to provide a 'baseline' for spotting any infections that they might contract during the final stages of preparation. After a countdown demonstration test that concluded with a simulated launch at 9.32 am local time on Wednesday, 3 July, they flew to Houston for the Fourth of July weekend. *Life* magazine published an issue

[7] In his debriefing after Apollo 11, Armstrong confirmed the fidelity of the LLTV, and thereafter each mission commander trained with it.

[8] The hypergolic propellants were nitrogen tetroxide oxidiser and a fuel comprising a 50:50 mix of hydrazine with monomethyl hydrazine. The RCS of the CSM required 300 pounds, the SPS of the CSM required 41,000 pounds, and the LM's propulsion systems required a total of 23,245 pounds.

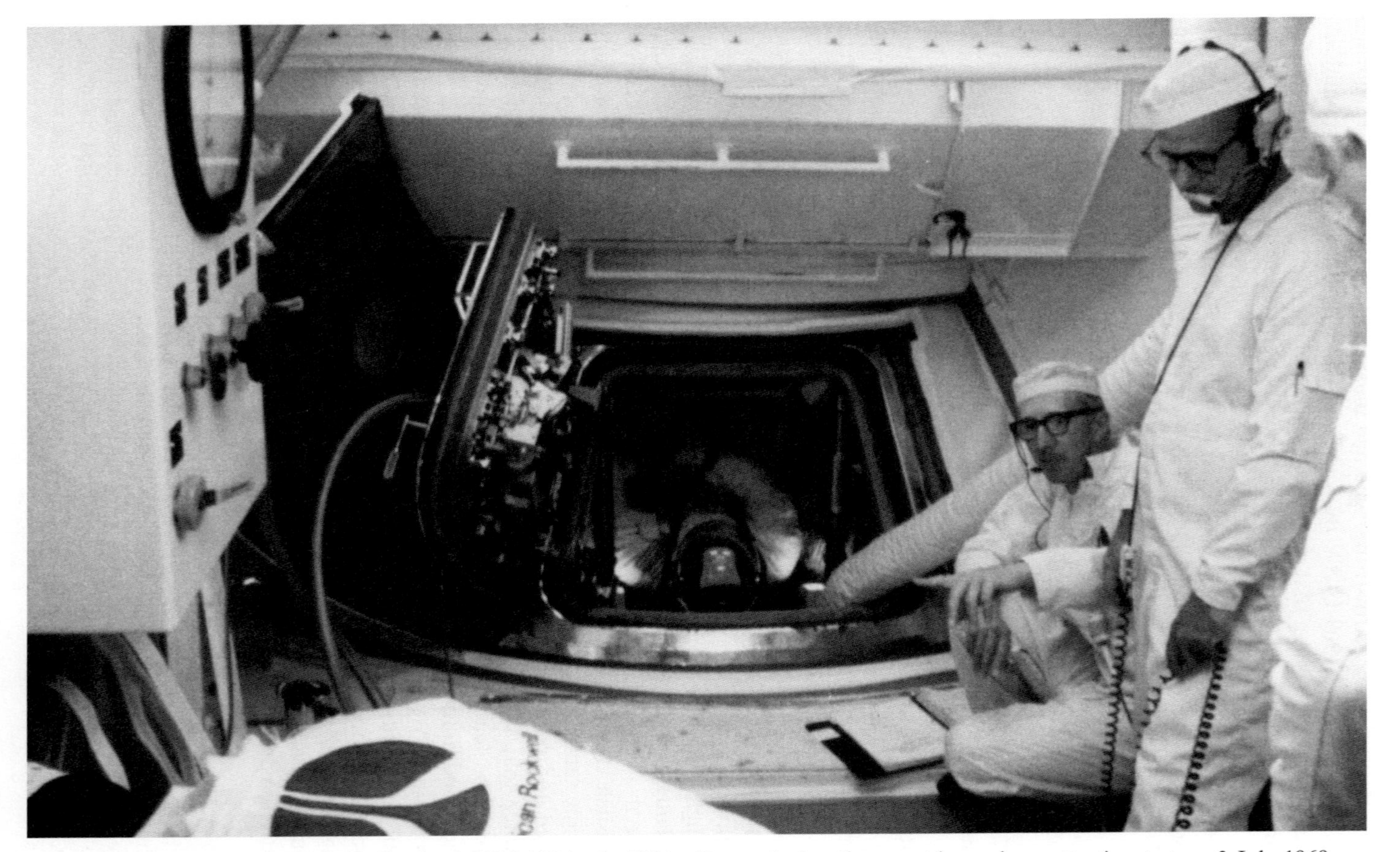

Guenter Wendt (seated) by the hatch of CSM-107 in the White Room during the countdown demonstration test on 3 July 1969.

with the cover 'Off To The Moon', with stories about their home lives. NASA would have loved to have scheduled the lunar landing for 4 July, but operational constraints did not permit this.

Gene Kranz's flight control team took 4 July off, but returned to work the next day for their 'graduation' simulations. As Armstrong and Aldrin were unavailable, Pete Conrad and Al Bean took their places as a welcome training opportunity for Apollo 12. The flight controllers successfully overcame six tough scenarios during the morning. The afternoon sessions were to be 'flown' by the Apollo 12 backup crew of Dave Scott and Jim Irwin, the rationale being that a less-experienced crew would increase the pressure on the flight controllers. Three minutes into the first run, Koos prompted the LM's computer to issue an alarm. A caution and warning light illuminated, and the computer flashed the numerical identifier for that particular problem. Computer alarms could result from a hardware fault, a software issue, out-of-tolerance data, or a procedural error either by the crew or the ground. Steve Bales, the guidance officer, was monitoring the LM's computer to ensure that it received the correct data from Earth and that its guidance, navigation and control tasks were being properly executed. In this case the alarm was a 12-01. Bales had previously seen it during functional tests of the computer on the ground, but never in a simulation, and certainly not in flight. While the LM crew awaited advice, he checked his manual: the 12-01 alarm was 'executive overflow', which meant that the computer was overloaded. The computer's executive was to repeatedly cycle through a list of tasks in a given interval of time, and evidently the time available was no longer sufficient to finish the tasks before it was obliged to begin the next cycle. Bales called Jack Garman, a support room colleague and software expert, and they agreed that the alarm was serious, especially since it was recurrent. With no mission rules to inform his decision-making, Bales called Kranz, told him that there was something amiss with the computer, although he could not say what, and recommended an abort. This call came out of the blue as Kranz had not been party to the discussion between Bales and Garman, but as a flight director must trust the judgement of his controllers – especially on abort calls – he confirmed it. Charlie Duke, serving as CapCom, relayed the abort to the crew, who performed the manoeuvre and made as if to rendezvous with their mother ship (which was not actually in the simulation). At the debriefing, Koos pointed out that the 12-01 had *not* necessitated an abort; in the absence of a *positive indication* that the computer was failing they should have continued. Shocked that he had made a bad call, Bales got together with the people from the Massachusetts Institute of Technology who had written the software, in order to investigate the alarm. Later that evening, he called Kranz and conceded there had been no need to abort. The next day, 6 July, Koos triggered a range of computer alarms to enable Bales' team to record data on the ability of the computer to continue to function. On 11 July Bales added a new mission rule listing the alarms that would require an immediate abort; in all other cases the powered descent was to continue pending a positive indication of a critical failure.

In 1966 Slayton had told George E. Mueller, Director of the Office of Manned Space Flight, that an Apollo crew would require 140 hours of training in the CSM simulator, with a lunar landing crew spending an additional 180 hours in the LM. In fact, as they completed their training, Collins had spent 400 hours in the CSM; Armstrong had spent 164 hours in the CSM, 383 hours in the LM, and a total of 34 hours in the Lunar Landing Research Facility at Langley and flying the LLTV; and Aldrin had spent 182 hours in the CSM and 411 hours in the LM, but had not used the other facilities. Training for lunar surface activities accounted for no more than 14 per cent of their time.

VEHICLE PREPARATION

Apollo spacecraft CSM-107 was built by North American Rockwell at its plant at Downey, California. The conical command module was 11 feet 5 inches high, 12 feet 10 inches in diameter, and provided a habitable volume of 210 cubic feet. The cylindrical service module was 12 feet 10 inches in diameter and 24 feet 7 inches tall. Radial beams divided it into a central tunnel, which contained tanks of helium pressurant, and six outer compartments, four of which held propellant tanks, one contained the fuel cell system and the sixth was unused.[9] The systems tests on the individual modules were completed on 12 October 1968, and the integrated tests on 6 December. The modules were flown to the Cape on 23 January 1969 by a 'Super Guppy' aircraft of Aerospace Lines. They were mated on 29 January, passed their combined systems testing on 17 February and altitude chamber tests on 24 March. At the Grumman Aircraft Engineering plant at Bethpage on Long Island, LM-5 completed its integrated test on 21 October 1968, and its factory acceptance test on 13 December. The ascent stage arrived at the Cape on 8 January 1969 and the descent stage on 12 January. After acceptance checks, the stages were mated on 14 February, passed their integrated systems tests on 17 February, and altitude chamber tests on 25 March. Overall, the vehicle stood 22 feet 11 inches tall. The descent stage was 10 feet 7 inches high and had a diagonal span of 31 feet across its foot pads. Two layers of parallel beams in a cruciform shape gave it a central cubic compartment (housing the descent engine), four cubic side compartments (each housing a propellant tank) and four triangular side compartments (carrying apparatus the astronauts would require during their moonwalk). The ascent stage comprised a pressurised crew compartment and midsection with a total volume of 235 cubic feet, and an unpressurised aft equipment bay.

The 138-foot-long, 33-foot-diameter S-IC first stage of the sixth launch vehicle in the Saturn V series was fabricated by Boeing at the Michoud Assembly Facility in Louisiana, and moved in a horizontal configuration by barge up the Intracoastal Waterway to the Mississippi Test Facility, arriving on 6 August 1968. It was then shipped around the southern tip of Florida, to the Kennedy Space Center. On arrival on 20 February 1969 the 24-wheeled trailer bearing the stage was offloaded by a

[9] The fuel cell system had three fuel cells, two tanks of cryogenic oxygen and two tanks of cryogenic hydrogen, and provided 28 volts.

prime mover and driven into the 'low bay' annex of the Vehicle Assembly Building. The S-II second stage had the same diameter as the S-IC, but was only 81 feet 6 inches in length. After assembly at the North American Rockwell plant at Seal Beach in California, it was shipped via the Panama Canal to the Mississippi Test Facility, where it was tested on 3 October 1968. On arriving at the Cape on 6 February 1969, the S-II, complete with its 18-foot-tall aft interstage 'skirt', was driven on its 12-wheeled trailer to the low bay. After tests at the Douglas Aircraft Corporation facility in Sacramento, California, the S-IVB third stage was flown to the Cape by 'Super Guppy' on 19 January 1969. In all, some 12,000 companies across America participated in the production of the launch vehicle.

The principal structure of the Vehicle Assembly Building was 718 feet long, 517 feet wide and 525 feet tall. Its internal volume of almost 130 million cubic feet required a 10,000-ton air-conditioning system to prevent a 'weather system' with its own rainfall developing. The cavernous interior provided four 'high bays' for simultaneous assembly of Saturn V vehicles. Each pair of bays shared a bridge crane located 462 feet above the floor. The operator was in walkie-talkie contact with his colleagues at the work sites, and used a computer to move loads of up to 250 tons with a tolerance of 1/228th of an inch. Mobile Launch Platform 1 was a two-level steel structure 160 feet long, 135 feet wide and 25 feet high. At one end was the Launch Umbilical Tower, which rose 398 feet above the deck, and offset towards the other end of the platform was a 45-foot-square hole to allow launch vehicle exhaust to pass through. On 21 February the S-IC was hoisted, turned to vertical, and clamped to the supporting arms, one on each side of the hole. The S-II was added on 4 March. The next day the 260-inch-diameter S-IVB, now with its flared aft skirt fitted, was added, and the Instrument Unit containing the guidance system for the launch vehicle (which had arrived on 27 February) was placed on top. The 28-foot-long truncated-cone to house the LM and support the 154-inch-diameter CSM was fabricated at the North American Rockwell plant in Tulsa, Oklahoma, and delivered on 10 January. The integrated CSM, LM, adapter and launch escape system tower was referred to as the 'spacecraft' because it was the payload of the three-stage launch vehicle. Its addition on 14 April completed the 'stack'. From the aperture of the F-1 engines of the first stage to the tip of the escape tower, the 'space vehicle', as the integrated launch vehicle and spacecraft was known, stood 363 feet tall. Nine hydraulically operated arms on the umbilical tower provided access to key sections of the vehicle.[10] The combined systems test of LM-5 was finished on 18 April. The integrated systems test of CSM-107 was completed on 22 April, and the spacecraft was electrically mated with the launch vehicle on 5 May. The overall test of the space vehicle was accomplished on 14 May.

The 6-million-pound transporter for the mobile launch system was 131 feet long,

[10] The swing arm numbers and their interface points are: 1, S-IC intertank; 2, S-IC forward; 3, S-II aft; 4, S-II intermediate; 5, S-II forward; 6, S-IVB aft; 7, S-IVB/IU forward; 8, SM; 9, crew access.

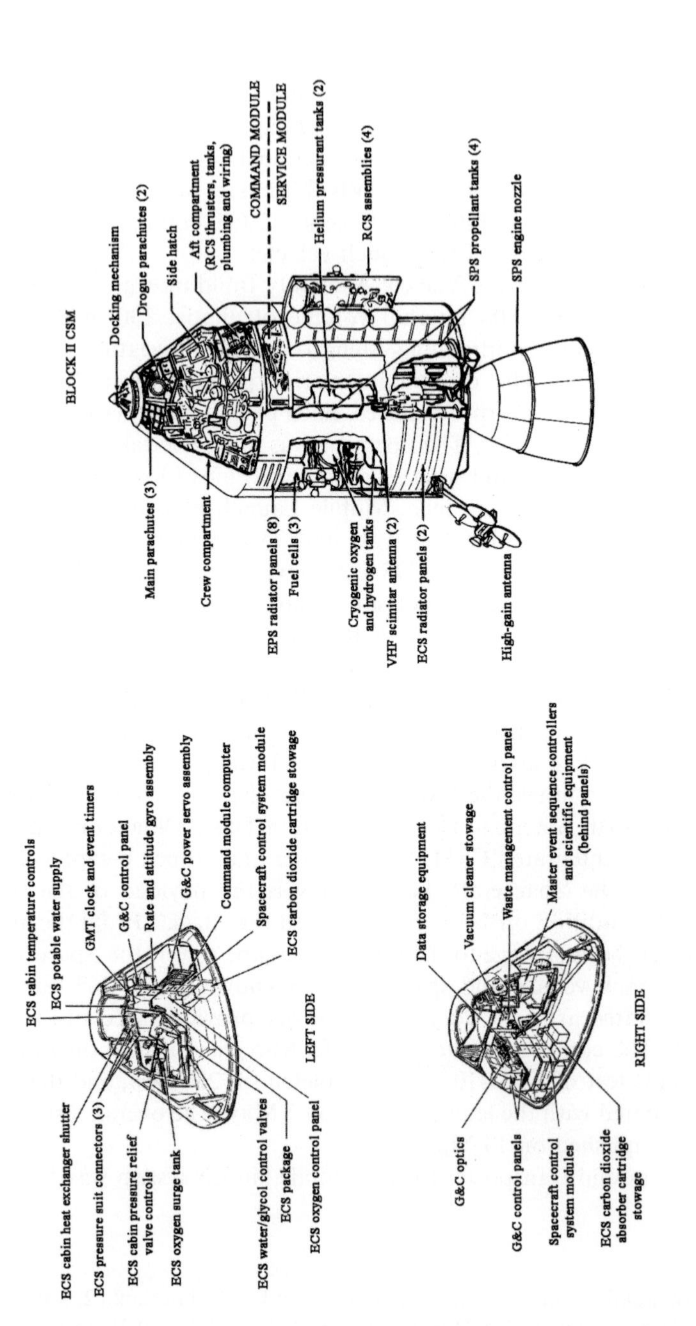

Cutaway diagrams of the Apollo CSM.

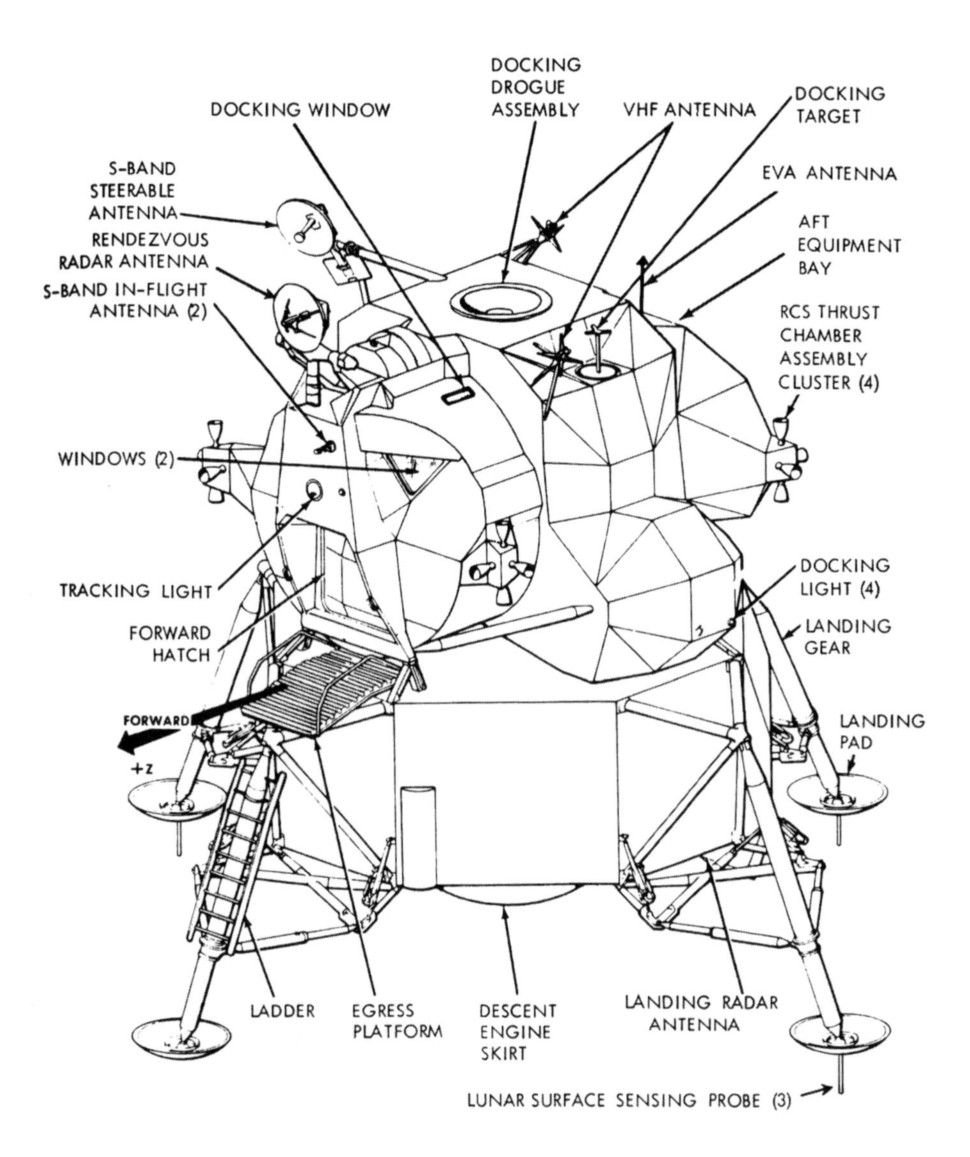

A diagram of the LM.

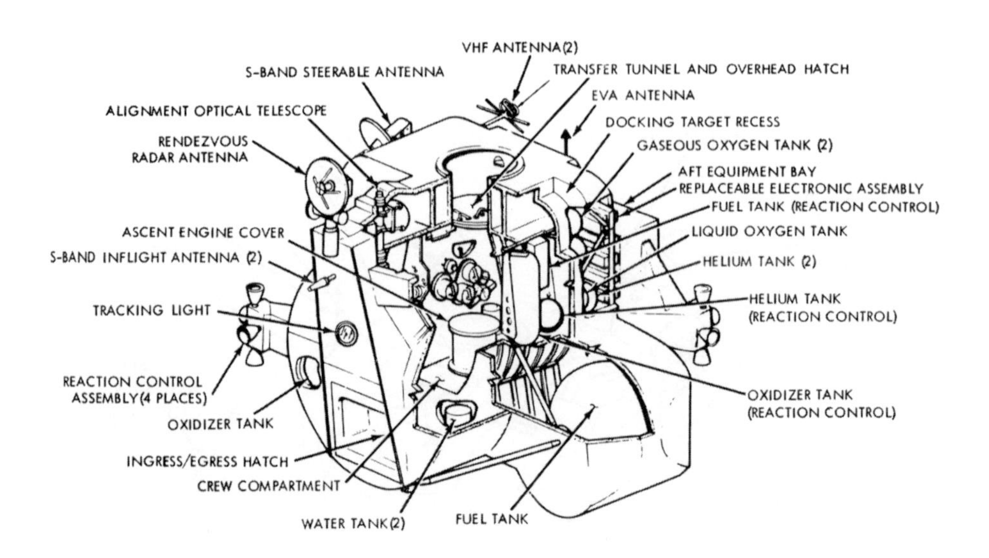

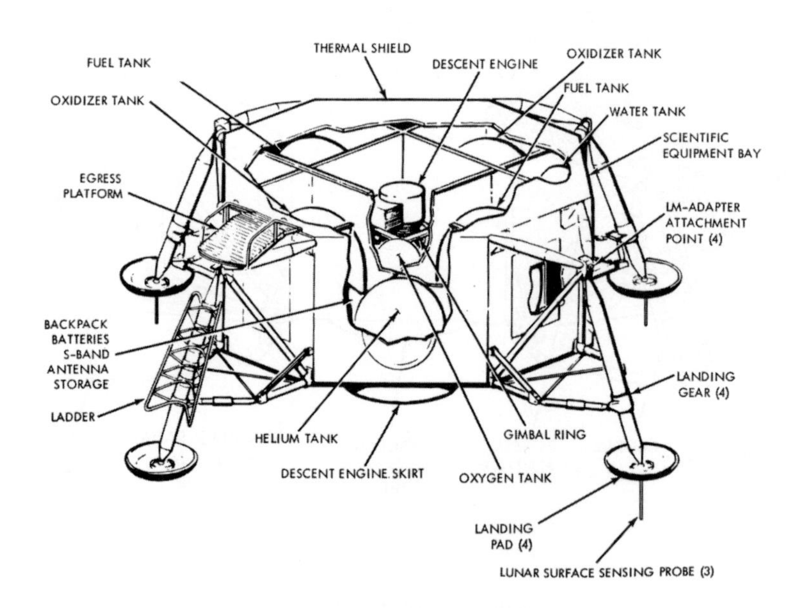

A cutaway diagram of the two LM stages.

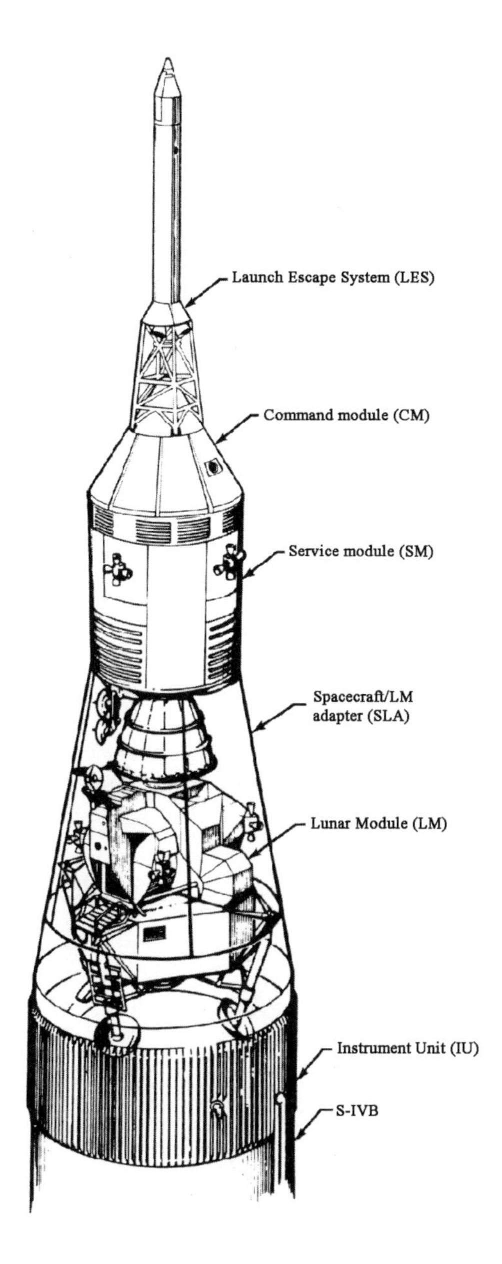

From the point of view of the Saturn V launch vehicle, the 'spacecraft' comprises the Launch Escape System, the CSM, and the LM contained within the adapter.

CSM-107 is hoisted from its work stand in the Manned Spacecraft Operations Building at the Kennedy Space Center on 11 April 1969.

CSM-107 is mated with the adapter of the Apollo 11 launch vehicle on 11 April 1969.

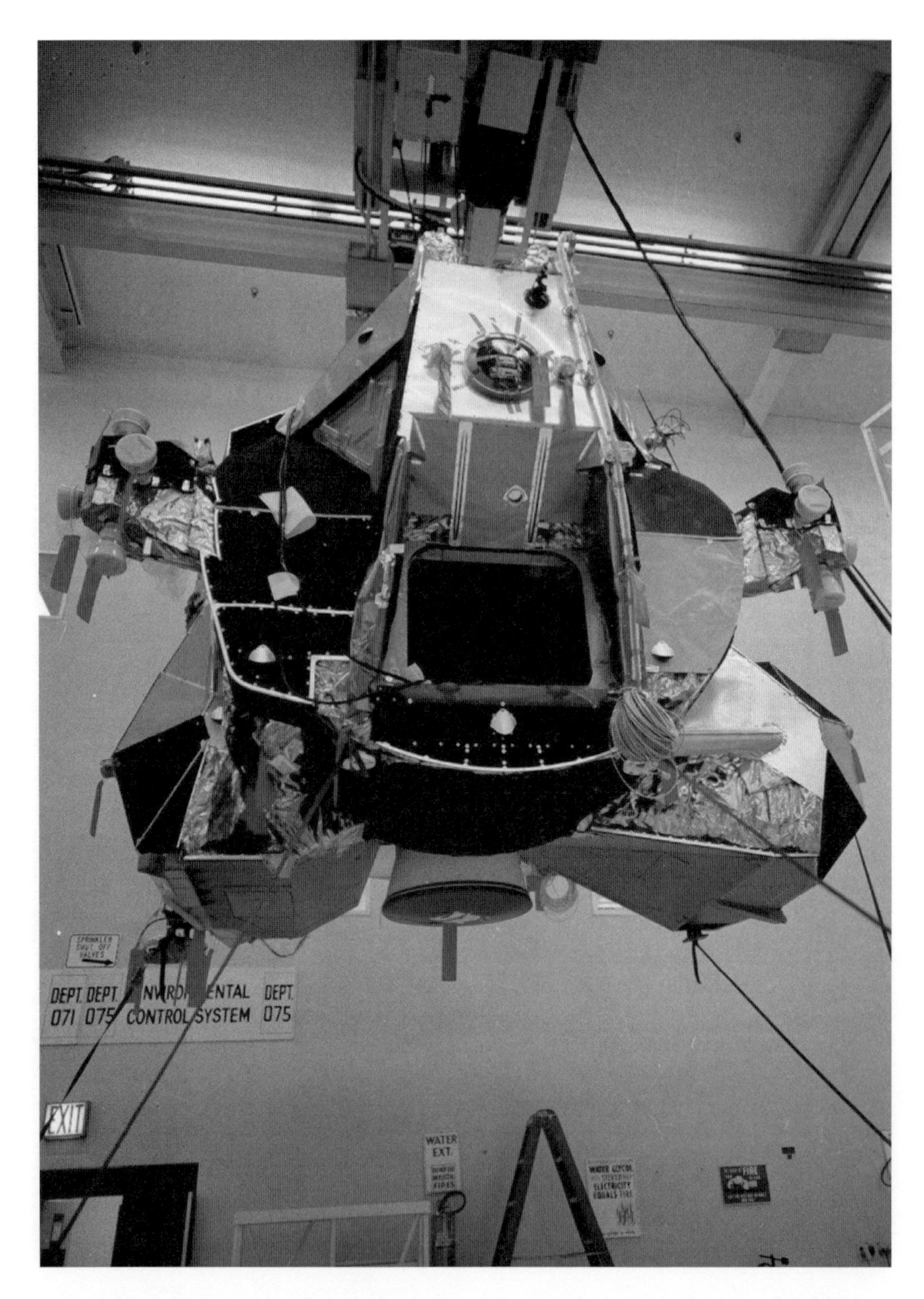

At the Grumman plant on 4 January 1969, the ascent stage of LM-5 is hoisted for shipment.

LM-5 undergoes final preparations in the Manned Spacecraft Operations Building at the Kennedy Space Center on 4 April 1969.

The S-IC first stage of the Apollo 11 launch vehicle on its trailer in the Vehicle Assembly Building on 21 February 1969.

The space vehicle for Apollo 11 is 'stacked' in the Vehicle Assembly Building (clockwise from top left): a crane hoists the S-IC on 21 February; the S-II is added on 4 March; the S-IVB is added on 5 March; and the spacecraft is added on 14 April 1969.

On 20 May 1969 the Apollo 11 space vehicle emerges from the Vehicle Assembly Building.

On 20 May 1969 the Apollo 11 space vehicle starts up the incline to Pad 39A.

On 20 May 1969 the Apollo 11 space vehicle arrives on Pad 39A.

On 22 May 1969 the Mobile Service Structure is driven up to Pad 39A.

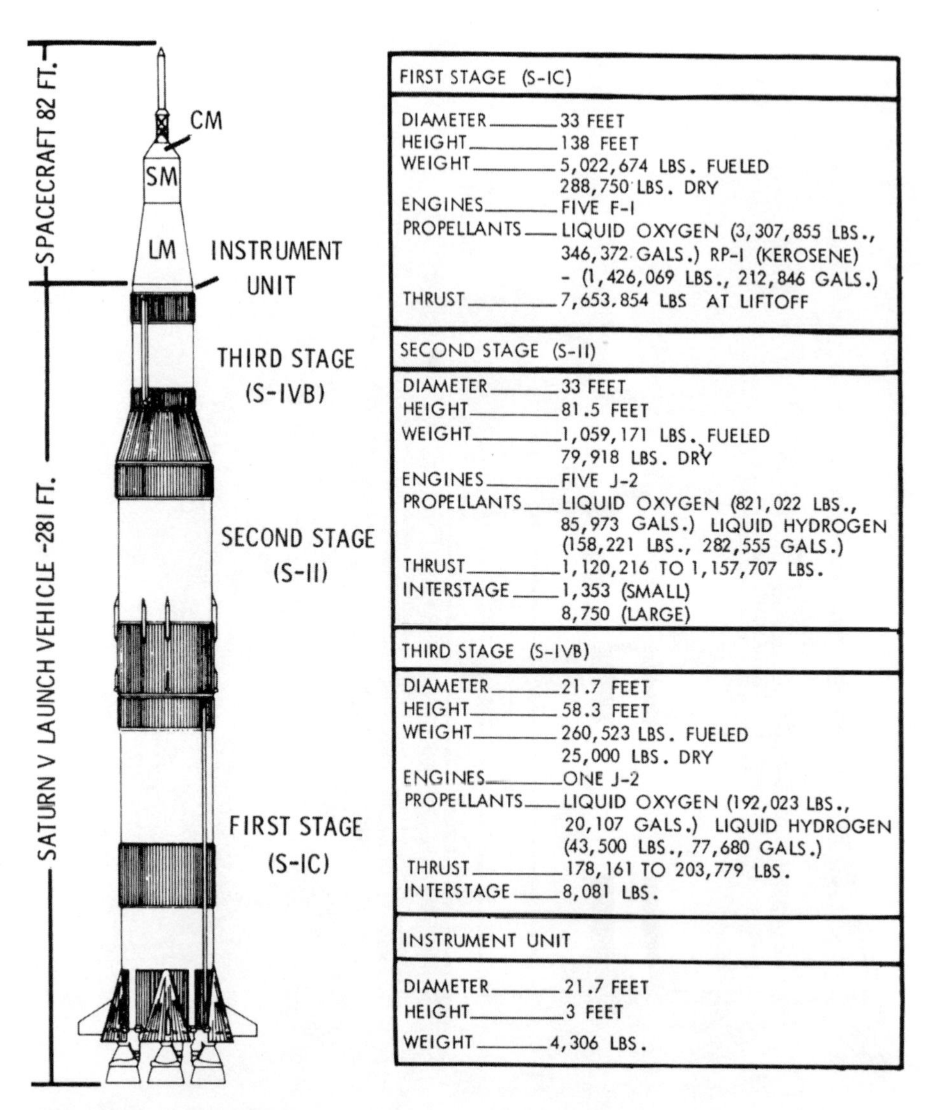

FIRST STAGE (S-IC)	
DIAMETER	33 FEET
HEIGHT	138 FEET
WEIGHT	5,022,674 LBS. FUELED 288,750 LBS. DRY
ENGINES	FIVE F-1
PROPELLANTS	LIQUID OXYGEN (3,307,855 LBS., 346,372 GALS.) RP-1 (KEROSENE) - (1,426,069 LBS., 212,846 GALS.)
THRUST	7,653,854 LBS AT LIFTOFF

SECOND STAGE (S-II)	
DIAMETER	33 FEET
HEIGHT	81.5 FEET
WEIGHT	1,059,171 LBS. FUELED 79,918 LBS. DRY
ENGINES	FIVE J-2
PROPELLANTS	LIQUID OXYGEN (821,022 LBS., 85,973 GALS.) LIQUID HYDROGEN (158,221 LBS., 282,555 GALS.)
THRUST	1,120,216 TO 1,157,707 LBS.
INTERSTAGE	1,353 (SMALL) 8,750 (LARGE)

THIRD STAGE (S-IVB)	
DIAMETER	21.7 FEET
HEIGHT	58.3 FEET
WEIGHT	260,523 LBS. FUELED 25,000 LBS. DRY
ENGINES	ONE J-2
PROPELLANTS	LIQUID OXYGEN (192,023 LBS., 20,107 GALS.) LIQUID HYDROGEN (43,500 LBS., 77,680 GALS.)
THRUST	178,161 TO 203,779 LBS.
INTERSTAGE	8,081 LBS.

INSTRUMENT UNIT	
DIAMETER	21.7 FEET
HEIGHT	3 FEET
WEIGHT	4,306 LBS.

NOTE: WEIGHTS AND MEASURES GIVEN ABOVE ARE FOR THE NOMINAL VEHICLE CONFIGURATION FOR APOLLO 11. THE FIGURES MAY VARY SLIGHTLY DUE TO CHANGES BEFORE LAUNCH TO MEET CHANGING CONDITIONS. WEIGHTS NOT INCLUDED IN ABOVE ARE FROST AND MISCELLANEOUS SMALLER ITEMS.

A page from the Apollo 11 Press Kit with information on its launch vehicle.

114 feet wide, and travelled on four independent double-tracked crawlers, each 'shoe' of which weighed about 1 ton. The access road was comparable in width to an 8-lane highway. It comprised three layers, averaging a total depth of 7 feet. The base was a 2-foot-6-inch-thick layer of hydraulic fill. Next was a 3-foot-thick layer of crushed rock. This was sealed by asphalt. On top was an 8-inch layer of river rock to reduce friction during steering. The vehicle was operated jointly by drivers in cabs located on opposite diagonals, who communicated by intercom. On 20 May the Apollo 11 space vehicle was driven to Pad A, the southernmost of the two launch sites of Launch Complex 39. Because the concrete pad was built above ground level to accommodate a 43-foot-tall flame deflector in the flame trench, the transporter had to climb a 5 per cent gradient while tilting the platform such that the tip of the launch escape system tower did not diverge more than 1 foot from the vertical alignment. Once in position, hydraulic jacks lowered the platform to emplace it on six 22-foot-high steel pedestals on the pad. In all, the 'roll out' lasted 6 hours. In its final orientation, the umbilical tower stood towards the north, with the axis of the central trench aligned north and south. After the transporter had withdrawn, the flame deflector was rolled in beneath the hole in the platform. On 22 May, the transporter collected the Mobile Service Structure from its parking place alongside the access road, and delivered it to the pad. The flight readiness test was completed on 6 June. The countdown demonstration test started on 27 June; the 'wet' phase was completed on 2 July, and the 'dry' phase on 3 July. As Kurt H. Debus, Director of the Kennedy Space Center, once said in jest, "When the weight of the paperwork equals the weight of the stack, it is time to launch!"

SMALL DETAILS

When NASA began to launch pairs of spacecraft during a single Apollo mission, it became necessary to introduce individual call signs while the vehicles were being operated independently. On seeing their CM arrive at the Cape tightly wrapped in a blue sheet, like a sweet, the Apollo 9 crew decided to name the CSM 'Gumdrop', and the LM was named 'Spider' for its arachnid appearance. In March 1969, after the Apollo 10 crew decided to name their vehicles 'Charlie Brown' and 'Snoopy' – characters in Charles L. Schulz's comic strip *Peanuts* – Julian Scheer, Assistant Administrator for Public Affairs, wrote to George M. Low, Manager of the Apollo Spacecraft Program Office in Houston, to suggest that the next mission, which was to try to land on the Moon, should use more dignified names. The Apollo 11 crew, of course, were fully aware of the historical significance of their mission. As Michael Collins recalled:[11]

[11] Based on accounts in *Carrying the Fire: An Astronaut's Journeys*, by Michael Collins, W.H. Allen, p. 332, 1975, and *'All we did was fly to the Moon': Astronaut Insignias and Call Signs*, by Richard L. Lattimer, The Whispering Eagle Press, Florida, p. 66, 1985.

"We had a variety of non-technical chores, such as thinking up names for our spacecraft and designing a mission emblem. We felt Apollo 11 was no ordinary flight, and we wanted no ordinary design, yet we were not professional designers. NASA offered to help us along these lines – wisely, I think. On Gemini 10, which [I flew with John Young, and] in my view has the best-looking insignia of the Gemini series, artistic Barbara Young had developed one of John's ideas and come up with a graceful design, an aerodynamic 'X' devoid of names and machines. This was the approach we wanted to take on Apollo 11. We wanted to keep our three names off it, because we wanted the design to be representative of everyone who had worked toward the lunar landing – and there were thousands who could take a proprietary interest in it, yet who would never see their names woven into the fabric of a patch. Further, we wanted the design to be symbolic rather than explicit. On Apollo 7, Wally Schirra's patch showed the Earth and an orbiting CSM trailing fire. On Apollo 9, Jim McDivitt produced a Saturn V, a CSM, and a LM. Apollo 10's was even busier! Apollo 8's was closer to our way of thinking, showing a figure of eight looping around Earth and Moon, on a command-module-shaped patch, but it had, like all the rest, three names printed on it. We needed something simpler, yet something which unmistakably indicated a peaceful lunar landing by the United States. Jim Lovell, Neil's backup, introduced an American eagle into the conversation. Of course! What better symbol – eagles landed, didn't they? At home I skimmed through my library and finally found what I wanted in a National Geographic book on birds: a bald eagle, landing gear extended, wings partially folded, coming in for a landing.[12] I traced it on a piece of tissue paper, and sketched in an oblique view of a pockmarked lunar surface. Thus the Apollo 11 patch was born – although it had a long way to go before final approval. I added a small Earth in the background and drew the sunshine coming from the wrong direction, so that to this day our official insignia shows the Earth [incorrectly oriented] over the lunar horizon. I pencilled 'APOLLO' around the top of my circular design and 'ELEVEN' around the bottom. Neil didn't like the 'ELEVEN' because it wouldn't be understandable to foreigners, so after trying 'XI' and '11', we settled on the latter, and put 'APOLLO 11' around the top. One day, outside the simulator, I was describing my efforts to Jim Lovell, and he and I both agreed that the eagle alone really didn't convey the entire message we wanted. The Americans were about to land, but so what? Thomas L. Wilson, our computer expert and simulator instructor, overheard us and said to add an olive branch as a symbol of our peaceful expedition.

[12] The eagle that attracted Collins's interest appeared on p. 236 of the book, *Water, Prey, and Game Birds of North America*, published by the National Geographic Society in 1965. In fact, the plate in the book was a mirror image of the original painting by Walter Alios Weber, which was published in the July 1950 issue of *National Geographic Magazine*. The eagle on the mission patch matches the orientation in the original.

> Beautiful! Where do eagles carry olive branches? In their beaks, naturally. So I sketched one in, and after a few discussions with Neil and Buzz over colour schemes, we were ready to go to press. The sky would be black, not blue, but absolute black, as in the real case. The eagle would be eagle-coloured, the Moon Moon-coloured, as described by Apollo 8, and the Earth also. So all we had left to play with, really, were the colours of the border and the lettering. We picked blue and gold, and then Stan Jacobsen in Houston assigned James R. Cooper, an illustrator at MSC, to do the artwork for us. We photographed the finished design and sent a copy through channels to Washington for approval. Washington usually rubber-stamped everything. Only this time they didn't, and our design came back disapproved. The reason? The eagle's landing gear – powerful talons extended stiffly below him – was unacceptable. It was too hostile, too warlike; it made the eagle appear to be swooping down on the Moon in a very menacing fashion – according to Bob Gilruth [Director of the Manned Spacecraft Center]. What to do? A gear-up approach was unthinkable. Perhaps the talons could be relaxed and softened a bit? Then someone had a brainstorm: just transfer the olive branch from beak to claw, and the menace disappeared. The eagle looked slightly uncomfortable clutching his branches tightly with both feet, but we resubmitted it anyway, and it greased on through channels and won final approval."

As regards the call signs, when it became apparent that Apollo 11 would be *the* mission, the crew began to receive suggestions for naming their spacecraft, some of which comprised pairs, others not. Names from mythology were dismissed for the simple reason that investigation invariably turned up something inappropriate. Romantic name pairings such as 'Romeo' and 'Juliet' were also rejected. 'Castor' and 'Pollux' were appealing, but were too suggestive of the Gemini program. Pat Collins argued for 'Owl' and 'Pussycat'. An important factor was that the names selected should have clarity in radio transmission. For Scott Carpenter's Mercury flight, his wife, Rene, had suggested 'Rampart', after the mountain range of his native Colorado, but he chose 'Aurora', which, lacking hard consonants, proved indistinct on the radio. It was decided that while the names must reflect American pride in the mission, they must do so with subtlety. To paraphrase Collins's account:

> "The choice of an eagle as a motif for the landing led swiftly to naming the landing craft Eagle. One day, I was chatting long-distance with Julian Scheer, Assistant Administrator for Public Affairs in Washington, who suggested the name Columbia for our CSM. It sounded a bit pompous to me, but it had a lot going for it – the close similarity of Jules Verne's mythical moon-ship cannon, the Columbiad, and the close relationship between the word 'Columbia' and our national origins: Columbia had almost become the name of our country. Finally, the lyrics 'Columbia, the Gem of the Ocean' kept popping into my mind and they argued well for the recovery of the spacecraft, which hopefully would float on the ocean. Since Neil and Buzz had no objections, and since I couldn't come up with anything better, Columbia it was."

The 'Apollo 11' call sign would be used until such time it became necessary to discriminate, whereupon the two vehicles would employ their own names. Prior to the mission, Armstrong and Aldrin had given some thought to whether they should continue to refer to themselves by the call sign 'Eagle' while on the lunar surface, or introduce some other name. As Aldrin recalls:[13]

> "It would be somewhat similar to a radio call sign, but we wanted to give it added significance. Moon One? Base Camp? Moon Base? When we made our choice, we told only Charlie Duke, who would be our Capsule Communicator back in Houston, who we felt should know the exact name in case transmission was garbled. I cannot remember which of us originated the selection, but once we had thought it over it was an obvious choice. We were landing in an area known as the Sea of Tranquility, and would call our landing site Tranquility Base."

Approval of the call signs was not forthcoming from headquarters until the beginning of July.

PRESS CONFERENCE

Although the astronauts were at home for the holiday weekend, Saturday, 5 July, was devoted to the media. It started with a press conference in the auditorium at the Manned Spacecraft Center. As the astronauts were in their 21-day prelaunch flight crew health stabilisation program, workmen had spent two days assembling a three-sided roofed-over box with a 12-foot-square base utilising 10-foot-tall transparent panels and fitted with fans to blow air outwards; smoke tests having been made to verify this forced ventilation. After Brian Duff, the Public Affairs Officer, had explained to the members of the press – many of whom represented foreign media – the requirement for the special precautions, the astronauts made their entrance wearing rubber masks. At that point, some of the local press, who had been alerted and had purchased surgical masks, donned up to poke fun. Once in the isolation area, the astronauts removed their masks and sat behind a large desk adorned with NASA's 'meatball' insignia and, now being revealed, the mission patch. A large Stars and Stripes formed the backdrop.

As mission commander, Armstrong spoke first. He reminded everybody that Apollo 11 would not have been possible without the achievements of the previous crews and of all the ground staff who assisted. Then Collins talked about how he would look after the CSM while his colleagues were on the lunar surface. Aldrin described how the descent would be conducted. There were then press questions, most of which were either directed at, or picked up by, Armstrong – although in some cases after he had said what he intended to say he invited one or other of his

[13] *Return to Earth*, by Buzz Aldrin with Wayne Warga. Bantam Press, p. 213, 1974.

On 5 July 1969 the Apollo 11 crew give their final preflight press conference at the Manned Spacecraft Center.

colleagues to continue the theme. After it was revealed that the radio call signs for the CSM and LM would be 'Columbia' and 'Eagle' respectively, Armstrong was asked whether he knew what he would say on stepping onto the lunar surface, and he replied that he had not yet decided. He did say that they intended to introduce an unofficial name for the landing site, but did not announce what this would be. When a foreign reporter asked about the plan to raise the Stars and Stripes on the lunar surface, he explained that Congress had directed that this be done. After the plaque that was to be affixed to the leg of the LM had been revealed, he pointed out, as the wording on the plaque proclaimed, that the landing was to be done for all mankind – the United States was not making a territorial claim on the Moon. Asked about the purpose of the mission, he explained that the primary objective was to demonstrate that it was possible to fly to the Moon, land, lift off and return safely to Earth – as President Kennedy had directed. When asked what would happen if the LM became stranded on the surface, he said that they would have supplies for a day or so, after which Collins would have no option but to return home alone. Asked what would be the most dangerous part of the flight, Collins replied, truthfully – though some took his remark to be flippant – that this would be the part they had overlooked in their preparations. As he would later explain, "In a test pilot's world, boring is good because it means that you have not been surprised, that your planning has been precise, and your expectations matched; conversely, excitement means surprise, and that is generally bad." Armstrong pointed out that as a result of Apollo 10 in particular, theirs would not be a mission into the unknown; only the act of landing would be new.

However, the press did not want to know about the technological challenge, they wanted to know how the astronauts *felt* about the mission.

Although, as George Low had surmised, the public expected its heroes to be cast from the same mould as Charles Lindbergh, attempts to coax the astronauts into a discussion of the philosophical implications of the mission were fruitless. Armstrong, Collins and Aldrin, all of whom were 38 years old, were less voluble in temperament than previous crews. Armstrong was one of the most taciturn of the astronauts, and shunned publicity. Aldrin was only marginally less reserved. Only Collins opened up to the reporters, but he was not to attempt the landing. As he would later point out, the task of a test pilot was to remember every aspect of a machine's behaviour, and hence he was trained to suppress his emotions lest these should interfere with a cold dispassionate analysis. If NASA wished its astronauts to emote (as the vernacular had it) it should form a crew comprising a philosopher, a priest and a poet; not three test pilots. On the other hand, such a crew would be unlikely to return to give a press conference because, having emoted all the way out and back, they would probably neglect to insert the circuit breaker to enable the parachutes to deploy. Nevertheless, Armstrong was not without humour. Asked what he would most like to take with him to the Moon, he replied, "More fuel."

The main conference was followed by another for the wire services, one for the magazines and filmed interviews with each television network – it was a long day. As for biological isolation, the crew were directed to leave, without masks, by the corridor through which the world's press had departed! They then went home to spend the rest of the weekend with their families. Furthermore, at the Cape they were in routine contact with secretarial staff, caretakers, suit technicians and simulator engineers. Collins would later compare flight surgeons to nervous old ladies who were convinced that their houses were haunted.

READY TO GO

On flying back to the Cape on Monday, 7 July, the astronauts returned to the semi-isolation of their quarters in the Manned Spacecraft Operations Building, going out only to use the simulators in a nearby building. On 10 July, having been medically checked, Tom Paine had a private dinner with the astronauts at which he implored of them, "If you get into trouble up there, do not hesitate to abort. Come on home. Don't get killed. If you do have to abort, I promise this crew will be slipped ahead in the mission sequence. You'll get another chance. Just don't get killed." Collins reflected that Paine's motivation was to eliminate "the obvious risk of letting our desire to be first on the Moon cloud our judgement in analysing the hazards". In fact, Paine had said the same to the crews of Apollo 8 and Apollo 10. Earlier that day, after tests had indicated an oxygen leak in the first stage of the launch vehicle, Walter Delle, a Boeing quality inspector, entered the tank and tracked the 'hiss' to the helium pressurant manifold. As this was such a delicate item, it was debatable which would be the least risky option: to accept the leak, or attempt to eliminate it. It was decided to try to stem the leak by applying torque to a nut using a wrench. If the manifold were to be damaged in the process, replacing it would take four days, which would require the launch to be postponed. But Delle was successful, and the final review cleared the mission to aim for launch on 16 July.

The week before launch, Charles Berry mentioned to a reporter that President Nixon had asked to have dinner with the crew on the night before launch, as Vice President Spiro T. Agnew had done with the Apollo 10 crew. Berry had stated that Nixon's presence at such a late stage would prejudice the crew health stabilisation program, since if the Apollo 11 crew were to return with an infection it would be essential to know whether this had been contracted prior to their leaving Earth in order to enable it to be dismissed as a potential lunar infection. In fact, considering that the launch of Apollo 9 had had to be postponed several days to allow its crew to recover from a mild upper-respiratory infection, it was remarkable that Agnew had been permitted to visit the Apollo 10 crew, since they would be in deep space by the time any symptoms that would have given rise to a postponement became manifest. However, NASA headquarters took the view that Berry's opinion was merely his recommendation; it was not for him to decide whether anybody could visit the crew. Frank Borman, assigned as space adviser to Nixon for Apollo 11, said the dinner should go ahead, since it would be a tremendous boost to crew morale. However, because the matter was now in the public domain, Nixon deferred. If an astronaut were to fall sick in space, Nixon would be open to the damning criticism of callously disregarding the professional advice of the chief flight surgeon. Although the press habitually referred to Berry as the astronauts' personal physician, he was Director of

Neil Armstrong and Buzz Aldrin reviewing procedures in a LM simulator on 11 July 1969.

Medical Research and Operations at the Manned Spacecraft Center. The final comprehensive medical examination of the astronauts was on Friday, 11 July by the physicians assigned to this mission: Al Harter, Jack Teegan and Bill Carpentier. The aim was to evaluate their biological state of heath by comparing the organisms in their systems with the 'baseline' established on 26 June. In the early days, nurse Dee O'Hara had taken blood and urine samples of departing astronauts, but as these tasks were now done by technicians she was responsible for the paperwork – which amounted to 18 pages per man. As a result of this examination, the crew was declared fit to fly. All that remained, medically speaking, was the basic check-up on the morning of launch.

Over the weekend, the pace slackened. They continued to use the simulators, but for proficiency rather than for training, and undertook a final review of the flight plan. Ted Guillory had supervised the writing of the 240-page flight plan for Apollo 11. It weighed 2 pounds, and addressed every aspect of the mission for a nominal duration of 195 hours 40 minutes. Two copies would be needed, one for the CSM and the other for the LM. In addition, there would be some 20 pounds of reference material on board. The general consensus was that the astronauts had reached their 'peak' right on time. As Aldrin observed: "We could spend another year trying to isolate [the open issues] one by one, and we'd never really get them all. We could spend too much time doing that, so much that we could forget what the mission is all about." On 16 May the first draft of the mission rules was issued. Updated weekly, by the time of launch it had expanded into a 350-page book that defined the actions to be taken in the event of a multitude of situations arising in flight. As such, it formalised the collective knowledge of all concerned.

Soon after Apollo 8 set off for the Moon Frank Borman had suffered a bout of 'space sickness', which came as a considerable surprise in view of the fact that he had he spent 14 days on board Gemini 7 in December 1965 with no ill effects. Rusty Schweickart suffered similarly on Apollo 9. It seemed that in the confined Gemini spacecraft astronauts had not been able to become disoriented, but by being able to move around in the much larger Apollo cabin they could develop motion sickness. In an effort to prime the vestibular mechanism of his inner ear for weightlessness, Collins drove down the coast to Patrick Air Force Base and flew aerobatically in a T-38 for an hour each day over the final weekend. Aldrin flew zero-gravity arcs in the KC-135 several days later, in the hope of doing the same.

The hot news on Sunday, 13 July, was the announcement by the Soviet Union that it had launched the unmanned spacecraft, Luna 15. The speculation was that this would land on the Moon, scoop a sample, and return this to Earth. There was some concern in NASA that the spacecraft's transmissions might interfere with Apollo 11. Since his Apollo 8 flight, Frank Borman had spent much of his time in Washington as Nixon's space adviser and undertaking goodwill tours. Three days earlier, he had returned from a 10-day trip to the Soviet Union, and was in fact, the first astronaut to visit that country. On a Washington stopover that afternoon he received a message from Chris Kraft in Houston requesting that he use his recent contacts to gain information on the Luna 15 mission. Borman called the office of Mstislav V. Keldysh, the 68-year-old leader of the Soviet Academy of Sciences.

Mike Collins and Deke Slayton at Patrick Air Force Base on 12 July 1969 after a flight in a T-38 aircraft.

On 13 July 1969 the Apollo 11 crew gives a final news conference by CCTV from the Manned Spacecraft Operations Building.

However, as it was 2 am in Moscow, Borman left a message and then returned to Houston. At 6 am the following morning he received a phone call from Keldysh's office reporting that the information was *en route*, and several hours later identical telegrams arrived at his home and at the White House specifying the orbit intended for Luna 15 and confirming that there would be no radio interference. In view of the rivalry between the space-faring nations, this was welcome cooperation. Even if Luna 15 returned only a few ounces of lunar material, the Soviets would be able to claim that they had 'beaten' America as long it returned ahead of Apollo 11. The stakes were therefore high. Wernher von Braun told a reporter that although the Soviets had lost the 'race' to send a man, their attempt "to soft land a spacecraft on the Moon and scoop up a sample of lunar soil and fly it back to Earth" represented a tremendous technical challenge, in which he wished them "full success" even though it would "take a little bit off our program". Then he made the telling point that such a mission could be no impromptu effort, because "to have the hardware ready, it would have had to have been planned ... years earlier".

In July, the afternoon temperature at the Cape regularly rose to 100°F, and the combination of heat and ocean humidity made conditions almost unbearable. The Sun blazed down day after day. The concrete was so bright that sunglasses were mandatory. But on Sunday, 13 July, it started to rain and the forecast was for several days of poor weather. Thus, even if everything else was ready, the weather might oblige the launch to be postponed, possibly to August.

On Monday, 14 July, Robert Gilruth, George Low, Deke Slayton, Chris Kraft and Max Faget flew to the Cape for the final flight readiness review, which, as expected, confirmed the launch date. At 5 pm the terminal countdown was picked up with the clock at T–28 hours, with two built-in 'holds' (the first to start at T–9 hours and last for 11 hours, and the second to start at T–3 hours 30 minutes and last for 1 hour 32 minutes) with a view to launching as the window opened for the primary landing site, which was 9.32 am local time on Wednesday, 16 July.

4

Setting off

AN EARLY START

Although Armstrong and Aldrin ran some LM simulations on Tuesday, 15 July, they spent the remainder of the day relaxing in the Manned Spacecraft Operations Building. In the evening, Lew Hartzell served a dinner of broiled sirloin steak and buttered asparagus for the crew, their backups, the members of their support crew, and Deke Slayton. The three astronauts then chatted with their wives by telephone, and retired at 10 pm. After clearing away the dinner, Hartzell went to the camper he kept in the nearby parking lot, but as it was too hot to sleep there he slept in a spare bedroom in the crew quarters, awakening at 2.30 am to prepare breakfast.

Guenter F. Wendt's job title was Pad Leader, but John Glenn had dubbed him *der pad fuehrer* on account of his Teutonic accent being as thick as the lenses of his spectacles. Although from Germany, he was not one of Wernher von Braun's rocket team; he had flown night-fighters for the Luftwaffe, as an engineer. After the war he emigrated to the USA, got citizenship, and joined McDonnell Aircraft. When the company won the contract to build the Mercury spacecraft, Wendt was given the task of ensuring that the spacecraft was ready for launch – supervising it from the moment that it arrived at the Cape, to the sealing of its hatch. When the company produced the Gemini spacecraft he continued at the Cape. However, the contract for the Apollo spacecraft was given to North American Aviation, which appointed its own pad crew.[1] After the loss of the Apollo 1 crew in a fire on the pad, Wally Schirra insisted that Wendt be rehired. Before Glenn's flight Wendt had told his wife Annie that while he could not guarantee her husband's safe return, he could promise that every effort would be made to ensure that the spacecraft was up to the job. This had remained his objective. Having spent most of 15 July methodically checking and rechecking, he went home at 6 pm, dozed until midnight, then rounded up three of the members of his team: NASA quality inspector 'Lucky' Chambers, North

[1] In 1967 North American Aviation merged with the Rockwell Standard Corporation, as North American Rockwell; in 1973 this became Rockwell International.

American Rockwell mechanical technician John Grissinger, and backup crew member Fred Haise.

Meanwhile, at 11 pm the chill-down process had begun, preparatory to loading cryogenic propellants into the launch vehicle. During the night, a communications issue on the ground delayed pumping liquid hydrogen into the S-II by 25 minutes, but this time was recovered during the scheduled hold at T–3 hours 30 minutes. A high-pressure cell over the ocean off North Carolina combined with a weak trough over the northeastern Gulf of Mexico to draw light southerly surface winds across the Cape, increasing humidity. The sky was heavily overcast and there was light rain, with occasional flashes of lightning off to the north. Nevertheless, the forecast was optimistic.

A full week before launch, people began to gather at the Cape communities of Titusville, Cocoa Beach, Satellite Beach and Melbourne. With four days to go, the Florida authorities were expecting 35,000 cars, 2,000 private aircraft and a flotilla of boats to converge on the Cape. People were drawn from all around the world to witness the launch and be able to tell their grandchildren that they had been present when men set off to make the first lunar landing. Jay Marks, a Houston car dealer and casual acquaintance of the Armstrong family, had arrived a week early, lived in his camper van, and spent the week fishing. He was not alone. As Marks put it, "Apollo 11 gave a lot of nice people a chance to get acquainted." By 15 July there was not a vacant room to hire. Hotels and motels allowed late-comers to set up their camp beds in lounges and lobbies, but most people spent the night on the beaches and roadsides, where vehicles were parked nose to tail for a 30-mile radius. Since it was to be a dawn launch, the countdown parties ran through the night. At one of the parties Wernher von Braun and his wife Maria met Hermann Oberth who, at 75 years of age, was the only one of the three pioneers of rocketry still alive to witness the great dream become reality. Konstantin Tsiolkovski had died in 1935 and Robert Goddard in 1945.

"It's a beautiful morning," said Slayton as he awakened Armstrong, Aldrin and Collins at 4.15 am local time on Wednesday, 16 July. The weather was clearing, as predicted. Once the astronauts had showered and shaved, they went to the exercise room where Dee O'Hara, wearing a crisp white uniform, short dark hair and vivid lipstick, gave them their final check-up. At 5 am they sat down for breakfast with Slayton and Bill Anders, a member of the backup crew, eating the traditional low-residue fayre of orange juice, toast, scrambled eggs and steak. In fact, Armstrong had earlier confided to his wife, "I'm sick of steak!" NASA artist Paul Calle sat in the corner of the room, unobtrusively sketching. After packing their possessions to be sent home, they made their way to the suit room.

Hamilton Standard of Windsor Locks, Connecticut, was prime contractor for the space suit, or pressure garment assembly. Earth's atmosphere has a sea-level pressure of about 15 psi and a gas mix of roughly 80 per cent nitrogen and 20 per cent oxygen. The International Latex Corporation of Dover, Delaware, was subcontracted to make an airtight bladder to hold pure oxygen at a differential pressure of 3.7 psi. Although contoured to the human shape, the extremely flexible material of the bladder would tend to 'balloon' when pressurised. It was therefore restrained by a

Campers settled on open ground around the Kennedy Space Center to wait for the launch of Apollo 11.

With accommodation fully booked, late arrivals sleep on the beach to be in place to watch the early morning launch.

The Apollo 11 crew, Bill Anders and Deke Slayton enjoy the traditional launch day breakfast of steak and eggs.

complex system of bellows, stiff fabric, inflexible tubes and sliding cables which, while maintaining the shape of the suit, impaired the mobility of the occupant. The design of the knee and elbow joints was simple, since these work in the manner of a hinge, but because the shoulder joint can rotate in several axes this was a greater challenge, so much so that at one point NASA briefly considered reassigning the contract. The bladder incorporated a network of ventilation tubes to cool the occupant and preclude the build-up of moisture. Two versions of the suit were required: one for use inside the spacecraft as protection against loss of cabin pressure, and the other to provide thermal and micrometeoroid protection plus other systems required when operating on the lunar surface.

The space suits varied in certain respects:

- Both suits shared a nomex inner layer, a neoprene-coated nylon pressure bladder, and a nylon restraint layer.
- The outer layers of the intravehicular suit comprised nomex and a double layer of teflon-coated beta cloth.
- The integral thermal and micrometeoroid protection for the extravehicular suit had a double-layer liner of neoprene-coated nylon, a number of layers of beta-kapton laminate and a teflon-coated beta cloth surface.
- The intravehicular suit had one pair of umbilical connectors installed on the chest to circulate oxygen from the cabin system.
- The extravehicular suit had two pairs of such connectors, one pair as on the intravehicular suit, and the other pair for use with the portable life-support system.
- The extravehicular suit also had a coolant water loop.
- Both suits had a connector for electrical power and communications.

The boots were part of the bladder, but the helmet and gloves used aluminium locking rings to maintain the integrity of the bladder. The helmet was a transparent polycarbonate 'bubble', with adequate air flow to prevent a build-up of carbon dioxide. The gloves were required to support a natural range of bending and rotating motions of the wrist, with a finger-covering material that was sufficiently thin and flexible to allow the manipulation of switches. Each astronaut had three individually tailored suits – a training suit for use in simulators and the low-gravity KC-135 aircraft, during which it was likely to suffer wear and tear, and two flight suits (one prime, the other backup) which, after integrity tests, were reserved for countdown demonstrations and the actual mission. Each suit had a US flag on the left shoulder, a NASA 'meatball' on the right breast and the mission patch on the left breast.[2] As his astronaut specialism, Collins had liaised between the crew systems division and the industrial teams to ensure that the suits were both fit for function and safe to use.

[2] The Stars and Stripes shoulder patch was introduced by Jim McDivitt and Ed White after being prohibited from naming their Gemini 4 spacecraft 'American Eagle'. In addition to retaining the flag, for their Gemini 5 flight Gordon Cooper and Pete Conrad introduced a mission patch. Both became standard adornments.

Joe Schmitt assists Mike Collins during suiting up.

Having already donned his 'Snoopy hat', Neil Armstrong lifts his 'bubble' helmet.

Neil Armstrong waves as he approaches the television crews outside the Manned Spaceflight Operations Building.

Carrying their ventilators, the Apollo 11 crew enter the transfer van for the drive to the pad.

Joseph W. Schmitt led a four-man team. He had supervised the suiting up of every American astronaut since Al Shepard in 1961. After the countdown demonstration test on 3 July, the three primary suits had been stripped, inspected for wear and tear, cleaned and reassembled – a four-day task. On arriving at 3.30 am, Schmitt had supervised the unbagging and inspection of the suits, and the astronauts arrived for simultaneous suiting at 5.30 am. This laborious process started with each man rubbing his posterior with salve prior to donning a diaper that would contain both fecal matter and associated odours. This was a precaution against a loss of pressure in the cabin when retrieving the LM from the final stage of the launch vehicle after translunar injection, in which event the astronauts might require to spend several days in their suits. Next was a prophylactic-style urine collector, with a collection bag worn around the waist. A connector on the thigh of the suit enabled the bag to be emptied while the astronaut was suited. Biosensors were attached to the chest, and linked to a signal-conditioning electronics pack that supplied telemetry through the electrical umbilical. After donning cotton long-johns, which NASA referred to as a constant-wear garment, each man was assisted into his one-piece pressure suit. Armstrong and Aldrin were to wear the 55-pound extravehicular suit, and Collins the lightweight 35-pound model for internal use. In the suiting-up procedure, the astronaut sat on a reclining couch, inserted his legs into the suit's open rear, inserted his arms, bent forward and eased his head through the rigid metal neck ring. He then had to stand and shuffle until the suit felt comfortable, whereupon a technician would seal the bladder and zipper. The next item was the brown-and-white soft communications carrier, dubbed a 'Snoopy hat', with its integrated earphones and microphones. Once the gloves were fastened to the wrist rings and the helmet was in place, the oxygen umbilicals were attached to the sockets on one or other side of the chest and the suit was pumped to above-ambient pressure in order to verify the integrity of the bladder, helmet and gloves. There was a pressure gauge on the right arm of the suit. The Omega watches on the suit wrists were set to Houston time, one hour behind the Cape. They would breathe pure oxygen at sea-level pressure to purge nitrogen from their blood stream, and thereby preclude 'the bends' when the pressure was reduced during the ascent to orbit. With the suit sealed, communication was by umbilical intercom.

At 6.20 am, after the astronauts had donned yellow rubber galoshes for the trip to the pad, suit technician Ron C. Woods led the procession from the suit room, with Schmitt bringing up the rear. At Guenter Wendt's request, Schmitt had put a sign on the corridor wall saying 'The Key To The Moon Is Located', the meaning of which was, as yet, obscure. As the astronauts made their way down to ground level, with each man carrying his ventilator like a suitcase, the corridors were lined with old friends and coworkers, but their good wishes were almost inaudible over the hiss of the oxygen circulation. Collins, by arrangement, was handed a brown paper shopping bag containing a surprise for Wendt. On emerging from the Manned Spacecraft Operations Building they waved at the television crews supervised by Charles Buckley, the head of security. Parked by the door were the two white transfer vans – one prime and the other a backup. Slayton checked the astronauts into the van, which had a large mission patch adorning its rear access door, wished

The White Room on Swing Arm 9 provided access to the spacecraft.

Neil Armstrong leads Michael Collins along Swing Arm 9.

them good luck, and then set off for Firing Room 1 of the Launch Control Center beside the Vehicle Assembly Building, where the 463 members of Rocco A. Petrone's launch team monitored consoles showing the status of the space vehicle, comprising the launch vehicle and the spacecraft. Schmitt and Woods joined the crew, and the two vans departed in a convoy, driving north to the Vehicle Assembly Building then swinging east over the Banana River causeway to Pad A of Launch Complex 39, a total distance of just over 8 miles. On the way, Armstrong had Schmitt extract a small card from his pocket and push it beneath his watchband. Just before he unplugged from the communications circuit, Schmitt wished the three astronauts "a real good flight", to which Aldrin replied, "You take yourself on a good vacation when you get us all off." As they arrived at the pad at 6.37, sunrise was imminent. The elevator of Mobile Launch Platform 1 was waiting. Once on the upper deck, as they crossed to the high-speed elevator of the Launch Umbilical Tower, Collins observed that on previous visits the site had been a hive of activity, but now it was utterly deserted.

On exiting the elevator at the 320-foot level, the astronauts were met by Wendt, wearing a white smock and cap. As they were not yet on intercom, he greeted each man with a pat on the shoulder. Because the White Room that provided access to the command module was so cramped, Aldrin remained on the tower while Wendt led Armstrong, Collins, Schmitt and Woods across Swing Arm 9. Wendt then handed to Armstrong the promised 'Key To The Moon'. Its shaft was a crescent Moon about 4 feet long made of styrofoam and covered by aluminium foil, with an oval loop on one end and a set of teeth on the other. Armstrong withdrew the card from his watchband and presented it to Wendt.[3] The card read: 'Space Taxi. Good Between Any Two Planets.' At 6.53 am Armstrong shed the galoshes that had protected the boots of his suit, stood in front of the hatch, which was set at floor level, grasped with both hands a bar that was located inside the cabin, inserted both of his feet, and slipped onto the centre couch. Haise, who had spent 90 minutes running through a 400-item checklist, setting switches and making checks, was already in the lower equipment bay to assist him to shuffle onto the left couch. For launch, the couch was adjusted to elevate the lower legs, and once in space it would be set flat. Schmitt entered to switch Armstrong's oxygen from the portable ventilator to the cabin's system and to plug in the communications umbilical. Armstrong checked in with Clarence 'Skip' Chauvin, the Spacecraft Test Conductor in the Firing Room. Jim Lovell, Armstrong's backup, came on the line. The previous evening Lovell had promised that if Armstrong did not feel up to the flight, he was ready to take his place; Lovell repeated his offer, but Armstrong assured Lovell that he was feeling just fine.

Meanwhile, because Wendt claimed to have caught an implausibly large trout, Collins had purchased the smallest trout available – just 7 inches in length – frozen it,

[3] As a treat, in his personal preference kit Armstrong had an opal that Wendt had supplied, which, upon its return to Earth, Wendt intended to give to his wife Herma.

and nailed it onto a wooden plaque with the inscriptions 'Guenter Wendt' and 'Trophy Trout'. It was in the brown paper bag. During the walk out to the van, Collins had dreaded dropping the bag in view of the television cameras, causing the world to wonder why a man bound for the Moon was carrying a dead fish. He presented it to Wendt, then entered the spacecraft.[4] As Aldrin had been CMP when backing up Apollo 8, and was familiar with the centre crewman's tasks during launch, it had been decided that he should retain this position, which placed Collins on the right.

Alongside the elevator, Aldrin enjoyed 15 minutes of solitude. He admired the view of sunrise and surf to the east, the cars and boats in the distance on the roads and rivers, and the monolith of the Vehicle Assembly Building to the west. Far to the south was 'Missile Row', with Pad 5 from which Al Shepard rode a Redstone on a suborbital flight in 1961; Pad 14 from which John Glenn rode an Atlas into orbit in 1962; Pad 19 from which Aldrin and his colleagues rode Titan II missiles on their Gemini missions in 1966; and Pad 34, where the Apollo 1 crew had been consumed by fire in 1967. After Schmitt escorted Aldrin across the access arm, Aldrin presented his fellow Presbyterian with a condensed version of the Bible entitled *Good News For Modern Man*, inscribed inside: 'On permanent loan to G. Wendt'.

At 7.22 am, having confirmed that there were no extraneous items in the cabin, Haise departed. As the couches were so closely spaced that the astronauts' elbows touched, he wriggled under the centre couch to reach the hatch. He could hear the crew on the intercom but could not speak to them to wish them luck, so once he was outside he leaned in and shook each man's hand. When Chauvin gave the go-ahead to close the hatch, Wendt tapped Aldrin's helmet and stepped aside; Grissinger then swung the big hatch closed and locked it. Once the hermetic integrity of the seal had been verified, Grissinger added that section of the boost-protective cover. At 7.52 am Wendt's team descended to ground level and drove to a nearby site in case their services should be required. Meanwhile, Swing Arm 9 with the White Room was rotated 5 feet from the spacecraft, ready to be either restored in an emergency, or swung completely clear just prior to launch.

In the spacecraft, the astronauts verified the switch settings to ensure that none had been disturbed, either by themselves ingressing or by Schmitt or Haise moving around in the capsule. Meanwhile, the cabin was purged. Following the loss of the Apollo 1 crew in a capsule fire, the practice of pressurising the cabin with oxygen for launch had been discontinued. The suited crew remained on pure oxygen, but the atmosphere in the cabin was replaced by 40 per cent nitrogen and 60 per cent oxygen. On being informed that elements of the count were 15 minutes ahead of time, Armstrong pointed out that he wanted them to wait for the launch window to open before starting the engines.

On Tuesday, 15 July, Reverend Ralph Abernathy, successor to the late Martin

[4] Wendt kept the trout in his deep freeze until having it remounted in a more conventional way.

Former President Lyndon Baines Johnson and current Vice President Spiro T. Agnew in the VIP stand.

Luther King as head of the Southern Christian Leadership Conference, had led a mule-drawn wagon and a small group of protestors to the Kennedy Space Center to decry "this foolish waste of money that could be used to feed the poor". NASA Administrator Thomas O. Paine had met them. After observing that to cancel the mission would yield no benefit to the poor, Paine had invited a delegation to watch the launch from the official guest area.

Overnight, seeing no one heading away from the Cape, drivers had switched lanes to get closer, generating the worst congestion in Florida's history. Even the residents of Cocoa Beach, to whom launches were routine, were caught up in the excitement. With the notable exception of alarm clocks, which had been sold out by Tuesday afternoon, local shopkeepers were able to supply the needs of the visitors. As dawn approached on 16 July, it was estimated that 1,000,000 people were on the roads, rivers and beaches, where 'Good Luck Apollo 11' had been etched in large letters in the sand. Worldwide, 1,000 times that number were watching the 'live' television coverage.

By the time that the countdown entered its final hour, the rain had stopped, the cloud cover was light cumulus topped by patches of cirrostratus, there was a 6-knot southerly breeze, the temperature was already 85°F, and the humidity was 73 per cent: it was going to be a scorcher of a day.

After seeing the astronauts off, Dee O'Hara went to watch the launch with her friend Lola Morrow, who had been hired by NASA in 1962 as a travel clerk and two years later had taken on the daunting challenge of organising the astronauts' office at the Cape.

Among the thousands of invitees in the VIP stand were Vice President Spiro T. Agnew, four cabinet ministers, 33 senators, 200 congressmen, 19 state governors, 40 city mayors, hundreds of ambassadors,[5] foreign ministers, ministers of science, military attachés, senior NASA employees, and representatives of the companies that built the launch vehicle and spacecraft. Also present were Lyndon B. Johnson and his wife Ladybird, and James E. Webb, NASA's former administrator. The nearby press enclosure contained 3,500 journalists, 812 of whom were drawn from 54 foreign countries, including 12 from Eastern Europe – but none from either the Soviet Union or the People's Republic of China. Each of the American television networks had its own team of commentators and consultants. Walter Cronkite, the CBS anchorman popularly regarded as 'the most-trusted man on television', was acutely aware that Apollo 11 was different from any previous mission. As he later recalled, "We knew darned good and well that this was real history in the making." If it succeeded, "this was the date that was going to be in all the history books" and "everything else that has happened in our time is going to be an asterisk". It became evident that there were three milestones in the space program in terms of press

[5] Britain's ambassador to Washington, John Freeman, having attended the launch of Apollo 10, declined his invitation to Apollo 11 on the basis that – as an embassy spokesman put it – "when you've seen one Apollo launch, you've seen them all".

presence at a launch, with the numbers increasing each time: John Glenn's orbital flight, Apollo 8's impromptu flight to orbit the Moon, and now, with luck, the accomplishment of John F. Kennedy's great challenge.

Meanwhile, in Wapakoneta, almost all of the 7,000 population were watching television. Armstrong had advised his parents not to attend the launch, in order to spare them press attention. Although NASA had dispatched Public Affairs Officer Thomas Andrews to fend them off, reporters were camped outside the house and there was an 80-foot-tall transmission tower in the driveway! On the other hand, on hearing that the house had only a black-and-white television, the networks had delivered a large colour set to enable the family to fully appreciate the coverage of the event.

Jan Armstrong had not attended the Gemini 8 launch because her husband had asked her not to, but for Apollo 11 she had insisted. To enable her to escape press attention, North American Rockwell arranged a corporate jet and moored a motor cruiser on the Banana River, several miles south of the pad. On Tuesday evening, Jan, sons Ricky and Mark, friends Pat Spann and Jeanette Chase, Dave Scott, his wife Lurton, and Dora Jane Hamblin representing *Life* magazine, flew to Patrick Air Force Base and were then driven to a friend's house on South Atlantic Avenue. At midnight, Jan drove to the Kennedy Space Center to look at the floodlit space vehicle from the astronauts' viewing area, 3 miles from the pad, then drove back to her hideaway. At 4 am the group boarded the boat. Listening to the commentary on a transistor radio, Jan hoped the launch would be on time because she was exhausted and needed some sleep.

Joan Aldrin set her alarm for 6 am Houston time, but when it sounded she cancelled it and slept for another 50 minutes. "I wish Buzz was a carpenter, a truck driver, a scientist – anything but what he is," she had confided on discovering that he was to make the first lunar landing. Her plan was to keep busy with housework to take her mind off the mission. Her first intended task had been to raise the US flag in the garden, but on seeing the reporters she left this to someone else. Among her guests was Jeannie Bassett, who once occupied the house beyond the backyard fence. After Charles Bassett's death in 1966, Jeannie had sold the property and taken the children to California, but had returned in order to keep Joan company during what promised to be a nerve-wracking mission. Pat Collins awoke about the same time. It had been a rough night in Nassau Bay, with a thunderstorm felling a tree on her lawn, and she arranged for its removal. When the television reported that the count was going exceptionally smoothly, she felt sure the launch would be on time.

Clifford Charlesworth's flight controller team was to handle the launch phase. When Chris Kraft, the Director of Flight Operations seated on Management Row behind the flight director's console, put a series of needless queries, Charlesworth turned around, smiled, and warned, "Chris, you're making me nervous!"

ASCENT

At T–3 minutes 45 seconds Paul Donnelly, the Launch Operations Manager in the Launch Control Center, wished the Apollo 11 crew, "Good luck, and Godspeed."

By T–2 minutes the 'boil off' of liquid oxygen had ceased, and pressurisation was underway in all three stages of the launch vehicle. With one minute remaining on the clock, Armstrong reported, "It's been a real smooth countdown." Ten seconds later, the launch vehicle went onto full internal power. Because the apparatus was much too complex for the final phase of the preparations to be managed manually, at T–20 seconds an automatic sequencer took over. At T–17 seconds, the guidance system in the Instrument Unit was released. The phased ignition sequence for the five F-1 engines was initiated at T–8.9 seconds, with the vehicle being held down by four clamps. Knowing that no Saturn V had lit its engines and then *not* lifted off, the astronauts turned their heads in their 'bubble' helmets and grinned at each other – they were going to fly! Jack King, the Public Affairs Officer at the Cape, counted down the remaining seconds, "3, 2, 1, 0. All engines running. Liftoff! We have a liftoff at 32 minutes past the hour. Liftoff on Apollo 11."

As the clock ran down through its final minute, Joan Aldrin had sat stiffly in a chair, close to tears, fidgeting nervously with a cigarette, twisting a handkerchief, flexing her hands. She watched in silence as the vehicle lifted off. Although in a room full of children, relatives and neighbours, she managed to capture a sense of solitude. In fact, everyone in the room was silent. In contrast, Pat Collins had been very focused, discussing aspects of the flight plan with Barbara Young, who had been through this on Apollo 10. At liftoff, Pat called out delightedly, "There it goes!"

As the vehicle began to rise from the pad, a plug drawn from its tail started the master event timer in the spacecraft. NASA specified the timing of mission events in terms of Ground Elapsed Time (GET), as measured from 'Range Zero', defined as the last integral second prior to liftoff – in this case 09:32:00 Eastern Daylight Time on 16 July 1969. Armstrong's heart rate was 110 beats per minute, Collins's 99 and Aldrin's only 88, in each case significantly lower than at this point in their Gemini flights. The first 12 seconds of a Saturn V launch were challenging, since the vehicle had to 'side step' away from the Launch Umbilical Tower, just in case a gust of wind pushed it towards the tower or one of the swing arms was tardy in rotating clear. As the vehicle gimballed its four outer engines to make this manoeuvre, it swayed this way and that; the effect being most pronounced at the top. As Collins observed later, "It was, I thought, quite a rough ride in the first 15 seconds or so. I don't mean the engines were rough, and I don't mean it was noisy, but it was very *busy* – that's the best word for it; it was steering like crazy." Once the vehicle had cleared the 400-foot-tall tower, operational control was handed to Houston. The Instrument Unit of the Saturn V now commanded an axial roll in order to align the vehicle with the flight azimuth. Armstrong was to report key events to Houston, and at an elapsed time of T + 13 seconds he called, "We've got a roll program." This was acknowledged by Bruce McCandless, a yet-to-fly astronaut serving as the Capsule Communicator (CapCom). Between T + 13.2 and T + 31.1 seconds, the vehicle rolled from a pad azimuth of 90°E to a flight azimuth of 72°E. Once aligned, the vehicle started to pitch over in order to arc out over the Atlantic on the desired ground track. "Roll is complete," called Armstrong, "and the pitch is programmed."

Apollo 11 lifts off.

A view of the launch of Apollo 11 from the Press site.

The F-1 engines of the 363-foot-long vehicle issue a tremendous exhaust plume.

Having done their job, members of the launch control team watch Apollo 11 depart.

Charles W. Mathews, Wernher von Braun, George E. Mueller and Samuel C. Phillips after the launch of Apollo 11.

For the crowds, the first indication that a launch was in progress was a light at the base of the vehicle. A jet of flame passed through a hole in the Mobile Launch Platform to a wedge-shaped deflector, which split and vented it horizontally north and south. Water had been pumped onto the pad to diminish the acoustic reflection from the concrete, and the water in the pit was vaporised and blasted out with the flame as a roiling white cloud. The space vehicle weighed 6.5 million pounds, 90 per cent of which was propellant. It was almost inconceivable that it could be raised off the ground, but the five F-1 engines, drawing propellants at the combined rate of 15 tons per second, yielded a total of 7.5 million pounds of thrust. As the vehicle slowly rose from the pad, exposing the flame, the intensity of the light rivalled the early morning Sun sufficiently to force observers not wearing sunglasses to squint their eyes. In the press stand, positioned some 3.5 miles from the pad because this was calculated to be as far as an exploding Saturn V could shoot a 100-pound fragment, some of the photographers, their cameras forgotten, simply stood and yelled 'Go!' again and again. No launch since John Glenn's had released such raw emotion in the press. At first it was like watching a silent movie, because the thunderous roar of ignition took 15 seconds to reach the official viewing sites. As the vehicle rose, the roar was overwhelmed by a staccato pop and crackle that was more felt than heard.[6] The ground shook sufficiently to register on remote seismic sensors. To some observers, it was debatable whether the Saturn V was rising, or its great thrust was pushing the Earth aside! Dee O'Hara and Lola Morrow had been joined at the astronauts' viewing site by Beth Williams, wife of C.C. Williams, an astronaut who had been lost in an aircraft accident in 1967. Tears of joy streamed down their faces. On her boat 5 miles away, Jan Armstrong did not have a very good view, but she preferred reality to a television screen showing the narrow view from a long-range camera.

The main screen in front of the Mission Operations Control Room displayed a plot of the trajectory of the Saturn V, which was exactly as programmed. If it were to suffer a guidance failure, Armstrong was ready to steer it himself, and was the first commander to have this facility. The vehicle passed through the region of maximum dynamic pressure at an altitude of 4 nautical miles,[7] while travelling at a speed of 2,195 feet per second. The slowly rising thrust from increasingly efficient engines, and the decreasing mass of the vehicle meant increasing acceleration. By design, the centre engine of the F-1 cluster shut down first to limit the acceleration. Once it had consumed some 4.5 million pounds of propellants, the S-IC was shut down. Its sustained thrust had compressed the vehicle lengthwise, and it snapped back to its true length when this force was suddenly removed, throwing the crew against their

[6] The Saturn V was so much more powerful than its predecessors that the sound of the first launch on 9 November 1967 took everyone by surprise. It not only rattled the tin roof of the VIP bleacher but also threatened to collapse the booth from which Walter Cronkite was providing his television commentary.

[7] NASA preferred to use nautical rather than statute miles for space missions. One nautical mile is 2,000 yards, or 6,000 feet; whereas a statute mile is only 1,760 yards or 5,280 feet.

harnesses; this 'eye-balls out' shock being particularly harsh immediately after the peak 'eye-balls in' load of 4 *g*.[8] As the 138-foot-long spent stage was released by pyrotechnic charges on its upper rim, small solid rockets in the fairings around its tail fired to retard it. Other such rockets on the exterior of the interstage pushed the remainder of the vehicle clear and gave ullage to settle the S-II's propellants prior to firing its five J-2 engines. Already on the edge of space, the S-II was to combine continuing to climb with building up horizontal velocity. The spent stage followed a ballistic arc, 357 nautical miles long, into the Atlantic. Although staging occurred at an altitude of 36 nautical miles and 50 nautical miles downrange, it was visible to viewers at the Cape through thin high-level cloud.

The S-II was rather quieter than the first stage, built up its *g*-load gently, and ran smoothly. The role of the interstage was to prevent the discarded S-IC stage from coming into contact with the engines of the second stage. However, because it represented 'dead weight', it was promptly jettisoned. If it had been necessary to abandon the launch vehicle, the launch escape system would have been fired to draw the command module clear. The main solid rocket motor had a thrust of 150,000 pounds, which was fully twice that of the Redstone missile that fired Al Shepard on his suborbital Mercury mission. At an altitude of 60 nautical miles the escape system was jettisoned by firing a secondary solid rocket motor. The tower took with it the conical cover that had protected the command module during the ascent through the atmosphere and, if an abort had been made, would have protected it from the escape rocket's exhaust. Up to this point, all windows except that in the hatch had been masked. With all five windows uncovered, the cabin brightened markedly.[9] As it was above the bulk of the atmosphere, the S-II could manoeuvre without enduring significant aerodynamic stress. It was tasked with correcting any trajectory errors inherited from the first stage. As with the S-IC, the middle engine was shut down first. On fuel depletion, the outer engines cut off. One second later, at an altitude of 101 nautical miles, 875 nautical miles from the Cape and far beyond the range of the television cameras, the S-II was jettisoned to fall into the Atlantic. Joan Aldrin, who had fidgeted throughout, now went into her bedroom to check the abbreviated flight plan that she had pinned on the wall.

On igniting its single J-2 engine, the S-IVB continued to navigate towards the 'keyhole in the sky' for orbit. As it pitched over, it presented the astronauts with a view of the curved horizon across the Atlantic; however, being veterans, they had seen it before. At a downrange distance of 1,461 nautical miles, the Instrument Unit of the vehicle noted that it had attained the required combination of altitude and

[8] A load of 1 *g* corresponds to an acceleration of 32.2 feet per second per second.

[9] The cabin had five windows, numbered 1 to 5 running left to right: outboard of the left couch, in front of that couch, in the main hatch by the centre couch, in front of the right couch and outboard of that couch, with the side windows being large and rectangular, the hatch window being circular and the forward-looking windows being small and wedge-shaped.

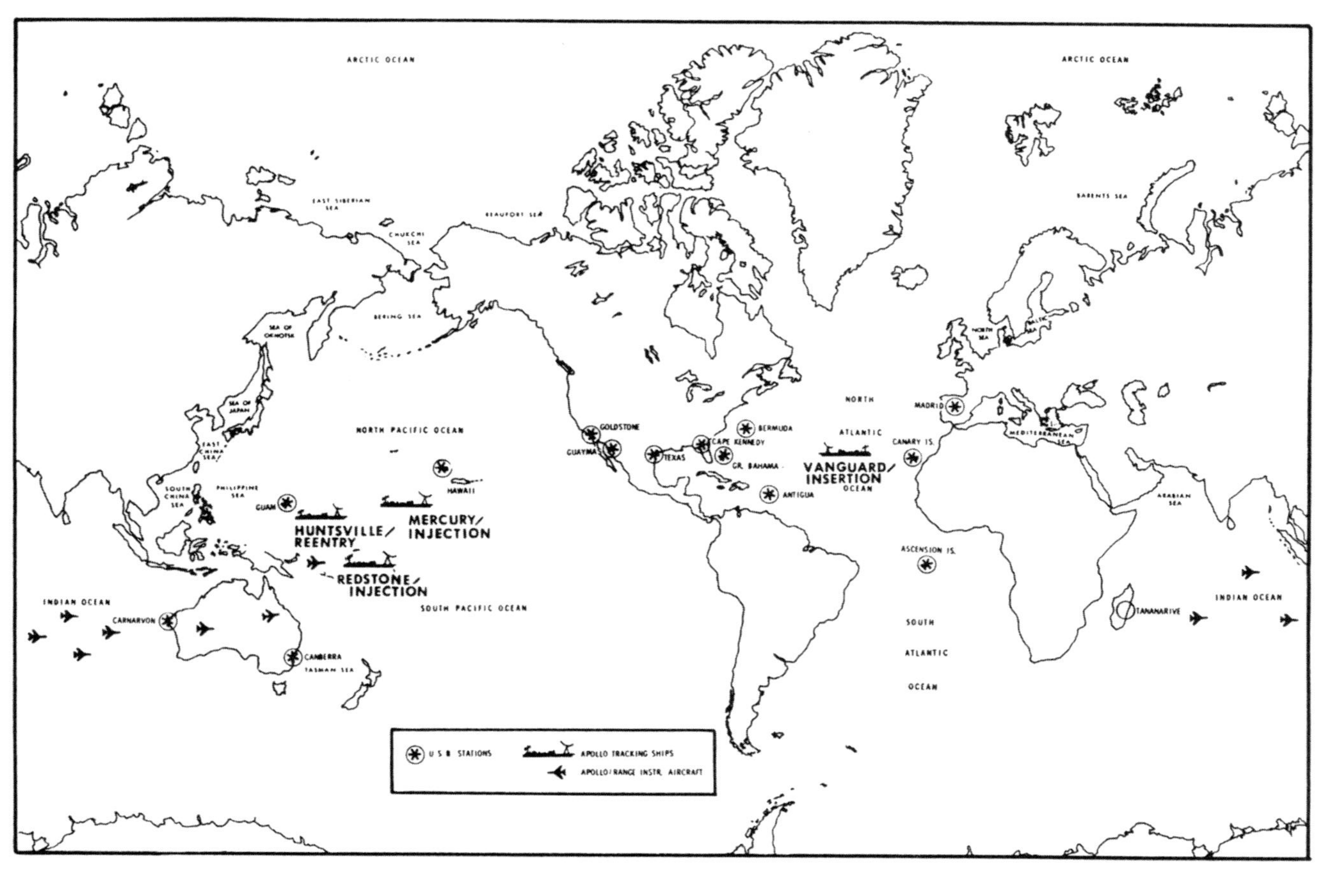

The Manned Space Flight Network as configured for the Apollo 11 mission.

velocity, and shut down the J-2 engine. At insertion,[10] the vehicle was travelling at 25,567.8 feet per second in a 'parking orbit' that ranged between a perigee of 98.9 nautical miles and an apogee of 100.4 nautical miles, was inclined at 32.521 degrees to the equator and had a period of 88.18 minutes – which was within 0.6 foot per second and 0.1 nautical mile of the specified velocity and altitudes respectively. Apollo 11 was off to an excellent start.

To communicate with Apollo spacecraft, NASA had established the Manned Space Flight Network (MSFN) using ground stations, ships and aircraft linked to the Goddard Space Flight Center in Greenbelt, Maryland, and then on to Mission Control in Houston. Although some of the stations were simply voice-relays for Mission Control, others had radars to provide the tracking data required to enable the mainframe computers in Houston's Real-Time Computer Complex to refine the parameters of the spacecraft's orbit in order to calculate the translunar injection (TLI) manoeuvre.

Once the space vehicle had disappeared from sight, Arthur C. Clarke remarked to the BBC's veteran space correspondent, Reginald Turnill, "At liftoff, I cried for the first time in 20 years – and prayed for the first time in 40 years." The protestor, Reverend Ralph Abernathy, having "succumbed to the awe inspiring launch" said, "I was one of the proudest Americans as I stood on this soil; I think it's really holy ground."

PARKING ORBIT

The USNS *Vanguard* had relayed communications as Apollo 11 entered orbit, then there was a hiatus until the spacecraft came within range of the tracking station on the island of Grand Canary, off northwest Africa. A valve in the command module had opened during the ascent through the atmosphere, and while its regulator had slowly allowed the cabin pressure to fall to 5 psi the environmental system started to increase the oxygen ratio and slowly purge the nitrogen that had been present to reduce the risk of a fire at sea-level pressure. Simultaneously, the pressure in the suits had also been reduced. For several hours prior to launch, the astronauts had been breathing oxygen at sea-level pressure to cleanse their bloodstreams of nitrogen, and on completing the post-insertion checklist they were able to doff their helmets and gloves, which were stowed in canvas bags. Despite the pre-breathing, Collins found that his left knee ached, as it had during his Gemini 10 mission, but in view of that experience he expected the discomfort to fade in a few hours.

"Apollo 11," McCandless called, "this is Houston through Canary."

"Post-insertion checklist is complete, and we have no abnormalities," reported Armstrong.

[10] Insertion occurred at T + 709.33 seconds; this being S-IVB cutoff plus 10 seconds to allow for engine tail-off and other transient effects.

As they left Canary behind, Collins unstrapped himself from his couch and went to the lower equipment bay to unstow the equipment that they would require. In a gravitational field the vestibular system of the inner ear facilitates a sense of balance, but in weightlessness the fluid that yields these cues sloshes about freely. Despite having performed aerobatics a few days previously to 'condition' himself, he avoided sudden head movements that might induce disorientation.

"Hey, Buzz?" Collins called.

"Yes?"

"How would you like the camera?"

"Okay."

"Here's a Hasselblad for you." The Hasselblad for internal use was fitted with an 80-millimetre lens.

"Just a second!"

"I'll just let go of it, Buzz; it'll be hanging over here in the air – it's occupying my couch."

A minute later, Collins asked, "Buzz, did you ever get that camera?"

"Yes."

"Are you ready for 16-millimetre?"

"Yes. How about a bracket?"

"Neil will give you the bracket."

Collins unstowed the 16-millimetre Maurer 'sequence camera', also known as a Data Acquisition Camera, and gave it to Aldrin, who mounted it on a bracket in the right-hand forward-facing window ready to film the retrieval of the LM following TLI. It could run at 1, 6 and 12 frames per second on automatic, and 24 frames on semi-automatic, with shutter speeds of 1/60th, 1/125th, 1/500th and 1/1000th of a second.

"I'm having a hell of a time maintaining my body position down here," Collins complained, "I keep floating up."

After briefly checking in with Houston through the relay site at Tananarive on the island of Madagascar in the Indian Ocean, they crossed the evening terminator. As the 'platform' of an inertial measurement unit tends to drift out of alignment, it requires frequent resetting. While they were in Earth's shadow, Collins was to run computer program 52 (referred to as 'P52') and take star sightings with the sextant in the wall of the lower equipment bay to realign the platform. At orbital insertion, the long axis of the vehicle was aligned with the velocity vector, with the sextant aimed towards space, and the S-IVB had established a pitch rate designed to maintain this 'orbital rate' attitude.

The platform used a nest of three gyroscopes and associated accelerometers to measure the spacecraft's attitude and velocity with respect to the orthogonal axes of a coordinate system in a specified frame of reference. Different frames were to be used at different times during the mission, each specified by a 'reference to the stable member matrix' (REFSMMAT). As the spacecraft rotated, the platform tended to remain fixed in inertial space and the gimbals rotated to compensate. The platform, aligned to a REFSMMAT – the 'stable member' against which the gimbals tracked the attitude of the spacecraft – was the one reference that remained in the same position relative to the stars irrespective of what the spacecraft did. For the start of the journey, the REFSMMAT was defined (in part) with respect to the line from the centre of the Earth through the launch site at the time of liftoff and also with respect to the flight azimuth. This frame would be retained through to TLI. The optical system had a telescope and a sextant. The telescope gave a 60-degree field of view, without magnification, and was intended to be used to identify a constellation and aim the coaxial sextant in the direction of a given star. The sextant's 1.8-degree field of view was magnified 28 times in order to make precise sightings. If the platform lost its sense of direction – which would occur if a pair of its three gimbals became co-aligned, locking them together (a condition appropriately known as 'gimbal lock') – Collins would run through this procedure to realign the platform from scratch. But when he merely wished to find out whether the system really knew its attitude, he would simply ask the computer to aim the sextant at a particular star, and check to see how far it was from the centre of the field of view, then mark its true position. With sightings on two stars, the computer could refine the alignment of the platform. As confirmation, he would then ask the computer to aim the sextant at a third star, which should then appear precisely centred in the field of view. With knowledge of its attitude and the 'state vector' specifying the spacecraft's position and velocity, the computer could, at least in principle, calculate all the manoeuvres necessary to conduct the mission independently of ground support.

Explorers have been navigating on Earth by telescope and sextant for centuries. The computer served essentially the same function as the chronometer. The inertial system not only acted as a compass, but also measured acceleration, which was not a quantity needed on the surface. The digital computer was the technological marvel – a decade earlier, navigating in space would have been a considerably more manual process. Indeed, it could be argued that a lunar landing mission would not have been very practicable prior to the 1960s. The display and keyboard (DSKY, pronounced 'disky') of the computer was supplied by the Raytheon Company. It had a power supply, decoder relay matrix, status and caution circuits, a 21-character display unit, and a 16-button key pad with the digits '0' through '9', 'VERB', 'NOUN', 'CLEAR', 'ENTER', 'PROCEED' and 'KEY RELEASE'. The executives, developed by the Instrumentation Laboratory of the Massachusetts Institute of Technology, were named 'Colossus IIA' for Columbia and 'Luminary' for Eagle, and used 'programs', 'nouns' and 'verbs'. Verbs were the instructions to do something in the context of a program, while nouns represented the structures that were to be operated upon. For example, if Verb 06 Noun 62 was entered, the verb meant "display in decimal" and the noun indicated what was to be displayed (in this case three numbers of the spacecraft's inertial velocity, vertical speed and altitude). Some programs had many verb and noun pairings. Numerical data took the form of five-digit quantities, scaled to fit and incorporating an implied decimal point. When a key was depressed, the appropriate item illuminated on the display. A verb would be flashed if the computer wished to attract attention when awaiting crew input. There were two DSKYs in Columbia – one on the main control panel, and the other in the lower equipment bay by the navigational instruments – and one in Eagle.

Star sightings were impractical while the spacecraft was in daylight because the particulates that the vehicle shed, and which floated alongside it, reflected sunlight and made it difficult to see the stars. But this material became invisible in Earth's shadow. After jettisoning the external cover of the optical system and inserting the eyepieces into the sextant and telescope, Collins peered though the telescope. The view was not encouraging. "I think I am seeing the horizon, but I'm far from being dark-adapted; it's hard to tell." He installed the handles that were to enable him to hold his weightless body in position to use the instruments. Finally, to cover his left eye while viewing with his right, Collins donned a small plastic patch attached to an elasticated string. The computer knew the celestial coordinates of 37 naked-eye stars that were widely distributed across the sky, each identified by an octal (i.e. base 8) number.

With the vehicle maintaining a fixed attitude with respect to Earth while they remained in low parking orbit, there were only a few reference stars available for this initial P52 check, and the mission planners had decided which ones he should use. Collins told the computer to aim the optics towards star 30, Menkent in the constellation of Centaurus. As the view through the telescope was poor, he went straight to the sextant. The star was slightly 'off'. He centred the cross hairs of the sextant on the star and instructed the computer to record its true alignment. Then he asked for star 37, Nunki, in Sagittarius, checked it, and marked it. The Apollo sextant was more practicable than the hand-held version he had tested on the first orbit of Gemini 10, when John Young, his commander, had dubbed him 'Magellan' after the captain of the first ship to circumnavigate the world. Having marked two stars and told the computer to update the platform, Collins instructed it to find star 34 as a final check. "Atria is there in the sextant, but it's not dab-smack in the middle." He asked the computer to display the alignment discrepancy as a fraction of a degree. "0.01, Goddammit! Now that's enough to piss a body off." Glenn Parker, an instructor at the Cape, had bet him he would not achieve an accuracy better than 0.02 degree, and Collins had been sure he could attain a perfect 0.00 alignment. Finished with the optical instruments, he removed and restowed the eyepieces.

Carnarvon on the southwestern coast of Australia was operating three tracking systems to ensure that the parameters of the orbit were accurately measured, to enable Dave Reed, the flight dynamics officer in Houston, to calculate the TLI burn. After McCandless confirmed that there were no issues involving the S-IVB that might threaten the continuation of the mission, Collins, judging his bet to have been a tie, asked McCandless to pass on a message, "Tell Glenn Parker at the Cape that he lucked out – he doesn't owe me a cup of coffee."

A black-and-white Westinghouse television camera with scan rate of 10 frames per second and 320 lines of resolution was carried on Apollo 7 and Apollo 8 with good results. A camera using the sequential colour technique was tested by Apollo 10, and issued to Apollo 11. Passing over Australia, Collins unstowed the camera and its ancillary apparatus and plugged in the power and signal cables. The camera was a rectangular box with a lens protruding from the front end. The monitor to enable them to see the field of view in order to verify the focus and lighting was the same width, about half the length, and slightly thicker, and had a small display

Table: Apollo navigation stars

Popular Name	Code (octal)	Official Name
Alpheratz	*01*	*alpha Andromedae*
Diphda	*02*	*beta Ceti*
*Navi**	*03*	*epsilon Cassiopeiae*
Achernar	*04*	*alpha Eridanus*
Polaris	*05*	*alpha Ursa Minor*
Acamar	*06*	*theta Eridani*
Menkar	*07*	*alpha Cetus*
Mirfak	*10*	*alpha Persei*
Aldebaran	*11*	*alpha Tauri*
Rigel	*12*	*beta Orionis*
Capella	*13*	*alpha Aurigae*
Canopus	*14*	*alpha Carinae*
Sirius	*15*	*alpha Canis Majoris*
Procyon	*16*	*alpha Canis Minoris*
*Regor**	*17*	*gamma Velorum*
*Dnoces**	*20*	*iota Ursae Majoris*
Alphard	*21*	*alpha Hydrae*
Regulus	*22*	*alpha Leonis*
Denebola	*23*	*beta Leonis*
Gienah	*24*	*gamma Corvi*
Acrux	*25*	*alpha Crux*
Spica	*26*	*alpha Virginis*
Alkaid	*27*	*eta Ursae Majoris*
Menkent	*30*	*theta Centaurus*
Arcturus	*31*	*alpha Boötis*
Alphecca	*32*	*alpha Corona Borealis*
Antares	*33*	*alpha Scorpii*
Atria	*34*	*alpha Trianguli Australis*
Rasalhague	*35*	*alpha Ophiuchi*
Vega	*36*	*alpha Lyrae*
Nunki	*37*	*sigma Sagittarius*
Altair	*40*	*alpha Aquilae*
Dabih	*41*	*beta Capricornus*
Peacock	*42*	*alpha Pavo*
Deneb	*43*	*alpha Cygni*
Enif	*44*	*epsilon Pegasi*
Fomalhaut	*45*	*alpha Piscis Austrinus*

* Three of these names were coined by Gus Grissom to celebrate his Apollo 1 crew ('Navi' was his middle name, 'Ivan', spelt in reverse; 'Dnoces' was the reverse spelling of 'second', as in Edward H. White II; and 'Regor' was the reverse spelling of 'Roger', as in Roger B. Chaffee) and, as far as the International Astronomical Union was concerned, they were unofficial.

screen on one end. The flight plan called for a test of the system by sending to the Goldstone station of the Manned Space Flight Network, situated on a dry lake in the Mojave Desert near the town of Barstow in California. Having had no in-flight anomalies, they set off across the Pacific Ocean with 20 minutes of free time.

"How does zero-*g* feel?" Armstrong enquired of Collins, who was the only one to have left his couch.

"I don't know," Collins mused. "It just feels like we're going around upside down."

Indeed, they were, because the S-IVB was maintaining them in a 'heads down' attitude, but Collins was referring to the sensation of 'full headedness' that arises from the accumulation of blood in the head with the onset of weightlessness.

Passing south of Hawaii, the eastern horizon began to glow.

"Stand by for sunrise," announced Collins.

"Neil hasn't seen many of those," pointed out Aldrin. Gemini 8 had performed an emergency return after only a few orbits, during which Armstrong had been too preoccupied to appreciate sunrise.

"We haven't got too many of them on this flight, so you might as well enjoy it while you can," Armstrong advised.

By now the view ahead was a spectacular arc of colour.

"Jesus Christ, look at that horizon!" Collins exclaimed.

"Isn't that something?" Armstrong agreed. "Get a picture of that!"

"Has anybody seen a Hasselblad floating by?" Collins asked. "It couldn't have gone far, a big son of a gun like that." He scrambled around looking for it. "Well, that pisses me off." As he searched, he spotted a ball-point pen that should not have been floating freely. "I've looked everywhere over here for that Hasselblad, and I just don't see it."

"It's too late for sunrise, anyway," Armstrong lamented.

"You want to get it before TLI," Aldrin noted, referring to the camera. Objects floating freely would have to be collected and stowed prior to the manoeuvre, to prevent them slamming into something else and causing damage.

"I know," agreed Collins. As he continued his search, he endeavoured to avoid making movements that might induce 'space sickness'. "Ah! Here it is, floating in the aft bulkhead." He belatedly snapped a few pictures of sunrise.

"How are we doing checklist-wise?" Armstrong prompted, several minutes later. "Let's make sure we don't screw up and forget something."

"After I extend the docking probe," Collins replied, "I have got to copy down a bunch of data, and you've got the RCS hot-fire."

As they approached Baja California, McCandless called via the relay station at Guaymas in Mexico to remind them that they were coming up on Goldstone, but the vehicle barely rose above Goldstone's horizon and the station received less than 1 minute's worth of the FM carrier signal without any image modulation. While this proved that the transmitter worked, it had yet to be demonstrated that the camera functioned. One point of progress, however, was that Collins decided that having the monitor float beside the camera was awkward, and that in future it should be taped on top of the camera.

Houston monitored the spacecraft's telemetry to verify that the docking probe extended properly, and that the RCS thrusters in the quads positioned at 90-degree intervals around the service module imparted pitch, yaw and roll impulses – using only small 'blips' in order not to disturb the S-IVB. The Instrument Unit was then updated by direct uplink with the parameters of its orbit and the time and duration of the TLI manoeuvre. This data, plus information on options for an abort if that burn did not go to plan, was also read up to Collins, who wrote it on the flight plan and read it back for confirmation. It had once been intended that there should be a teleprinter to simplify the provision of such information, but this had been deleted in an effort to control the mass of the spacecraft. Because Goldstone had not been able to verify the television camera, as they passed over Florida they transmitted to the Merritt Island Launch Area where, although there was no longer any apparatus to process an image, it was possible to confirm that the carrier signal was modulated. McCandless reported via Canary Island on revolution 2 that the telemetry from the hot-firing indicated that one of the RCS quads was cold. On checking, Armstrong reported that the switch for the heater in that unit had been set incorrectly. "It was Off; it's On now. Thank you." With the heater operating, the unit rapidly warmed up.

As Apollo 11 approached Earth's shadow for the second time, Collins resumed his couch and they put their helmets and gloves back on in preparation for the TLI burn. McCandless attempted to call through Tananarive but received no response; it transpired that the remote station was not uplinking. Unconcerned, the astronauts worked on through the checklist.

TRANSLUNAR INJECTION

The use of a parking orbit provided a full revolution of Earth in which to confirm that the S-IVB stage and the spacecraft were fully operational prior to attempting the TLI manoeuvre. The alternative would have been to time the launch to enable the S-IVB to undertake a single long burn directly onto a translunar trajectory. Such a manoeuvre was viable, but offered fewer contingency options. Specifically, if the spacecraft were to be found to have a fault that would require an abort and return to Earth, it would already be outbound by the time this became evident. However, the time spent in parking orbit posed a thermal issue for the S-IVB, which held liquid hydrogen at –423°F and liquid oxygen at –293°F. Nevertheless, the designers had cleverly exploited this fact, because as heat leaked in and caused the hydrogen to boil, the gas was vented through two small aft-facing nozzles for ongoing ullage to maintain the propellants settled in their tanks.

The burn was to begin at 002:44:14 and nominally last 5 minutes 47 seconds, but the Instrument Unit was programmed to terminate the manoeuvre when the desired velocity had been attained, and would therefore cut off early if the engine were to overperform and extend the burn if the engine underperformed. Armstrong was to allow it an additional 6 seconds before intervening, and asked Collins to monitor the duration and yell out at 5 minutes 53 seconds.

The configuration of Apollo 11 at the time of the TLI manoeuvre.

When Apollo 11 again came into range of Carnarvon, McCandless called with good news, "You are Go for TLI."

"Thank you," Collins acknowledged.

The S-IVB terminated its hydrogen venting, pressurised its propellant tanks, and ran through the pre-start sequence for the J-2 engine.

NASA had equipped several KC-135 aircraft as Apollo Range Instrumentation Aircraft, and stationed them in a line across the Pacific between Australia and the Hawaiian Islands. "They're going to try uplinking both on S-Band and on VHF," McCandless advised. "So if you make sure your S-Band volume is turned up, we'd appreciate it. We should have continuous coverage from now on, right through the TLI burn."

"Very good," acknowledged Armstrong.

McCandless established a relay through ARIA 4 but the signal was noisy, so he switched to ARIA 3, which was much clearer. The aircraft relayed telemetry to Houston. "We just got telemetry back down on your booster," McCandless advised, "and it is looking good."

"Everything looks good here," Armstrong confirmed.

Although Apollo 11 would be the third mission to perform the TLI manoeuvre, it was by no means routine, as any major engine burn represented a potential point of catastrophic failure. The manoeuvre began in the vicinity of the Gilbert Islands, about half-way between Australia and Hawaii. The Instrument Unit was in control, but it provided cues to the spacecraft, and illuminated a lamp immediately prior to igniting the J-2 engine.

"Whew!" exclaimed Armstrong on the intercom circuit when the engine lit.

"We confirm ignition," called McCandless, "and the thrust is Go."

"Pressures look good," Armstrong noted.

"About 2 degrees off in the pitch," Aldrin pointed out.

"I wouldn't worry too much about that," Armstrong advised.

The Real-Time Computer Complex was using the ARIA data to calculate the S-IVB's departure trajectory, and any emergency course correction that the spacecraft would have to make in the event a shortfall in velocity.

"One minute," called McCandless. "Trajectory and guidance are looking good, and the stage is good."

For the crew, the burn was silent. Although the engine was 110 feet behind and directed its plume aft, Collins noticed intermittent flashes of light through his side window. "Don't look out of window 1," he chuckled, "because if it looks like what I'm seeing out of window 5, you don't want to know."

"Oh, I see a little flashing out there, yes," Armstrong said.

From the centre couch Aldrin's field of view was limited, but he looked across Collins, "Damn. Kind of sparks flying out there."

There was a lurch when, as programmed, the S-IVB adjusted the ratio of fuel to oxidiser being fed to the engine.

McCandless provided a reassuring update, "Thrust is good."

As the S-IVB consumed propellant and its structural dynamics altered, it began to vibrate. On Apollo 8 the amplitude of the vibration had caused Frank Borman to

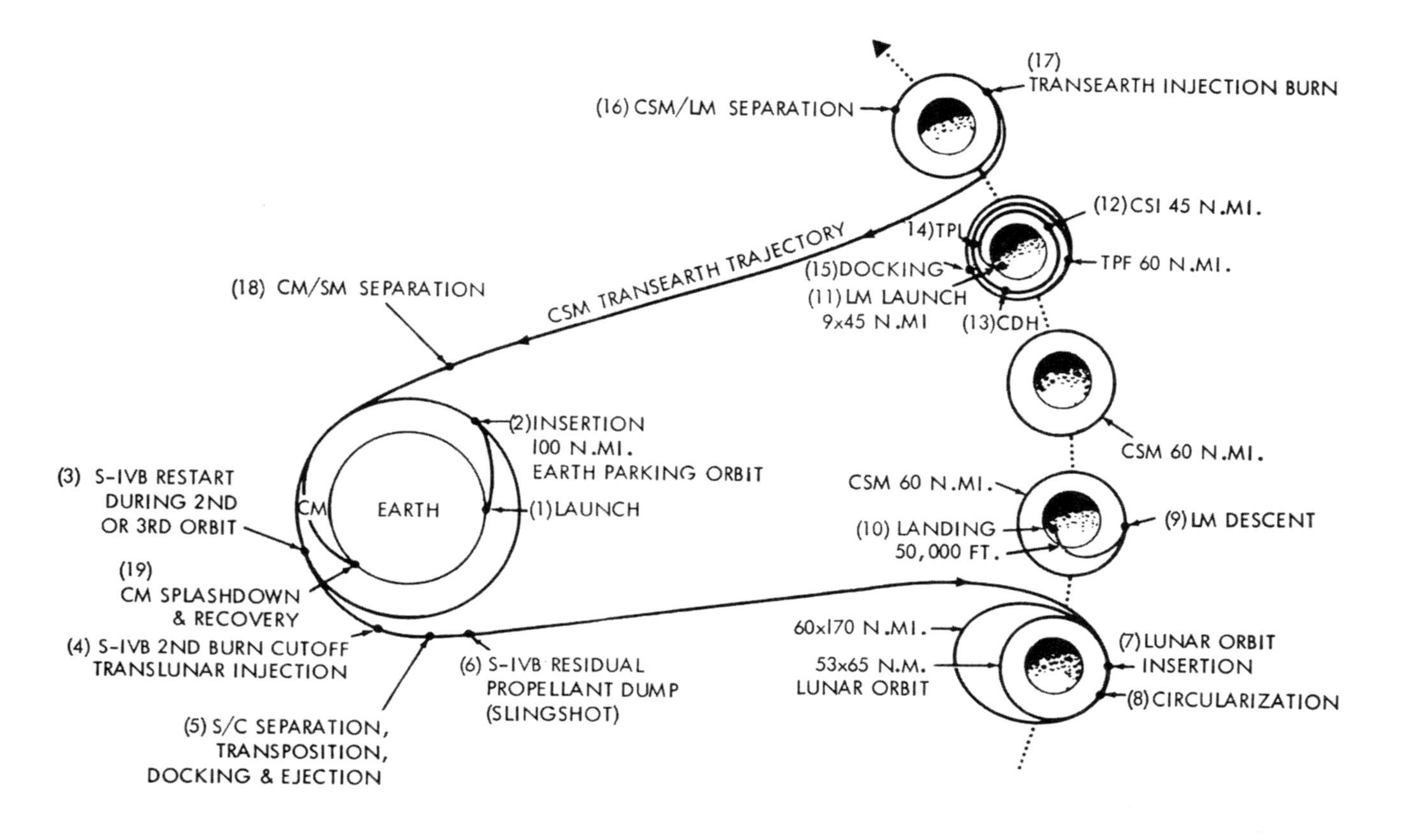

The main manoeuvres and trajectory planned for Apollo 11.

consider aborting the burn. As the rattle built up, Aldrin became concerned that it might shake loose the Maurer camera that he had mounted in window 4. Collins, not wanting the camera to fall on his helmet, checked that it was secure.

As they climbed, they emerged from Earth's shadow rather earlier than if they had remained in low orbit. The sunlight that suddenly flooded in made it difficult to read the instruments, but Armstrong had had the foresight to mount a card in his window to serve as a shade. "I'm glad I got my card up!"

"Neil, that was a hell of a good idea," congratulated Collins. "I can't see very much."

"You are Go at 5 minutes," McCandless advised.

As long as the J-2's burn continued to within about 45 seconds of the intended duration, the mission would be able to continue by having the CSM later employ its main engine to make up the velocity shortfall. A greater shortfall would require an abort in which the CSM cancelled the result of the interrupted TLI and pursued a trajectory designed to end with a landing in the primary recovery zone, where the US Navy had ships already on station.

The S-IVB's acceleration built up as it consumed its fuel and became lighter.

"What kind of *g* we pulling?" Aldrin asked.

"1.2, or 1.3 *g*, maybe," Armstrong replied.

"Gee, it feels a lot more than that."

"Here we go," announced Collins, as the nominal time for cutoff approached.

Two or three extra seconds passed, then the engine shut down.

"Shutdown!" called Armstrong.

Prior to the TLI manoeuvre, the vehicle had been in a nearly circular orbit at an altitude of about 100 nautical miles. The burn was to accelerate by 10,435 feet per second, to a velocity of 35,575 feet per second at an altitude of 174 nautical miles. At cutoff, it had attained 35,579 feet per second and 177 nautical miles, which was excellent by any measure. On the ground, the mass of the space vehicle had been 6.5 million pounds; at insertion into orbit it was slightly less than 300,000 pounds; it was now 138,893 pounds.

"Houston, that Saturn gave us a magnificent ride," Armstrong announced. "We have no complaints with any of the three stages." It was another great success for Wernher von Braun's rocket team at the Marshall Space Flight Center.

The TLI manoeuvre inserted the vehicle into an elliptical orbit with an apogee beyond the orbit of the Moon. It did not shoot the spacecraft directly towards the Moon, but some 40 degrees ahead of the current position of the Moon in its nearly circular orbit of Earth. However, as the spacecraft neared apogee in three days' time, the gravitational field of the now-present Moon would deflect its trajectory. If the spacecraft were to do nothing, it would pass around the 'leading limb' on a figure-of-eight 'free return' path that would send it back to Earth. However, the plan was to fire the main engine while behind the Moon in order to slow the spacecraft and enter lunar orbit.

Meanwhile, at home

Jan Armstrong remained on North American Rockwell's boat until TLI, which she refused to mark with champagne, preferring instead to defer celebration until the crew was safely home, then she returned to Patrick Air Force Base for a flight to Houston. After TLI, Pat Collins went onto her front lawn to give her first in-flight interview and, adhering to the wives' formula, told the reporters she was "thrilled, proud and pleased".

5

Translunar coast

RETRIEVING EAGLE

Immediately after the TLI manoeuvre, the S-IVB adopted an orientation calculated to yield favourable illumination for the separation of the CSM and its subsequent transposition, docking and extraction (TD&E) of the LM. Armstrong yielded the left-hand couch to Collins who, as CMP, had trained for this delicate operation. The reaction control system (RCS) comprised four units at 90-degree intervals on the side of the service module, each with a cluster of four rocket thrusters that could be fired separately or in various combinations to control roll, pitch and yaw motions. During the climb through the atmosphere, the conical adapter mounted on top the S-IVB was allowed to vent to ensure that there would be no sudden release of air when the spacecraft separated.

"You're Go for separation," called Bruce McCandless, 3 hours 15 minutes into the mission.

After using the left-hand translational controller to start the aft-facing thrusters, Collins threw a switch to detonate pyrotechnic charges around the rear rim of the service module to detach that module from the adapter. Once a display indicated that he was moving at about 1 foot per second, he ceased to thrust. Meanwhile, the four panels of the adapter hinged open like petals and then detached to drift away. After 15 seconds, he used the right-hand rotational controller to initiate a 2-degree-per-second pitch motion. When so instructed, the digital autopilot was to maintain this angular rate but, to his frustration, it cancelled the rotation and adopted a fixed attitude. He had to repeat the procedure several times before the autopilot accepted the rotation. In consequence, the rotation used rather more propellant than planned. The issue was later determined to be procedural in nature. Once the spacecraft was facing the S-IVB, Collins terminated the rotation and fired the aft-facing thrusters again, this time to halt 100 feet out, whereupon he fired them a third time in order to move back in.

"I hope you're getting some pictures, Buzz," Armstrong said.

"I've got the 16-millimetre going at 16 frames per second," confirmed Aldrin, referring to the Maurer camera he had mounted in window 4. The LM was clearly visible, supported within the annular ring near the base of the adapter by fixtures on its folded legs.

"Be sure that your RCS is working," Armstrong prompted Collins, to confirm that the forward-facing thrusters to be used for braking were functional, because if they failed the CSM would smash into the LM. When Collins did so, he noted that the efflux rippled the aluminium thermal shielding of the ascent stage, and hoped he wouldn't damage it.

The extended probe on the apex of the spacecraft was not visible through the small forward-facing window, but Collins did not need to see it as there was a 'stand off' target on the roof of the LM which, when correctly viewed, meant that the probe was centred on the conical drogue. With the Sun over his shoulder, the roof of the LM was nicely illuminated. In easing the probe into the drogue, Collins was conscious that his vehicle had a mass of 65,000 pounds, and that the 33,000-pound LM was attached to the 'dead weight' of the spent S-IVB. As the probe penetrated the socket at the apex of the cone, three small capture latches around the tip of the probe automatically engaged for a 'soft docking'. When he was sure that the vehicles were lined up, Collins threw a switch and a discharge of nitrogen gas pneumatically retracted the probe, in the process drawing the two collars together and triggering 12 spring-loaded latches that established a rigid connection, or 'hard docking'.

Although the docking was accomplished, Collins was dissatisfied. "That wasn't the smoothest docking I've ever done."

"Well, it felt good from here," Armstrong complimented.

"I mean the gas consumption," explained Collins. He had used rather more fuel during the transposition manoeuvre than expected.[1]

While manoeuvring, the quality of signal using the omnidirectional antennas had degraded to the point at which communication became impractical. Aldrin operated the controls to slew the high-gain antenna mounted on a boom on the rear of the service module to point its beam towards Earth to restore communications; once locked on, the system would steer itself to maintain maximum signal strength as the line of sight evolved.

The 30-inch-diameter tunnel to the LM ran through the apex of the command module. As per the plan, an open valve in the LM's overhead hatch had allowed its cabin to vent. Having already raised the pressure in the command module, Collins opened a valve to allow oxygen to pressurise the tunnel, and thence the LM. On opening his hatch, he noted an odour reminiscent of charred electrical insulation, but all the exposed wiring in the tunnel appeared to be factory fresh and he pressed on with his checklist, jiggling each of the docking latches by hand to confirm that it was properly engaged.

In 1958 America's first satellite, Explorer 1, discovered that there is an intense belt of charged-particle radiation present within the Earth's magnetic field. In fact, as subsequent satellites revealed, there is an inner belt of high-energy protons and an outer belt of electrons. These 'radiation belts' were named after the scientist who

[1] Telemetry showed the RCS propellant supply to be about 20 pounds below nominal following the transposition manoeuvre.

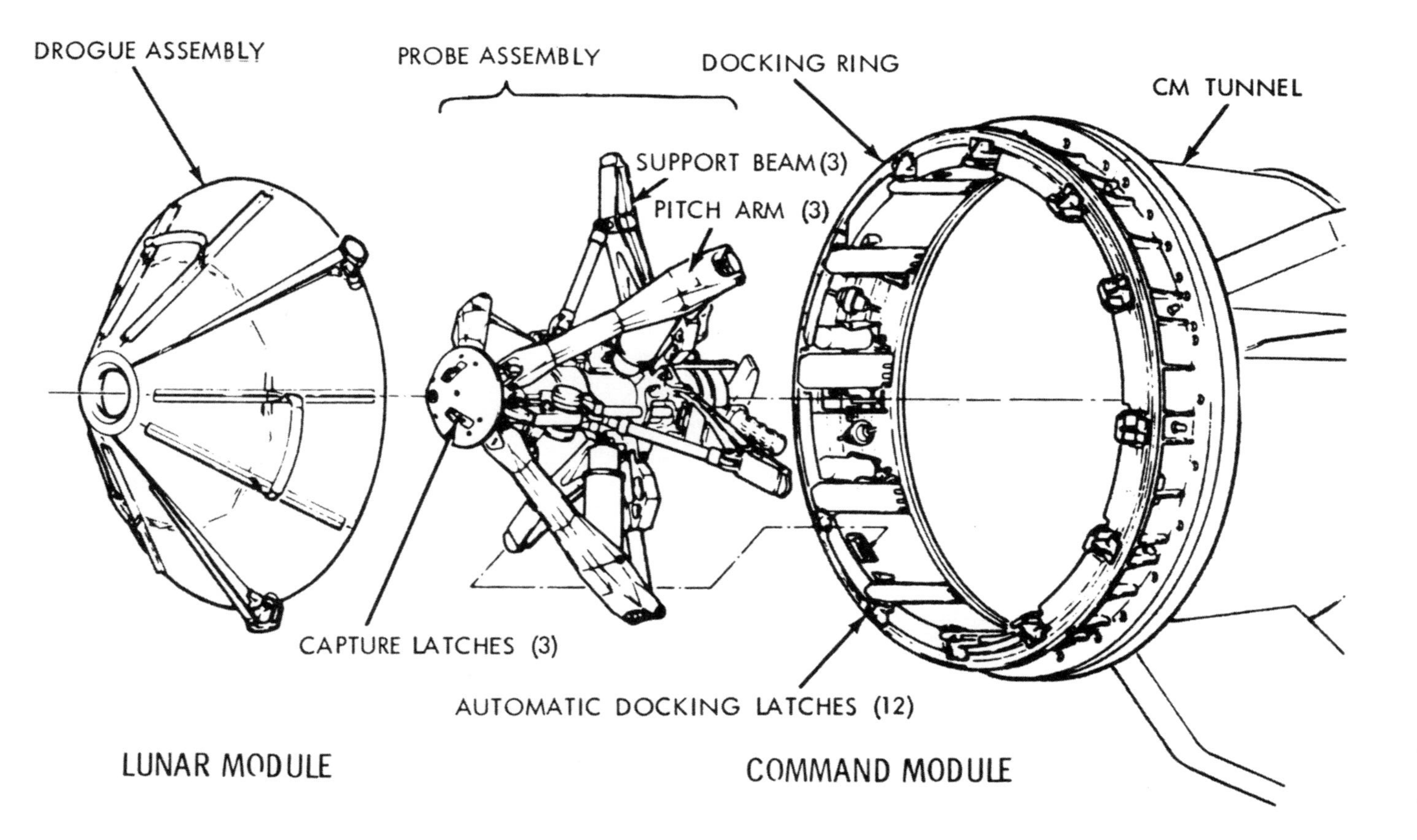

A diagram of the probe and drogue docking mechanism.

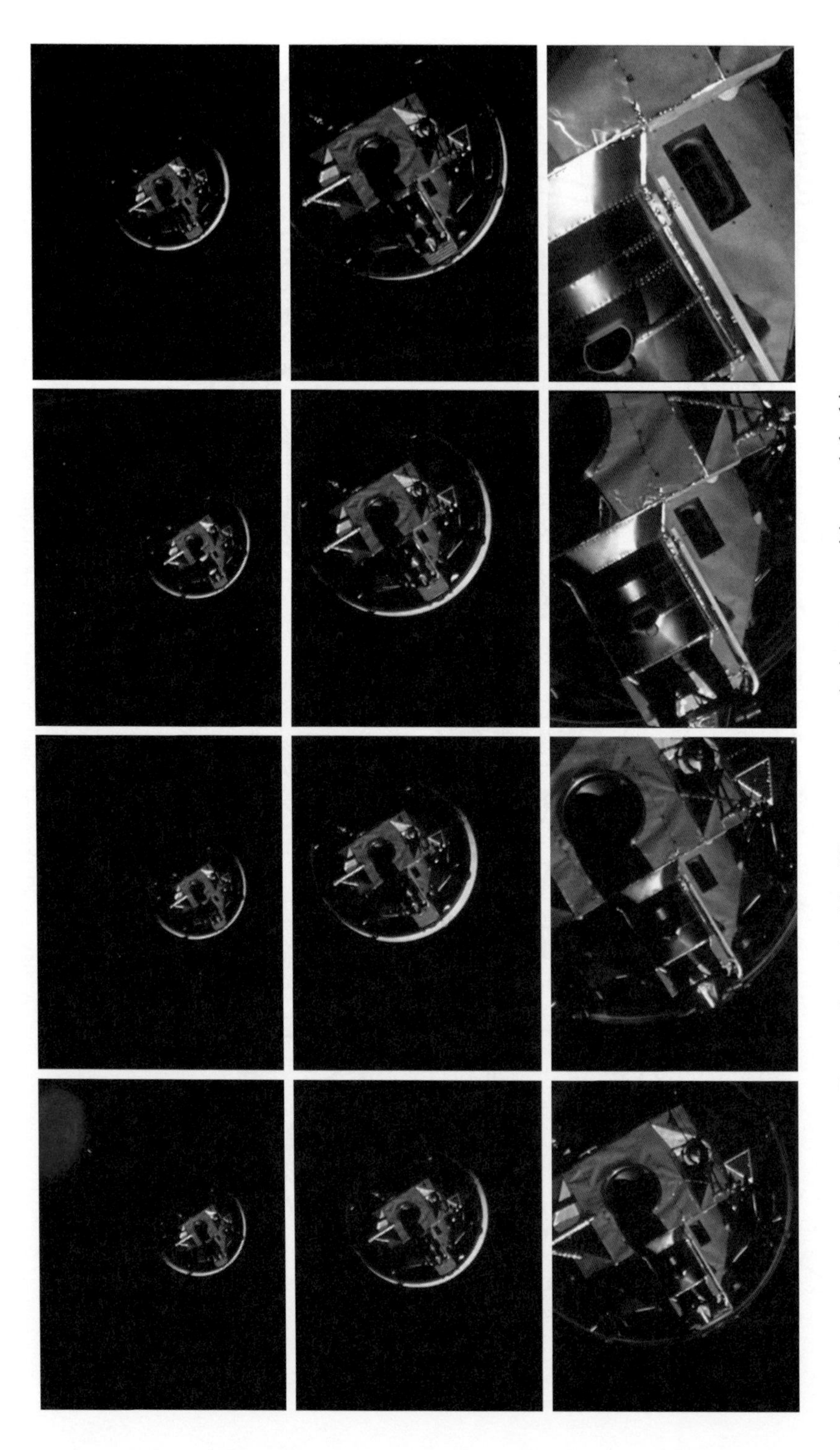

Frames from the 16-millimetre camera during transposition and docking.

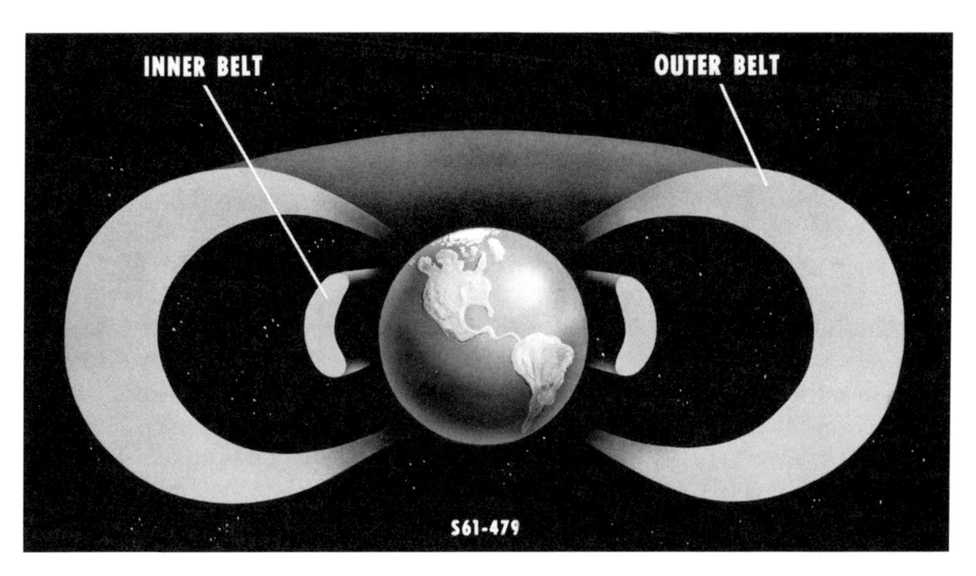

A depiction of the inner and outer van Allen radiation belts.

supplied the instrument that made the initial discovery – James van Allen of Iowa University. The proton belt would pose a severe health risk to any astronaut unwise enough to linger within it, but fortunately it was concentrated above the equatorial zone and extended out only a few thousand miles, and because the Moon orbits Earth at an angle to the equator Apollo 11 was able to avoid the most intense part of this belt. Also, as with any object that is thrown upwards in Earth's gravity, as soon as the S-IVB shut down its engine it began to slow as it pursued a ballistic trajectory. At that time it was at an altitude of 177 nautical miles and was climbing at 21,345 nautical miles per hour. When Collins separated the CSM from the S-IVB, their altitude was over 3,000 nautical miles, and this distance had doubled by the time he completed the transposition and docking 10 minutes later. The radiation received was therefore no worse than that involved in having a chest X-ray. The radiation within the Earth's magnetic field originates from the 'solar wind', which is a flow of charged particles emanating from the Sun. This was discovered in 1962 by Mariner 2, the first spacecraft to report on conditions beyond the realm dominated by the Earth's magnetic field. Cislunar space – the environment between Earth and the Moon – is comparatively benign, unless a 'solar flare' directs a blast of charged-particle radiation in our direction. In terms of sun spots, the intensity of solar activity varies with an 11-year periodicity and, in fact, 1969 was a risky time to be heading for the Moon!

Looking back at Earth, which was rapidly shrinking, Armstrong displayed his keen sense of geography – a childhood passion. Their departure trajectory offered a better view of the northern hemisphere than of southern latitudes. "Houston," he called, "you might be interested that out of my left-hand window right now, I can observe the entire continent of North America, Alaska and over the pole, down to the Yucatán peninsula, Cuba, the northern part of South America – and then I run out of window."

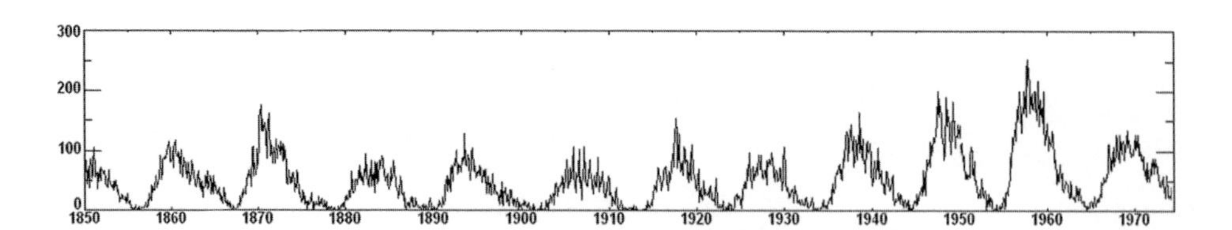

The number of sunspots follows a roughly 11-year cycle.

Meanwhile, Collins had inserted electrical plugs to feed power through to the LM's heaters and to pyrotechnics in the adapter. McCandless gave them the go-ahead to retrieve the LM. Precisely one hour after initiating the transposition and docking sequence, and now 12,600 nautical miles from Earth, Collins fired the charges to release the LM, and springs eased the 98,000-pound docked spacecraft away from the spent stage. If the charges had failed, the CSM would have undocked and gone on to fly an Apollo-8-style mission in lunar orbit, but this would have been a poor consolation.

The next task, some 20 minutes later, was to perform a manoeuvre to increase the rate of separation from the spent S-IVB. The cutoff velocity at TLI had been set marginally faster than that of the ideal trajectory, with the intention of using the service propulsion system (SPS) of the CSM to cancel the excess. This engine had been designed in such a manner that it was unlikely *not* to fire, but it was wise to test it early on. Delivering 20,500 pounds of thrust, it would be able to impart the desired velocity change (delta-V) of 19.7 feet per second in 3 seconds, which was just long enough for the telemetry to verify the propellant supply, the combustion chamber pressure and the stability of the gimbal. There was no need for an igniter because the hydrazine-based fuel and nitrogen tetroxide, being hypergolic, ignited immediately on coming into contact after being fed through a pattern of holes in the injector into the combustion chamber.

After running through the post-burn checklist, Armstrong returned to the view of Earth. "We didn't have much time to talk about the view out the window when we were preparing for LM ejection, but up to that time we had the entire northern part of the day hemisphere visible – including North America, the North Atlantic, and Europe and the northern region of Africa. There was a cyclonic depression over northern Canada, east of Athabasca. Greenland was clear, and it appeared that we were seeing just the icecap. All of the North Atlantic was pretty good. And Europe and northern Africa seemed to be clear. Most of the United States was clear. There was a low, it looked like a front, stretching from the centre of the country, up north of the Great Lakes and on into Newfoundland."

"I didn't know what I was looking at," added Collins, "but I sure did like it!"

"The view must be pretty good from up there," McCandless agreed. "We show you about 19,000 nautical miles out now." Although this was the altitude at which satellites in geostationary orbits reside, Apollo 11 was far above the plane in which such satellites operate.

If the S-IVB were to pursue the trajectory established by TLI, then it, like the spacecraft, would pass just in front of the leading limb of the Moon. To eliminate any possibility of a collision in transit, once Apollo 11 was clear of it, the ground commanded the S-IVB to execute a venting process designed to deflect its path to pass close by the *trailing* limb for a gravitational 'sling shot' into solar orbit. This venting would involve several phases over the period of about 40 minutes, and the planners had worked out how the spacecraft should be oriented to enable the crew to observe it.

"We've completed our manoeuvre to observe-the-slingshot attitude," Collins reported, "but we don't see anything – no Earth, and no S-IVB." Houston provided a revised attitude, and Collins started the manoeuvre, but because he had not only used more propellant than expected during the transposition and docking, but had also undertaken a fruitless manoeuvre, he initiated this new manoeuvre at a slow rate.

The venting could not be postponed because, once commanded, the Instrument Unit could not be interrupted. "It doesn't look to us like you will be able to make it around to the observation attitude in time," McCandless warned. "We recommend that you save the fuel."

"Our manoeuvre has already begun," pointed out Collins, "so it's going to cost us about the same amount of fuel to stop it no matter where we stop it, and we may as well keep going."

"We've got the S-IVB in sight at what I would estimate to be a couple of miles away," announced Aldrin several minutes later. "The dump appears to be coming out at radially opposite directions." As this preliminary part of the sequence was to be non-propulsive, the stage was venting oxidiser from opposite sides at the rear. About 25 minutes later, the venting was switched to occur through the J-2 engine, but by then the S-IVB was no longer visible. Finally, the two auxiliary propulsion system modules on the rear of the stage were fired to complete the manoeuvre.

As McCandless read up instructions to reconfigure the fuel cells that generated electrical power, the astronauts started their first meal. "If we're late in answering you," Collins apologised, "it's because we're munching sandwiches."

"I wish I could do the same here," said McCandless.

"No. Don't leave the console!" Collins warned. "Flight doesn't like it. How is he today?" The flight director was Clifford Charlesworth, and Collins had served as his CapCom during the Apollo 8 mission.

"Oh, he's doing quite well," McCandless laughed.

"I think today is the birthday of California," Armstrong called. "I believe they are 200 years old, and we send them a happy birthday. I think it is also Dr George Mueller's birthday, but I don't think he's that old!" George E. Mueller, Director of the Office of Manned Space Flight, was celebrating his 51st birthday.

Six hours into the mission, Charlesworth's Green Team handed over to Gene Kranz's White Team, and Charlie Duke took over as CapCom. After crossing the road to brief the press in the News Center, Charlesworth joined his controllers in the 'Singing Wheel', a red-painted wooden barn of a building that housed the bar that served as a 'watering hole' for off-shift flight controllers heading home.

Having taken sextant sightings to realign the inertial platform, Collins initiated a deep-space navigation test. This P23 exercise required him to measure the angles of five stars relative to Earth's horizon. In each case, the computer was to orient the vehicle to enable the sextant to measure from the 'substellar point'. However, the system was not aligning properly, and a measurement along a line that was not perpendicular to the horizon would yield an inaccurate result. The technique was valid, it was simply that something had gone amiss in planning. But he persevered because the data would help the engineers to determine the problem. The test was scheduled at this time in order to confirm the ability to make such measurements before the spacecraft was so far away that Houston could not provide a check. This technique was intended primarily for navigating the approach to re-entry, in order to provide a contingency against loss of communications on the way home.

Meanwhile, first Aldrin and then Armstrong doffed their pressure suits. It was a dynamic process involving a colleague unzipping the rear of one man's suit and holding him stable while he squeezed out. After finishing his navigation exercise, Collins joined them. With the various parts of the three suits adrift, there was, as Aldrin later put it, "a great deal of confusion, with parts and pieces floating about the cabin as we tried to keep the logistics under control". The suits were carefully folded, stuffed into storage bags and stowed beneath a couch. It was a great relief to remove the urine-collection and fecal-containment devices, which were stowed in a locker in the lower equipment bay. There was no toilet. Urination would be through a tube into a plastic bag that would be periodically vented overboard into vacuum. Defecation would involve the use of a plastic bag, part of which fitted over the hand like a glove. When finished, a germicide pill would be inserted into the bag to prevent bacteria generating gas. This 'glove bag' had proved effective on Gemini flights, when, because of the pill, it was euphemistically referred to as a 'blue bag'. Used bags were stowed in a special container in the lower equipment bay. It typically took 45 minutes to defecate. The use of the bags did not obviate the associated odours, however.[2] Over their constant-wear garments the men wore two-piece teflon-fabric flight suits. With the crew in 'shirt sleeves' and the centre couch removed, the cabin appeared considerably larger. The décor was 'battleship grey'. At an early stage a psychologist had recommended that, in order not to cause disorientation, there ought to be a two-colour scheme with brown below a certain horizontal level and blue above it, which sounded very reasonable until it was pointed out that in weightlessness there was no unique sense of 'up', and, by the psychologist's own logic, if the colour scheme really did matter, then an inappropriate one would be upsetting! Aldrin, who had mastered weightlessness spacewalking on Gemini 12, found the freedom of movement in the command module enjoyable because, as he put it, he was spacewalking indoors! Fortunately, due to the care taken during their first few hours in space, none of the Apollo 11 crew developed any symptoms of 'space sickness'.

[2] In some ways, the most unfortunate person involved in the mission was the man who opened the hatch immediately following splashdown!

PASSIVE THERMAL CONTROL

The REFSMMAT for the translunar coast used the ecliptic, which is the plane in which Earth orbits the Sun: two of the three axes of the REFSMMAT were relative to the ecliptic, and the third was aligned along the Earth–Moon line at the time of TLI. To prevent one side of the spacecraft roasting and the opposite side freezing in the constant sunlight of cislunar space, it was to be oriented with its principal axis perpendicular to the ecliptic, then set rolling on a 20-minute cycle for passive thermal control (PTC) in a regime more popularly known as the barbecue mode. This had to be established before the astronauts could retire for the night. There were four options in the flight plan for midcourse corrections to refine the trajectory for lunar orbit insertion. The first was scheduled 9 hours after TLI, but since the delta-V was only 17 feet per second the burn was deleted, and the initiation of PTC advanced. Ten hours into the mission, after orienting the spacecraft perpendicular to the ecliptic, Collins gave the computer 20 minutes in which to fire the RCS thrusters as necessary to eliminate any axial wobble.

While waiting, Aldrin enquired about their altitude, which was 50,000 nautical miles. "It's a beautiful sight," he enthused, referring to Earth. "I can see snow on the mountains of California. It looks like LA doesn't have much of a smog problem today. With the monocular, I can discern a definite green cast to the San Fernando Valley."

"How's Baja California look, Buzz?" Duke asked.

"It has got some clouds up and down it, and there is a pretty good circulation system a couple of hundred miles off the west coast of California."

"Your rates look really great, now," Duke called a few minutes later. "You can start your PTC."

"If you would like to delay PTC for 10 minutes or so, we can shoot you some television of a seven-eighths-phase Earth," Armstrong offered.

Duke checked whether Goldstone was configured to receive television; it was. "It will be recorded at Goldstone and then replayed over here, Neil. Any time you want to turn her on, we're ready."

"We're sending a picture of Earth right now. Let us know if they're receiving at Goldstone."

"Goldstone says it looks great," Duke confirmed.

Armstrong was pointing the camera through the window beside his couch, and they had put up shades to prevent the Sun entering the other windows and causing internal reflections. He zoomed in until Earth filled the screen, and was pleased to observe that the automatic gain control adjusted the aperture to compensate.

"We'd like 10 minutes' worth of television," Duke requested. "And we'd like a narrative on the exterior shots. We also suggest you might try an interior position."

"We're seeing the eastern Pacific Ocean," replied Armstrong. "We've not been able to visually pick up the Hawaiian Island chain, but we can clearly see the western coast of North America – the United States, the San Joaquin Valley, the High Sierras, Baja California, and Mexico down as far as Acapulco and the Yucatán peninsula. And you can see on through Central America to the northern coast of

South America, Venezuela and Colombia. I'm not sure you'll be able to see all that on your screen down there."

"Roger, Neil," acknowledged Duke. "We wanted a narrative so that when we get the playback we can correlate what we're seeing."

Although Armstrong and Aldrin had each remarked on the diminishing view of Earth, Collins had been too busy sighting on stars and operating the computer. "I haven't seen anything but the DSKY so far," he pointed out.

"It looks like they're hogging the windows," Duke consoled.

"You're right," Collins confirmed.

There was a distinct sense of distance – Earth and the Moon were both far off – but the distances were so immense that, in the absence of points of reference along the way, there was no sensation of movement despite the spacecraft's high speed.

After 15 minutes, Duke announced, "You can terminate the television at your convenience, and then initiate PTC." The impromptu telecast had verified that the television system was working beautifully.

Collins initiated the roll. With the Sun, Earth and Moon passing in procession by the windows, the illumination in the cabin became very dynamic.

As the camera was packed away, Armstrong gave a crew status report, reading their individual radiation monitors, noting that they had not taken any medications, and saying that in his opinion they were "fit as a fiddle", by which he meant they had not developed any symptoms of 'space sickness'.

"As far as we can see," Duke agreed, "you're cleared for some zzzz's." In fact, as a result of having deleted midcourse correction 1, they were two hours ahead on the time line.

When testing the CSM with Apollo 7, Wally Schirra's crew had maintained a staggered sleeping cycle in which there was always one man awake to monitor the spacecraft. On Apollo 8 Frank Borman's crew had done the same, with the result that they slept poorly and were exhausted on entering lunar orbit. It was therefore decided that future crews should adopt a single sleep cycle. Armstrong was eager for his crew to conserve their energies in transit, to ensure that they would be alert on arrival in lunar orbit. In fact, it had been agreed that he and Aldrin would act as passengers because their mission would not start until they were in lunar orbit, and Collins would act as the chauffeur for the translunar coast. Although Duke signed off early, the crew did not retire until 10.30 pm on their Houston-time watches. Having arisen very early to prepare for launch, it had been a long day, and in their last few hours they were feeling distinctly drowsy. They had their supper, placed shades over the windows to block out the Sun and dimmed the internal lamps. Armstrong and Aldrin then disconnected their communications links and snuggled into enclosed hammocks – in effect mesh sleeping bags slung beneath the side couches – while Collins, the 'watch keeper', donned a lightweight headset with its earpiece on low volume, in case of a call from Mission Control, and strapped into the left-hand couch using a lap belt; as he rested, he realised that he no longer had any discomfort in his knee.

Cliff Charlesworth, Gene Kranz and Glynn Lunney were to rotate 8-hour shifts during the translunar coast. The 'graveyard' shift was handled by Lunney's Black

Team. For the first 24 hours of a mission, flight controllers maintain a particularly close eye on the telemetry in order to identify the quirks of the new craft early on, and to establish norms and consumable rates.

Goldstone sent the taped telecast to Houston by a land line, where the signal was fed through conversion equipment for display on the wall screen and release to the media. On arriving home from the Cape and finding the press on her lawn, Jan Armstrong informed them that it had been a long day and she really needed to sleep, but happened to switch on her television as the networks started to run the telecast; the other wives missed it.

FLIGHT DAY 2

While Apollo 11 had been in Earth orbit it had been able to communicate through the standard facilities of the Manned Space Flight Network, but once it had set off for the Moon NASA added in the large antennas of the Deep Space Network at Goldstone in California, Madrid in Spain and Canberra in Australia, which, being located at 120-degree intervals in longitude, provided continuous communications. Ron Evans, the CapCom on the 'graveyard' shift, never had occasion to talk to the crew. Biomedical telemetry indicated that the astronauts had been active for some time when McCandless made contact. "Apollo 11, this is Houston."

"Good morning, Houston," replied Armstrong immediately.

As flight day 2 began, almost 23 hours into the mission and with Houston time approaching 8 am on Thursday, 17 July, Apollo 11 was 93,085 nautical miles from Earth and its velocity had slowed to 5,638 feet per second.

McCandless immediately read up the flight plan updates and the status of the spacecraft's consumables. As part of the post-sleep checklist, Collins gave a crew status report: he and Armstrong had slept for 7 hours, but Aldrin had gained only 5.5 hours. While the crew freshened up and prepared breakfast, McCandless read a selection of lighthearted news provided by the Public Affairs Office. "From Jodrell Bank, England, via Associated Press: The big Jodrell Bank radio telescope stopped receiving signals from the Soviet Union's unmanned moonshot at 5.49 Eastern Daylight Time today. A spokesman said it appeared Luna 15 'had gone beyond the Moon'. Another quote from a spokesman for Sir Bernard Lovell, Director of the Observatory: 'We don't think it's landed.' Washington, United Press International: Vice President Spiro T. Agnew has called for putting a man on Mars by the year 2000, but Democratic leaders replied that priority must go to needs here on Earth. Agnew, the ranking government official at your launch, apparently was speaking for himself, and not necessarily for the Nixon administration. Laredo, Texas, AP: Immigration officials in Nuevo Laredo announced, Wednesday, that hippies will be refused tourist cards to enter Mexico unless they take a bath and get haircuts. Huberto Cazaras, Chief of Mexican Immigration in Nuevo Laredo, said authorities in popular tourist spots had registered complaints about the hippies. Next is from UPI in Washington: The initial reaction to President Nixon's granting of a holiday, Monday, to Federal employees so they can observe a national day of participation in your lunar landing was one of surprise. Rodney Bidner, AP, London: Europe is Moon-struck by your mission. Newspapers throughout the continent filled their pages with pictures of the Saturn V rocket lifting off to forge Earth's first link with its natural satellite. The headline-writers taxed their imaginations for words to hail the feat. 'The greatest adventure in the history of humanity has started', declared the French newspaper *Le Figaro*. It devoted 4 pages to reports from the Cape and has diagrams of the mission. The tabloid *Paris Jour* proclaimed, 'The whole world tells them, bravo'. The communist daily *L'Humanite* led with the launch picture, and devoted its entire back page to an enthusiastic report describing the countdown and launch, the astronauts' wives and families, and some background for the lunar activities. Hempstead, NY: Joe Namath officially reported to the New York Jets training camp at Hofstra University, Wednesday, after a closed-door meeting with his team mates to discuss his differences with the pro-football commissioner, Peter Rozelle. London, UPI: The House of Lords was assured, Wednesday, that a midget American submarine would not 'damage or assault' the Loch Ness monster. Lord Nomay said he wanted to be sure anyone operating a submarine in the Loch 'would not subject any creatures that might inhabit it to damage or assault'. He asked that the plan to take a tissue-sample with a retrievable dart from any monster be done without damage and disturbance. He was told that it was impossible to say if the 1876 Cruelty to Animals Act would be violated unless and until the monster was found."

And with that, it was back to work. McCandless provided feedback to Collins regarding the difficulties he had encountered in locating the substellar point during the P23 sextant sightings of stars during the previous day's deep-space navigation. At the 24-hour point, Collins halted the PTC, Houston uplinked a new state vector based on tracking by the Manned Space Flight Network, and Collins performed a P52 to realign the platform prior to recalibrating the sextant as a preliminary to resuming the P23 exercise. Now at a distance of 102,436 nautical miles, Earth had a much smaller angular diameter and its horizon was sharper. He started with the star Alpheratz because it was near the horizon, simplifying the task of checking whether the automatic alignment correctly identified the star's substellar point. Although the axis of measurement was clearly perpendicular to the horizon, the star itself was not apparent. "Everything looks beautiful except there is no star in sight. It is just not visible."

McCandless announced the belated realisation that the star was occulted by the body of the LM, and recommended a different attitude.

While Collins performed the manoeuvre, Jim Lovell called, "Is the commander aboard?"

"This is the commander," replied Armstrong.

"I was a little worried. This is the backup commander still standing by. You haven't given me the word yet. Are you Go?" Lovell was reminding Armstrong that if he was not feeling up to the mission, he was willing to take his place – in jest, of course, as an exchange was impossible following launch.

"You've lost your chance to take this one, Jim."

"Okay. I concede."

Collins now announced that in the second attitude the axis of measurement was rotated 90 degrees from that needed to measure the elevation of the star. "I'm going to hold right here for your next suggestion."

Charlesworth decided that they should curtail the P23 exercise, and prepare for midcourse correction 2. Initiated at 026:44:57.92 at a distance of 109,245 nautical miles from Earth while travelling at 5,033 feet per second, the 2.91-second burn of the SPS engine slowed the spacecraft by 20.9 feet per second to reduce the closest approach to the Moon from the initial 175 nautical miles to the desired 60 nautical miles.

"We saw about 87 or 88 psi chamber pressure," reported Armstrong, referring to the SPS. This was rather low. "I'd like you to look at that on the ground."

McCandless said he would pass on the result of an engineering analysis of the telemetry. Two hours later, he confirmed that the chamber pressure had been stable at 94 psi, which was acceptable. Further analysis established that the performance of the engine during both burns matched that of the acceptance trials, and that the discrepancy was merely inaccurate calibration of the onboard gauge.[3] As the crew worked through the post-burn checklist, McCandless said, "we played the recorded television back last night, after you all turned in for your rest period, and the pictures came out quite well".

Having abandoned the P23 exercise, Collins began to re-establish PTC. He was impressed that the rotational axis had remained stable through their sleep period. While the computer worked to damp out oscillations, Lovell provided angles for the high-gain antenna, which was one of Aldrin's tasks. "You may have to repeat some of that, James," Collins warned. "We have got a LM guy taking care of the high-gain right now, and he's got his head out the window."

"I understand," said Lovell. "I had trouble on Gemini 12 with him, too."

"Hey, Jim," Aldrin called, "I'm looking through the monocular now and, to coin an expression, the view is just beautiful. It's out of this world. I can see all the islands in the Mediterranean – Majorca, Sardinia and Corsica. There is a little haze over the upper Italian peninsula, some cumulus clouds out over Greece. The Sun is setting on the eastern Mediterranean now. The British Isles are definitely greener in colour than the brownish green on the Iberian peninsula."

"Do you find that the monocular is any good to you, Buzz?" Lovell asked.

"It would be nicer if it had another order of magnitude of power on it. Of course, it has a tendency to jiggle around a little bit, and you might want to have some sort of a bracket. There is an anticyclone down in the southern hemisphere, southeast of Brazil, and the diameter of it must be over 2,000 miles across." Then he reported something he had not expected. "I've got a comment about the point on the Earth where the Sun's rays reflect back up toward us. In general, the colour of the oceans is mostly uniform and it's bright and darker blue except for this circular area that's about one-eighth of an Earth's radius in diameter in which the blue of the water turns a greyish colour."

[3] Houston would revisit this issue at the start of flight day 4.

"We noticed the same thing on Apollo 8. It's very similar to looking at a light shining on something like a billiard ball or a bowling ball; you get this bright spot in the blue of the water, and that turns it to sort of a greyish colour." The technical term for this phenomenon was specular reflection.

Aldrin had also been experimenting. "Hey, Jim, the best way to get a steady view through the monocular is to steady it out and set it close in front of your eye, and then you kind of float up next to it so that you're not touching it at all."

"How does it feel to be airborne again, Buzz?" Lovell asked.

"I've been having a ball floating around inside here – it's like being outside, except more comfortable."

"It's a lot bigger than our last vehicle," Lovell observed, referring to Gemini.

Collins, the chauffeur, joined in the banter. "Oh, yes. It's nice. I've been very busy so far. I'm looking forward to taking the afternoon off. I've been cooking, and sweeping, and almost sewing, and you know, the usual little housekeeping things."

"It's very convenient the way they put the food preparation system right next to the navigation station," Lovell said, implying that Collins could do everything from his position in the lower equipment bay.

"Everything is right next to everything else in this vehicle," Armstrong noted, meaning that the designers had done well to squeeze everything into such a small volume.

With the spacecraft's attitude stabilised perpendicular to the ecliptic, Collins initiated the PTC roll. As he did so, they crossed the point where they were equidistant between Earth and Moon, 112,386 nautical miles from each. However, as they were still slowing down, they were by no means half-way in terms of time. As Earth drifted by, Collins took the opportunity to take a look for himself. "I've got the world in my window for a change, and looking at it through the monocular it's really something. I wish I could describe it properly. The weather is very good. South America is coming around into view, and I can see all the way down to the southern tip of Tierra del Fuego." With that, Earth drifted from his field of view. Aldrin promptly pointed out that he was waiting to pick it up in the sextant for a magnified view.

"It sounds like one of those rotating restaurants!" McCandless said.

The mention of a restaurant prompted them to start lunch.

"Is that music I hear in the background?" McCandless asked.

"Buzz in singing," Collins replied.

"Houston," Armstrong called, "we're just looking at you out our window here, and it looks like there's a circulation of cloud that just moved east of Houston over the Gulf and Florida area. Did you have any rain this morning?"

"Our report from outside says it's raining now. It looks like you've got a pretty good eye for the weather there!"

"Well, it looks like it ought to clear up pretty soon," Armstrong advised.

After lunch, Armstrong and Aldrin reviewed LM activities, and Collins did the routine chores of purging contaminants from the fuel cells, topping up the charges of the batteries, dumping waste water and exchanging the lithium hydroxide filter that absorbed the carbon dioxide of the astronauts' exhalation.

A telecast was scheduled for later in the afternoon, and the astronauts decided to conduct another system test without announcing the fact. After several minutes, McCandless called, "Eleven, Houston. Goldstone reports that they are receiving a television picture coming down. It's a little snowy, but a good television picture."

"We're just testing the equipment up here," Armstrong explained.

"Ask if they can read the numbers," Collins prompted. The view was showing the DSKY, and he wanted to know if the display was readable, as a measure of the image quality.

"What numbers are you referring to?" McCandless asked.

"Well," Collins sighed, "I guess if they can't see any numbers, it's kind of a lost cause!"

"We want to know what numbers, before we ask them," McCandless said.

Collins realised the misunderstanding. "I'm showing them the DSKY."

After conferring, McCandless reported, "They can read the numbers, 'VERB', 'NOUN', 'PROGRAM', and the 'COMPUTER ACTIVITY' light is flashing.

"Very good. Thank you."

Without providing commentary, the crew, handing the camera back and forth, aimed it through window 4 at the overhead window of the LM, through window 2 at the docking target, and then provided close-up views of the main control panel of the command module.

The picture was snowy because it was transmitted through the omnidirectional antennas, which could not provide the signal strength of the high-gain antenna, and the spacecraft was just about at the limit for transmitting television in this manner. For the telecast due later in the day they were to halt PTC in order to maintain the high-gain antenna pointing at Earth.

As the flight control teams switched shifts, McCandless handed over to Duke.

"How's the White Team today?" Collins enquired.

"Bright-eyed and bushy-tailed."

Collins, on the left side, braced his arms against the bulkhead 'above' his head to maintain his feet against the wall of the lower equipment bay. "Have you got any medics down there watching telemetry? I'm trying to do some running in place here, and I'm wondering just out of curiosity whether it brings my heart rate up."

"We see your heart beating."

Armstrong, in the centre, joined in.

"Look at Armstrong's and Collins's and see if they go up any," Collins called. "We're running in place up here. You wouldn't believe it."

"I'd like to see that sight," Duke replied. "Why don't you give us a television picture of that?"

"I think Buzz is trying," Collins said. "Have you got it?

"It's coming in at Goldstone, but we don't have it here in the Control Center."

"I'm afraid this isn't going to help out the PTC very much!" Collins reflected. "I don't know if it's a vibration or what it is, but it makes the pitch and yaw rate needles on the FDAI oscillate a little bit where we jump up and down."

"Goldstone say they see you running there, Mike. You're about a 96 heart beat now."

"That's about all that's reasonable, without getting hot and sweaty." They had only one change of clothes, one for the trip out and the other for the trip home.

Aldrin then aimed the camera through window 5 at Earth. At 31 hours into the mission, Apollo 11 was 121,158 nautical miles out, and travelling at 4,613 feet per second.

"For this television program coming, you might give some thought to how you want us to stop PTC for the best high-gain angle," Collins suggested. Also, it'd be nice if you could stop us at such an attitude that we'll have Earth out of one of our windows."

"Boy," announced Aldrin, "you sure get a different perspective of the cabin in zero-*g*. Right now, Neil has got his feet on the forward hatch, and not only can he reach all five windows with his arms but he can also reach down into the lower equipment bay."

"Sounds like Plastic Man to me," laughed Duke.

"I'm hiding under the left-hand couch, trying to stay out of his way," Collins chipped in.

"Good idea, Mike," Duke agreed. He then specified an attitude for the telecast that would both provide a good line of sight for the high-gain antenna and position Earth in the left-hand window. In fact, since the recommended attitude would hold the primary axis perpendicular to the ecliptic, Collins would be able to establish it simply by halting the roll.

When Duke said the weather over the Houston area had cleared, as Armstrong had predicted, Collins took a look through the sextant. "I can see that the coastline is clear; those clouds have moved inland. It looks like the southeastern part of the country is socked in, but California looks nice. The San Joaquin Valley shows up as a real dark spot with a lighter brown on either side of it. You can't tell that it's green; it looks just sort of dark grey, or maybe even real dark blue."

"How does the Mojave look?" Duke asked. "Is it clear?"

"Yes – as usual."

"Can you pick out Edwards in the sextant?"

"I can see an F-104 taxiing out for takeoff on the runway," Collins joked.

"That's super!"

"They almost always have a 104 taxiing!" Collins explained.

"Hey, Charlie," Aldrin called, "what's the latest on Luna 15?"

"*TASS* reported this morning that it has entered orbit close to the lunar surface, and everything seems to be functioning normally." Nevertheless, the Russians had not confirmed that Luna 15 was to attempt to land, let alone that it was to scoop up a sample for return to Earth; they never announced missions in advance.

The flight plan called for a telecast starting at an elapsed time of 34 hours (i.e. 6.32 pm in Houston) and lasting for 15 minutes. After terminating PTC, they began to transmit, and this time Mission Control was configured to view 'live'. The telecast began with a long shot of Earth. "This is Apollo 11, calling in from about 130,000 miles out," Armstrong announced. "We'll zoom our camera in slowly and get the most magnification we can." Although the quality of the image using the high-gain antenna was excellent, Duke requested a description. "We're looking at the eastern

Pacific Ocean, and in the top half of the screen we can see North America, Alaska, United States, Canada, Mexico and Central America, but South America becomes invisible just off beyond the terminator, or inside the shadow. We can see the oceans with a definite blue cast; white bands of major cloud formations across the Earth; the coastline of the western US; the San Joaquin Valley; the Sierra range; the peninsula of Baja California; and some cloud formations over the southeastern US. There's one definite mild storm about 500 to 1,000 miles southwest of Alaska, and another very minor storm at the south part of the screen, at probably 45 degrees or more southern latitude. We can pick out the browns in the landforms pretty well. Green doesn't show up very well, but there is some showing along the northwestern coast of the United States and Canada."

"The whites are distinct, but on this monitor the landmasses appear to be just a darker greyish colour rather than a brown," Duke said.

"It's true that we don't have the depth of colour at this range that we enjoyed at 50,000 nautical miles out," Armstrong said. "But the oceans still are a definite blue and the continents are generally brownish in cast, although they're tending more towards grey now than they were at the closer range."

"I've just looked at another monitor," Duke announced, "and sure enough, the browns are coming in a lot more distinctly on the Eidophor that we have up on our screen in the Control Center."

On the front wall of the Mission Operations Control Room there was a 10-foot by 20-foot main screen in the centre, with two 10-foot by 10-foot screens on each side. Television could be displayed using an Eidophor projector, a technology that was developed in the 1950s to create theatre-sized television images. Its optical system was similar to a conventional movie projector, but instead of spooling film it had a slowly rotating disk covered with a thick oil, and a scanning electron beam created electrostatic charges on the oil, deforming its surface in such a way that when light was passed through the oil it produced an image with light and dark areas. As the disk rotated, a blade discharged and smoothed the surface, readying the oil for the next cycle. The modern Eidophors could show colour television using sequential red, green and blue projections.

At this point, Collins, who was holding the camera while Armstrong provided commentary, interjected, "Okay, world, hold on to your hat. I'm going to turn you upside-down." He slowly rotated the camera through 180 degrees.

"That's a pretty good roll," Duke complimented.

"I'm making myself seasick, Charlie. I'll just put you back rightside-up where you belong." He completed the circle.

"We would like to see some smiling faces up there, if you could give us some interior views," Duke prompted. "I'm sure everybody would like to see you." The astronauts' wives especially.

The interior lights had been switched off in order to prevent reflections on the window through which the camera was viewing. Once the lights were reactivated, the camera was swung inside to display Collins. "Hello there sports fans. Buzz is doing the camera work now. I would have put on a coat and tie if I'd known about this ahead of time."

"Is Buzz holding your cue cards for you?" Duke enquired.

"We have no intention of competing with the professionals, believe me. We're very comfortable up here, though. We have a happy home. There's plenty of room for the three of us, and I think we're all learning to find our favourite little corner to sit in. Zero-*g* is very comfortable, but after a while you get to the point where you sort of get tired of rattling around and banging off the ceiling and the floor and the side, so you find a corner somewhere and put your knees up or something like that to wedge yourself in." Armstrong tended to remain in a couch and Aldrin spent much of his time in the lower equipment bay, but Collins, who was operating the CSM, was for ever flitting about.

The view switched to Armstrong. "It's a real good picture we're getting here of Commander Armstrong," Duke complimented.

"Neil's standing on his head again, trying to make me nervous," said Collins. Armstrong had his feet up by the apex. "Directly behind his head are our optical instruments, the sextant and the telescope that we use to take sightings with."

"It's a beautiful picture," Duke enthused. "The clarity is outstanding." In fact, the quality was even better than on Apollo 10. Unfortunately, this led the audience to expect to view the moonwalk in similar style.

Aldrin aimed the camera at one of his star charts, which he had taped up over window 5 to serve as a sunshade.

"He doesn't really need the charts," Collins noted. "He's got them memorised. They're just for show." Aldrin had even brought a slide rule in case the computer should fail!

Aldrin explained that with Earth visible in one side window and the Sun in the other, the spacecraft's current attitude was perpendicular to the ecliptic. From their viewpoint, the Moon was approaching the Sun in the sky and facing its darkened hemisphere towards them, which was why they could not show the audience their destination. He handed the camera to Armstrong, grasped wall handles using both hands, and gave a demonstration of weightless exercises.

Pointing out that it was approaching dinner time in Houston, Collins went to a locker in the lower equipment bay. "We'll show you our food cabinet here."

"We see a box full of goodies there," said Duke.

"We've got all kinds of good stuff," Collins said as he pointed to the groups of packets. "We've got coffee up here at the upper left; and various breakfast items – like bacon in small bites; beverages like fruit drink; and over in the centre part we have all kinds of things. Let me pull one out here, and see what it is." He read the label. "Would you believe you're looking at chicken stew? All you have to do is add 3 ounces of hot water and wait for 5 or 10 minutes. Now we get our hot water out of a little spigot up here with a filter on it that filters out any gases that may be in the drinking water, and we just stick the end of this little tube in the end of the spigot and pull the trigger three times for 3 ounces of hot water and then mush it up and slice the end off it and there you go, beautiful chicken stew."

"Sounds delicious," Duke said.

"The food so far has been very good," noted Collins. "We couldn't be happier with it." Because that part of the cabin was dimly illuminated, he had been using a small torch to show the food packs. He showed how he could leave the torch in a stable orientation to illuminate the locker in order to have both hands free to sort through the packets. "The problem is that no matter how carefully you let go, you bump it just a tiny little bit and set it in motion." He tapped one end of the slowly rotating torch with his finger, and set it spinning as a demonstration of rotation in weightlessness.

Aldrin pointed out that in preparing for the telecast, they had realised the cable for the television camera (which they had not seen in training) was probably long enough to run through the tunnel into the LM, and said that they intended to try to televise the LM inspection scheduled for the third day. Duke said that would be a great idea.

Armstrong pointed the camera back at Earth and operated the zoom to make it diminish. "This is Apollo 11 signing off."

"Thank you for the show," said Duke. At about 35 minutes, it had lasted twice the scheduled duration.

As they had not departed from an attitude perpendicular to the ecliptic, all that Collins needed to do to reinstate PTC was to resume the roll.

Meanwhile, at home

Joan Aldrin began Thursday, 17 July, by raising the flag, thereby allowing the photographers on the front lawn their first 'snap' of the day. Following lunch, she sneaked out of the back yard in order to shop with a friend, Mary Campbell. Although she thought she was incognito, in the mall a salesperson who processed her purchases recognised her and said, "Thank you, Mrs Aldrin. We all wish you the best of luck." Jan Armstrong whiled away the afternoon in the yard clearing the swimming pool of leaves that had been shed during the storm when she was at the Cape. Meanwhile, in the Collins' yard, the tree that had blown down was being removed. Jan Armstrong watched the telecast sitting on the floor, with sons Ricky and Mark, although Mark was bored. Joan Aldrin, wearing the new outfit she had purchased at the mall, watched from the couch with son Andy, Robert and Audrey Moon, Jeannie Bassett and Dee O'Hara. Joan was eager to see Buzz, but because he was operating the camera initially she had to wait to catch sight of him, and then she laughed at his weightless exercises. Pat Collins watched with her children, but Michael was more concerned with his rabbit, and his sisters fidgeted. Pat grew frustrated with the view of Earth; she wanted to see the crew. When Collins finally appeared, she exclaimed, "Mike's growing a moustache!"

Back in space

An hour after the telecast, Duke asked the crew to participate in an experiment using a laser. "If you have got Earth through any of your windows, or in the telescope, would you so advise?"

"I've got you in the telescope," Collins replied.

"We have a blue-green laser that is to flash at a frequency of on-for-a-second, off-for-a-second. It's from the McDonald Observatory near El Paso, Texas, which should be just on the dark side of the terminator. Would you take a look and see if you can see it." Collins agreed. "McDonald reports there is a break in the clouds," Duke announced a few minutes later, "and they're beaming it through." Observing nothing through the telescope, Collins switched to the magnifying sextant, with no better result. Armstrong joined him. Aldrin reported, "Neither Neil nor Mike can see it. We did identify the El Paso area, and there did appear to us to be a break in the clouds there." However, at their distance from Earth, the beam would be just a few thousand feet across, and was very likely not illuminating the vehicle.

Half an hour later Duke signed off, "The White Team bids you good night."

"You earned your pay today, Charlie," Aldrin said.

As the Black Team began the 'graveyard' shift, with Gerry Griffin standing in for Glynn Lunney as flight director, the astronauts finished miscellaneous chores, had supper, and settled down for their second sleep period, this time with Aldrin as the watch-keeper. By this time, Apollo 11 was 137,219 nautical miles from Earth, and travelling at 4,132 feet per second.

FLIGHT DAY 3

As the scheduled end of the sleep period approached, Cliff Charlesworth, now in charge, decided that since the crew were sleeping soundly and there was nothing on the flight plan requiring urgent attention they should be left in peace. An hour later, telemetry indicated that the astronauts were stirring. Armstrong and Aldrin had slept for 8 hours, and Collins for 9 hours – consistent with the early retirement and late awakening.

"Apollo 11, this is Houston," McCandless called.

"Good morning, Houston," Aldrin replied promptly. "How do all our systems look?"

"They're looking great, and as far as we can tell everything is good from down here."

"It looks like the attitude held up really well during PTC last night," observed Aldrin. The spin axis had remained within 10 degrees of the ideal perpendicular to the ecliptic, providing excellent thermal control.

"How's the Green Team this morning?" Collins enquired.

"It was a very quiet night. The Black Team is complaining that they didn't get a chance to make any transmissions. Ron Evans is getting to be known as the silent CapCom."

"That's the best kind, Bruce," Collins teased.

"Okay," chuckled McCandless. The banter over, he launched straight into the flight plan updates, the most important of which was the cancellation of midcourse correction 3 at 54 hours. "At 53 hours we have a P52. We're requesting that you do this while in PTC, and we plan to continue PTC throughout the day." However, one of the first chores of the day – dumping the unwanted water produced by the fuel cells – imparted an impulse that perturbed the roll. "We're showing you about 20 degrees out in pitch right now and about 6 degrees in yaw," McCandless called 30 minutes later. "That's a little more than twice as much as the deviation you had prior to the waste-water dump. We'll watch it down here, and let you know if we believe any corrective action is required."

"Maybe next time we ought to split that in half," Aldrin suggested. "Put half of it on one side and half on the other, or something like that."

"We could do that," McCandless agreed. "We're actually interested in seeing what the effect on PTC is of this waste-water dump. We don't recall ever having performed a waste-water dump during PTC on previous missions." In other words, a chore had transformed into an engineering experiment, with the data being filed for future reference. "We've been working under the assumption that it would take an hour for a water dump to dissipate to the point where you could reasonably take star sightings for platform alignment, navigation, or something of this sort. If you have a spare minute or two, could you comment on observation conditions, now?"

"My guess," Collins replied after looking, "would be the telescope's probably pretty useless, but you can differentiate in the sextant between water droplets and stars by the difference in their motions." But he had not been using the telescope very often. "With the LM attached, the telescope is just about useless because the Sun bounces off its structure. Those star charts that the mission planning and analysis division people provided us, I think, would be most useful if for some reason we had to mark through the telescope – we could use those as a guide for what we're looking at and say, 'Well, that bright blob over there *must* be such and such a star because that's the position we're in'. But so far we haven't been able to pick out any decent star patterns [using the telescope]."

After two hours of sparse interaction between Apollo 11 and Mission Control, McCandless announced, "I've got the morning news here if you're interested."

"Yes, we sure are," Collins replied.

"Interest in Apollo 11 continues at a high level," McCandless assured, "but a competing interest in the Houston area is the easing of watering rules. Mayor Louie Welch has promised to lift lawn-watering restrictions if the rains continue. Today is partially cloudy, with a 30-per-cent chance of thunderstorms in the afternoon. In Washington, the Senate Finance Committee has approved extension of the income tax surtax, but a Senate vote on the bill currently seems remote. In Austin, State Representative Ray Lemmon of Houston has been nominated as the National Director of the American Society for Oceanography. Lemmon has proposed a study of the possibility of establishing an institute of oceanography in Texas. This would be the first such institute on the western Gulf of Mexico. In Minneapolis, Minnesota, the weather bureau, after recapping today's weather showing a high of 88 and a low of 72, has noted 'snowfall: none'. From St Petersburg, Florida, comes a radio report from the Norwegian explorer, Thor Heyerdahl, which points out that the crew of his papyrus boat, the *Ra*, will sail into Bridgetown, Barbados, despite damage from heavy seas. The crew, however, are sleeping on their escort vessel. Norman Baker, navigator of the expedition, said the crew was aboard the *Ra* today repairing damage from storms this week that split the footing of the mast. Part of the broken mast was jettisoned overboard. The vessel is now 725 miles east of the Barbados. 'It is possible but uncomfortable to sleep aboard the *Ra*,' Baker said. 'But the purpose of our voyage is not a test of strength or human endurance.' That's the reason the crew was spending nights on the escort vessel *Shenandoah*, which rendezvoused with the *Ra* on Tuesday." In his sports roundup, McCandless related the story of an Irishman, John Coyle, who won the world's porridge-eating championship by consuming 23 bowls of instant oatmeal in a 10-minute time period from a field of 35 other competitors.

"I'd like to enter Aldrin in the oatmeal eating contest next time," Collins said.

"Is he pretty good at that?" McCandless asked.

"He's doing his share up here," Collins confirmed.

"Let's see. You all just finished a meal not long ago, too, didn't you?"

"I'm still eating," Aldrin pointed out.

"He's on his 19th bowl!" Collins joked.

"Are you having any difficulties with gas in the food bags, like the Apollo 10 crew reported?"

"That's intermittently affirmative, Bruce," Collins replied. "We have these two hydrogen gas filters that work fine as long as you don't actually hook them up to a food bag. But the entry way into the food bag gives enough back pressure to cause the filters to lose efficiency. A couple of times, I have been tempted to go through that drying out procedure, but we found that simply by leaving the filters alone for several hours their efficiency seems to be restored – it ranges anywhere from darn near perfect to terrible, just depending on the individual characteristics of the food bags. Some bags are so crimped near the entry that there is no way to work them loose to prevent back pressure." The gas separator comprised two stainless steel cylinders about 5 inches long and a little over 1 inch in diameter, attached to the water dispenser. These contained two filters, one to attract water and the other to repel it, in the process removing the gas. The design had been modified after the Apollo 10 crew had reported problems, but it evidently required additional work. The ingested hydrogen gave rise to what Collins would later describe as "gross flatulence in the lower bowel, resulting in a not-so-subtle and pervasive aroma" reminiscent of "a mixture of wet dog and marsh gas". Aldrin would later jest that by the time they were on their way home they were suffering so badly that if the RCS thrusters were to have failed they would have been able to provide manual attitude control!

The first version of the flight plan had envisaged the tunnel remaining sealed until in lunar orbit. Aldrin, however, had successfully argued for an inspection of the LM during the translunar coast, since if the rigours of launch had so damaged that vehicle as to render it unusable it would be best to discover this sooner rather than later. However, as a result of the mass limit imposed on the design of the LM, it could accommodate only six chemical storage batteries, which in turn limited the total electrical power supply, with the result that at this point in the flight it would not be feasible to power it up to transmit telemetry to enable Houston to check its systems. Nevertheless, an early entry would enable Aldrin to make a start on chores such as removing and stowing protective covers.

Six hours into the day, McCandless asked whether they were still intending to take the camera into the LM to televise this inspection.

"If the cord lengths work out all right," Armstrong confirmed.

In view of the growing instability of the roll axis, and the fact that PTC would have to be halted for the telecast, Charlesworth decided that they should go ahead and adopt an attitude in which the high-gain antenna could readily be maintained facing Earth.

"When you work up an attitude for the high-gain, is there any way we could get partial Sun in one of the two LM front windows?" Aldrin asked.

"We'll have a look at it," McCandless promised. Several minutes later he relayed, "We recommend stopping PTC at 054:45:00; this should put you at just about the right roll angle to give you Earth in window 1 of the command module, aim the high-gain antenna for television, and put the Sun on the forward hatch of the LM. If you take down the window shades, you should get some sunlight in."

While Collins made the manoeuvre, the Green Team handed over to the White Team, and Charlie Duke took over as CapCom.

The telecast was not expected to start until 056:20, or 4.52 pm in Houston, but at 055:10 Apollo 11 began to transmit. "They're getting television at Goldstone," Duke announced. "We're not quite configured for it here, but we should be up in a couple of minutes."

"This is just for free," Collins explained. "This isn't what we had in mind."

They had decided to enable Houston to watch the tunnel being opened. Aldrin was operating the camera. Collins had just removed the apex hatch and was in the process of stowing it in a bag beneath the left-hand couch. After Armstrong made a preliminary inspection of the probe assembly using a torch, Collins entered the tunnel to release the mechanism by repeatedly cycling its ratchet handle. If it had failed to release, there was a toolkit with which to dismantle it. As the astronauts were not providing running commentary, Duke made occasional observations, but because his remarks ran 12 seconds 'late' owing to the time it took to convert the picture this sometimes gave rise to confusion. "It's a pretty good show here," he began. "It looks like you've almost got the probe out."

"Can you see that?" Armstrong asked. "There isn't much light up in there, just the tunnel lights."

"Roger, Neil. It's really good."

"It's coming down," Armstrong said, as Collins pulled the bulky mechanism of interconnected rods from the tunnel.

"It looks like it's a little bit easier than doing that in the chamber," noted Duke. The mechanism was heavy on Earth, and as part of their training they had removed the probe in an altitude chamber.

"You have to take it easy," Collins observed. The probe was weightless, but it still had inertia.

"Mike must have done a smooth job in that docking," Armstrong announced, "because there isn't a dent or a mark on the probe." They used elasticated cords to stow the mechanism by the wall at the foot of the couch. The conical drogue was stowed alongside the probe. With the picture lagging so far behind the audio, Duke experienced a sense of déjà vu in which he listened to the astronauts describing an action in real-time, and then waited to watch them do it.

Mike Collins in a CM similator prepares to open the apex hatch.

Having removed the apex hatch, Mike Collins prepares to remove the probe assembly.

Collins re-entered the tunnel and checked an indicator that showed the angular offset between the two docking collars; the fact that this was only two degrees was a tribute to his skill in performing the docking.

"It looks like we'll be ready to go into the LM early, if that's okay with y'all down there," Armstrong said. They were about 40 minutes ahead of the time line.

"Go ahead any time you wish," Duke replied.

Aldrin passed the camera to Armstrong, entered the tunnel, turned the handle on the hatch and hinged it inwards into the LM, activating the cabin lights. For this inspection, the LM would draw power from the CSM. Armstrong returned the camera to Aldrin, who pointed it into the other craft to display items of equipment in stowage on the floor of the cabin 'above' him.

"Buzz, are you already in?" Duke asked.

"I'm halfway in." Having never seen the LM from this perspective in training, Aldrin was momentarily disoriented. "I'd better turn around, I guess." By making a half somersault to restore his frame of reference, he immediately felt at home in the cramped cabin. He would later report this to have been the strangest sensation of the mission. Although in the LM, he was on a communications umbilical running back through the tunnel into the command module. Because the mass limit on the LM precluded the use of panelling, the wire bundles and plumbing were largely exposed. The hull interior had been sprayed with a dull-grey fire-resistant coating. The front and sides carried a mass of switches, circuit breakers and instruments. The walls were very thin in places, but were not required to carry structural loads – they were only a pressure shell against the vacuum of space. Although the shades were over the main windows, sunlight diffused through, providing a low level of illumination. Aldrin pointed the camera back down the tunnel to show Armstrong at the far end, with Collins behind him, watching.

"Hey, that's a great shot," Duke said. "I guess that's Neil and Mike – it better be, anyway!"

Armstrong entered the tunnel to hold the camera, to enable Aldrin to make a start on his inspection. "I'll open up the windows to see what the lighting's going to be like," Aldrin announced. He pulled the shades, first from the right and then the left window, then donned his sunglasses.

"The lighting is superb!" Duke exclaimed.

Armstrong made his first contribution. "Yes, the lighting in the LM is very nice now, just like completely daylight; and everything is visible."

"The vehicle is surprisingly free of any debris floating around," said Aldrin. "It's very clean." In fact, during his inspection he would discover only one 'lost' washer floating adrift. After inspecting the miscellaneous stowed equipment, he tested the LM's telescope, mounted near the roof in the centre of the front panel, noting that when he looked 'up' he could see the shiny surface of the command module. Meanwhile, Armstrong rotated the camera to point down the tunnel to show Collins poking through an oxygen umbilical with which to ventilate the LM's cabin. Then Aldrin took the camera and pointed it through the narrow overhead window to show one of the forward-looking windows of the command module. "Charlie, can you see Mike staring out the window?" The view was indistinct because there were so many

Buzz Aldrin during the LM inspection.

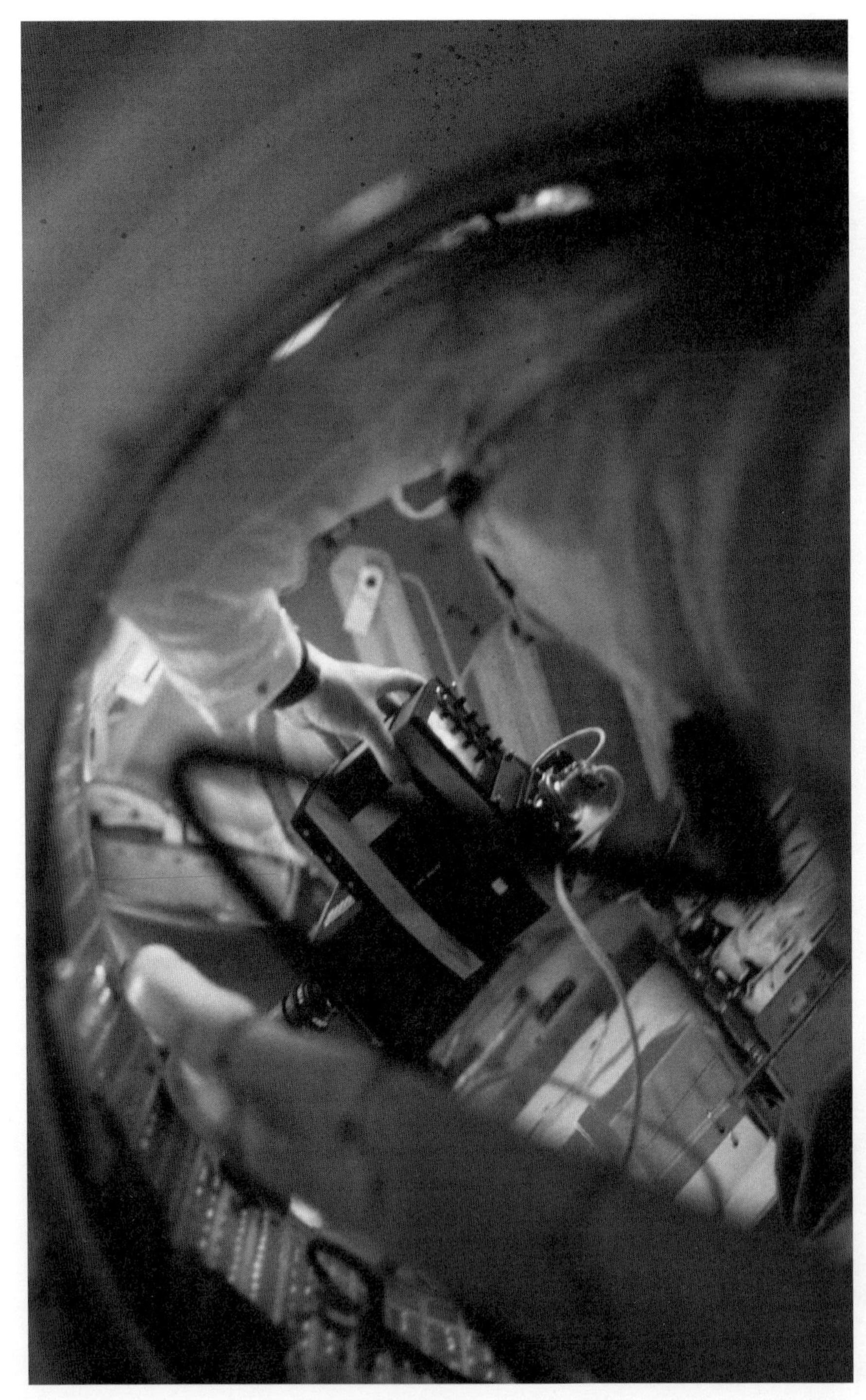

From his position in the tunnel, Neil Armstrong documented the LM inspection using the television camera.

layers of glass, but when Collins put his head up close to the window his face became apparent.

"We see him staring back at us," Duke confirmed.

At the scheduled start time for the telecast, the networks picked up the feed. "Your show is going out to the US now," Duke announced. "We're about to get the satellite up and then it will go 'live' to Japan, western Europe, and much of South America." Aldrin was installing the bracket to the top-right corner of his window on which he was to mount a Maurer camera to document Armstrong's descent of the ladder at the start of the moonwalk. Next, he installed another bracket midway along a horizontal bar across the window for later photography.

While Aldrin worked, Armstrong zoomed in to show the instrument panel to the estimated audience of 200 million people. "That's real good camera work," Duke complimented.

"It's got to be the most unusual position a cameraman's ever had, hanging by his toes from a tunnel and taking the picture upside-down," said Aldrin, referring to Armstrong. Aldrin unstowed an assembly designed to fit over a 'bubble' helmet for extravehicular activity on the lunar surface, and demonstrated how its twin visors operated.

Armstrong then aimed the camera back down the tunnel. "Ah, that's a good view of Mister Collins down there," Duke said. "We finally see him again!"

"Hello there, Earthlings," said Collins.

"It's like old home week, Charlie, to get back in the LM again," Aldrin said to Duke.

"I can imagine," Duke agreed. "Is Collins going to go in and look around?"

"We're willing to let him," Armstrong replied, "but he hasn't come up with the price of the ticket yet."

"I'd advise him to keep his hands off the switches," Duke warned Armstrong.

"If I can get him to keep his hands off my DSKY," Collins retorted, "it would be a fair swap."

On returning to the command module, Armstrong aimed the camera at Earth, now 177,000 nautical miles away. "We're going to turn our television off now for a short bit while we do some other work. Apollo 11 signing off."

"That was one of the greatest shows we've ever seen," Duke complimented. It had lasted about 1 hour 36 minutes, during which time the spacecraft had travelled over 2,000 nautical miles.

Meanwhile, at home

Joan Aldrin had hosted an afternoon pool party. Pat Collins attended with her sister Ellie Golden. Jan Armstrong brought her sister Carolyn Trude. After the wives had appeared together for the press on the front lawn, they retreated to the swimming pool, joining Jeannie Bassett. The Collins children had been sent to a day camp, but the Armstrong and Aldrin children were present and played in the pool with Kurt Henize, son of Karl and Caroline Henize. Valerie Anders made a brief visit. Audrey Moon prepared snacks and Bob Moon served as a drinks waiter. At 4.30 pm Jan Armstrong and her sister set off for home. When Jan switched on the car radio she was surprised to hear that the telecast was already in progress. On reaching home, they found the house to be even busier, because their mother, Mrs C.G. Shearon of Pasadena, California, and their sister Nan Theissen and her husband Scotty had arrived. Jan watched what remained of the telecast. When Joan switched on her television she was delighted to see Buzz hogging the show. She noted, wryly, that he had gained more air time in this one transmission than she, a trained actress, had managed in her entire career! Pat Collins, who left the Aldrins' at the same time as Jan Armstrong but did not promptly switch on her television, missed most of the telecast. All three wives were frustrated that their NASA minders had not alerted them to the change in schedule – if they had been told that the crew had started transmitting over an hour earlier than planned, they could have watched from the viewing gallery of the Mission Operations Control Room. Pat Collins successfully eluded the reporters to eat in a favourite restaurant named *Rendezvous*, but with so many guests Jan Armstrong phoned for a bulk pizza delivery.

Back in space

"Our recommendations on the activities for the next hour or so," said Duke at 57 hours, "are to continue your LM familiarisation as desired until 58 hours, then ingress to the CSM, close the hatch, and set up PTC shortly thereafter." In view of the extent to which a water dump had upset the roll axis, it was decided that the tank should be totally vented prior to re-establishing PTC. As the spacecraft had been oriented perpendicular to the ecliptic for the telecast and was stable, it would be necessary only to initiate the roll manoeuvre.

When Collins said that he wanted to do a P52 in order to correct the platform prior to setting up PTC, he found that Owen Garriott, a member of the small group of scientist–astronauts that NASA recruited in 1965, was on the CapCom console. "Are the Maroons on now?"

"Not permanently, Mike," Garriott explained. "The Maroon Team will be on tomorrow." Once Apollo 11 was in lunar orbit, Kranz's team of flight controllers were to take a rest in preparation for handling the powered descent, and their slot in the shift cycle would be worked by Milton Windler's team.

With the tunnel restored and PTC re-established, the crew had their supper to the accompaniment of music. They had a cassette player with a variety of music tapes. Armstrong had selected the *New World Symphony* by Antonin Dvorák and *Music Out Of The Moon* by Samuel J. Hoffman, whose titles seemed appropriate. Collins had a variety of 'easy listening'. Aldrin was content with his colleagues' choices, and rarely played his own selections.

"This PTC looks sort of weird to us," Duke called after they had eaten. "We've already drifted out to 70 degrees in pitch, and we're wondering if you might have done anything that could have caused us to pick up these rates." When Collins said he knew of nothing that could have upset the rotation, Duke continued: "When we started off, it looked real fine to us. Now it's drifting off with a funny pattern that we have not seen previously on a flight. We're trying to figure out." Five minutes later he was back, "We'd like to terminate this PTC, as we have no assurance that we're going to get through the sleep period with this funny pattern." On previous missions the automatic thruster firings designed to damp out oscillations had kept the crew awake. Collins halted the motion, re-established the proper attitude, gave the computer time to cancel out any tendency to drift, and restarted the roll. This time, the axis remained stable, and the astronauts retired for the night.

INTO THE MOON'S SPHERE OF INFLUENCE

At 061:40, Apollo 11 entered the Moon's sphere of influence. The strength of a gravitational field is directly proportional to the mass of the gravitating body and inversely proportional to the square of the range from the body. With Earth fully 81 times more massive than the Moon, the 'neutral point' was roughly 90 per cent of the way to the Moon; more specifically, 186,437 nautical miles from Earth and 33,822 nautical miles from the Moon. By this milestone, the velocity of the spacecraft had slowed to a relative crawl of 2,990 feet per second with respect to Earth and 3,272 feet per second with respect to the Moon. When Apollo 8 first blazed this trail, the less technically minded members of the press expressed surprise that the crew did not feel a jolt. The origin of this misunderstanding was a comment at a press conference by flight dynamics officer Philip C. Shaffer. After pointing out that Apollo 8 had *slowed* as it climbed from Earth, and beyond the neutral point *accelerated* towards the Moon, he added that as the spacecraft crossed the neutral point, the computers in the Real-Time Computer Complex, which constantly calculated its position and velocity, switched from a terrestrial to a lunar frame of reference and the way the numbers were crunched made it appear that the position of the ship had jumped several miles. There was, in fact, no physical manifestation of passing from one gravitational field to the other. Although Apollo 11 entered the Moon's sphere of influence, it was not committed to it because, being on a free-return trajectory, if it were to do nothing it would pass around the back of the Moon and be deflected back to Earth.

Meanwhile, at home

During the overnight shift, those members of Glynn Lunney's Black Team who had been asleep during the day and had missed the telecast of the LM inspection, watched a replay.

6

Into lunar orbit

FLIGHT DAY 4

As Apollo 11's trajectory was predicted to produce a closest approach to the Moon of 62 (± 2) nautical miles, as against the nominal 60, Glynn Lunney, leading the overnight Black Team, decided to cancel midcourse correction 4 and, as the crew would not now require to devote time to a P52 platform realignment and the other chores for the manoeuvre, he also decided to extend the sleep period either by two hours or until the astronauts awoke of their own accord.

When the scheduled wake-up time passed without word from Houston, Aldrin, who was standing watch, put in a call, "Houston, Apollo 11." Getting no response, he repeated the call twice more.

It was approaching 6 am in Houston on Saturday, 19 July. "Good morning," replied Ron Evans, making his first communication of the mission as 'graveyard' shift CapCom.

"Are you planning midcourse correction 4?" Aldrin asked.

"That's negative," said Evans. "We were going to let you sleep in until about 71 hours, if you'd like to turn over."

"Okay. That'll be fine."

Several minutes later, however, Armstrong and Collins also stirred, and it was decided to make an early start while remaining off the air until Houston made the formal call. Armstrong and Collins had slept 7.5 hours; Aldrin for only 6.5 hours. During this hiatus, Cliff Charlesworth's Green Team began their shift, with Bruce McCandless as CapCom.

The astronauts' wives had an early start, too. Jan Armstrong went out onto the lawn to give the photographers the first of their daily pictures. Pat Collins went to have her hair done, only to find that three female journalists also just happened to have appointments. On returning home, she took delivery of flowers from Annie Glenn, wife of John Glenn, with a card that stated: 'May God watch over you and your family.' Meanwhile, Joan Aldrin welcomed Michael and Rosalind Archer, her father and step-mother, who had just arrived from Pensacola, Florida. Rusty Schweickart and his wife Clare dropped off an enormous casserole of cold roast turkey with which to help feed the growing crowd.

"Good morning *again*," said Aldrin when McCandless put in the formal wake-up call at 71 hours.

After relaying the flight plan updates, McCandless, with lunar orbit insertion imminent, returned to the low-pressure indication in the combustion chamber of the main engine during midcourse correction 2. "I've got a few words for you on the SPS engine performance."

"We're ready to listen," Armstrong replied.

"The onboard combustion chamber pressure reading is due to a known gauge calibration factor between what is in the chamber and what you're reading on the gauge. We expect a single-bank operation to be 90 psi on the gauge for an actual chamber pressure of 95 psi; in dual-bank operation, a chamber pressure of 94 psi on the gauge is an actual of 99 psi." The redundant propellant-feed valves of the engine were actuated by nitrogen and could be used individually (single bank) or together (dual bank). The lunar orbit insertion burn was to be done in dual-bank mode. A mission rule required the crew to cut short the manoeuvre if the chamber pressure fell below a given value, and McCandless clarified this. "Similarly, 80 psi on the gauge correlates to 83 psi actual, and we recommend that you stick to an LOI-termination-cue of 80 psi on the gauge – that is, no change to mission rules."

"Roger," replied Armstrong. "We got all that."

Although Apollo 11 was now only 10,000 nautical miles from the Moon, the crew had not been able to see it since flight day 2 because the geometry of their trajectory meant that when they looked in that direction they were blinded by the Sun. With the spacecraft still in PTC, sunlight flooded in through a succession of windows, but while they were having breakfast the spacecraft passed into the Moon's shadow and the cabin was suddenly plunged into darkness. Earth orbits the Sun at an average distance of 93 million statute miles. The visible surface of the Sun, known as the photosphere, is 860,000 miles in diameter, but this is surrounded by a very hot but tenuous atmosphere. This glowing corona is normally seen by Earthlings only when the Moon's shadow falls on Earth during a total solar eclipse – having flown into that shadow, Apollo 11 had arranged its own eclipse. On looking out, the astronauts saw the corona projecting out behind the now large, but dark, lunar disk.

Deciding to take a picture, Collins asked McCandless to suggest an exposure for the corona. He got a typically comprehensive response. "We recommend that you use high-speed black-and-white film, interior lights off, the Hasselblad with the 80-millimetre lens and, as you'll be hand-holding the camera, an f-stop of 2.8, with a sequence of time exposures at one-eighth of a second, one-half of a second, and, if you think you can steady the camera against something, exposures of 2, 4 and 8 seconds."

"It's quite an eerie sight," said Aldrin. "There's a marked 3-dimensional aspect of having the corona coming from behind the Moon the way that it is." From their perspective, with the Moon occulting the Sun, most of the lunar disk was revealed by 'Earthshine'. The phases of the Moon and Earth are opposed, in the sense that when the Moon is a crescent, Earth is gibbous – which was the case at this point in the mission. Also, owing to their actual sizes, the angular diameter of Earth's disk is fully four times that of the Moon as viewed from Earth. Furthermore, Earth has a

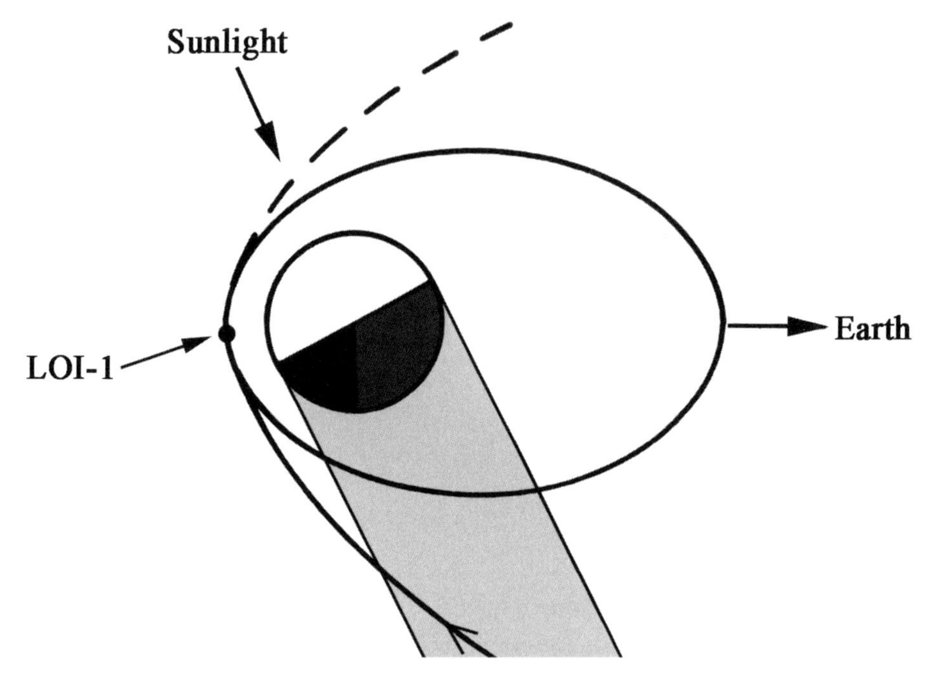

A diagram of the 'Earthshine' illumination as Apollo 11 approached the Moon to perform the LOI-1 manoeuvre.

greater albedo, meaning that it reflects a higher percentage of the sunlight that falls on it. Thus, Earthshine illuminates the lunar nightscape more brightly than a 'full' Moon does a terrestrial night. "I guess what's giving it that 3-dimensional effect is the Earthshine," Aldrin mused. "I can see the crater Tycho fairly clearly – at least, if I'm right-side-up, I believe it's Tycho. And, of course, the sky is lit all the way around the Moon." From Apollo 11's vantage point, it was possible to see a dark crescent of the hemisphere that faces away from Earth, with the limb silhouetted against the glowing corona.

"The Earthshine coming through the window is so bright you can read a book by it," Collins noted.

"How far out can you see the corona extending?" McCandless asked.

"I would suggest that along the ecliptic we can see the bright part of the corona out to about two lunar diameters; and perpendicular to that line, only one-eighth to one-quarter of the lunar radius," Armstrong ventured. As their eyes adapted to the darkness, they saw the stars. "It's been a real change for us. Now we're able to see stars again and recognise constellations for the first time on the trip. The sky's full of stars. It's just like the night-side of Earth. All the way here, we have only been able to see stars occasionally, and perhaps through the monocular, but we haven't been able to recognise any star patterns."

"I guess it's turned into night up there really, hasn't it?" said McCandless.

"It really has," Armstrong agreed.

"If you have a minute or so free, we can read you up the morning news here," McCandless offered a few minutes later.

"Let's hear it," Armstrong replied.

"Hot from the wires of the Manned Spacecraft Center's Public Affairs Office, especially prepared for the crew of Apollo 11: First off, it looks like it's going to be impossible to get away from the fact that you guys are dominating all the news back here. Even *Pravda* in Russia is headlining the mission, and refers to Neil as 'The Czar of the Ship'; I think maybe they got the wrong mission."

Fred Haise read the next item, "West Germany has declared Monday to be 'Apollo Day'. School children in Bavaria have been given the day off. Post Office clerks have been encouraged to bring radios to work, and Frankfurt is installing television sets in public places."

McCandless again, "The BBC in London is considering a special radio alarm system, to call people to their television sets in case there is a change in the EVA time on the Moon."

Haise, "And in Italy, Pope Paul VI has arranged for a special colour television circuit at his summer residence in order to watch you – that's even though Italian television is still black-and-white."

McCandless, "Here in Houston, your wives and children got together for lunch yesterday at Buzz's house and, according to Pat, it turned out to be a gabfest. The children swam and did some high jumping over Buzz's bamboo pole."

Haise, "In Moscow, space engineer Anatoli Koritsky, quoted by *TASS*, stated that Luna 15 could accomplish everything that has been done by the earlier Luna spacecraft. This was taken by the press to mean it could investigate gravitational fields, photograph the Moon, and then go down onto the surface to scoop up a bit for analysis."

McCandless, "Mike, even your kids at camp got into the news. Michael Junior was quoted as replying 'Yeah' when somebody asked him if his daddy was going to be in history. 'What's history?' he asked, after a short pause."

After a sports round-up, Haise concluded with a novelty item, "You might be interested in knowing, since you are already on the way, that a Houston astrologer, Ruby Graham, says that the signs are right for your trip to the Moon. She says that Neil's clever, Mike has good judgement and Buzz can work out intricate problems. She also says Neil tends to see the world through rose-coloured glasses, but he is always ready to help the afflicted or distressed. Neil, you are also supposed to have intuition that enables you to interpret life with feeling. Buzz is supposed to be very sociable, cannot bear to be alone and has excellent critical ability. Since she didn't know at what hour you were born Mike, she has decided that you either have the same attributes as Neil or you're inventive with an unconventional attitude that might seem eccentric to the unimaginative."

"Who said all that?" Collins asked.

Haise laughed, "Ruby Graham, an astrologer here in Houston."

"Tell Michael Junior that history or no history, he'd better behave himself," Collins requested.

"We'll pass that along," McCandless promised.

A few minutes later, Collins announced that he was going to halt PTC because the translunar coast was drawing to a close. By design, their trajectory would take the spacecraft in front of the Moon's leading limb. However, because the Moon is 2,160 statute miles in diameter, and pursues its orbit around Earth at 2,287 miles per hour, this rendezvous required precise navigation through cislunar space.[1] Although it was possible to enter lunar orbit by executing a single manoeuvre, if the engine were to fire for too long the spacecraft might dip so low thereafter as to crash. As the lunar orbit insertion (LOI) burn was required to occur on the far side of the Moon, out of communication, in planning Apollo 8 it had been decided to guard against this outcome by splitting the manoeuvre. The initial burn (LOI-1) would produce an orbit with a high apolune on the near side of the Moon, and the Manned Space Flight Network would track the spacecraft during the first two revolutions to make a precise calculation of the delta-V for the burn (LOI-2) that would circularise the orbit at the desired altitude, and this strategy had been retained by the later missions. Three hours before LOI-1 McCandless read up the details: time of ignition would be 075:49:49, the duration of the burn was to be 6 minutes 2 seconds, and the objective of the retrograde delta-V of 2917.3 feet per second was to achieve a lunar orbit with a perilune of 62 nautical miles and an apolune of 170 nautical miles. The loss of signal (LOS) as the spacecraft passed behind the Moon's leading limb in the run-up to this manoeuvre was predicted at 075:41:23. If the burn was as intended, then the acquisition of signal (AOS) upon appearing around the trailing limb would be at 076:15:29. If for some reason the engine did not fire, acquisition would be 10 minutes earlier, and the gravitational 'slingshot' resulting from the encounter would have deflected the spacecraft's trajectory back to Earth. If the engine were to cut off prematurely, then Apollo 11's fate would depend upon whether the delta-V attained was sufficient to enable the Moon to capture it. If it attained an orbit with a very high apolune, this might possibly be lowered to continue the lunar phase of the mission. If, however, the burn did not last long enough to enable the Moon to capture the spacecraft, then it would emerge from the trailing limb heading into deep space and, in the event of the SPS being deemed unusable, their fate would rest upon whether it was possible to use the descent propulsion system of the LM to *attempt* a return to Earth; there was no guarantee.

At 075:30, after Charlesworth had polled his team, McCandless passed on the decision, "You're Go for LOI."

"Roger," acknowledged Aldrin. "Go for LOI."

"All your systems are looking good going around the corner, and we'll see you on the other side."

[1] In the event, only one of the four midcourse correction options had been necessary.

OVER THE HILL

Jack Riley was the Public Affairs Officer in Mission Control. "We've just had loss of signal as Apollo 11 passed behind the Moon. At that time we were showing its distance from the Moon as 309 nautical miles and its velocity with respect to the Moon as 7,664 feet per second. Here in the Control Center, two members of the backup crew, Bill Anders and Jim Lovell, have joined Bruce McCandless at the CapCom console. Fred Haise, the third member of the backup crew, just came in, too. And Deke Slayton, Director of Flight Crew Operations, is also present. The viewing room is filling up: among those on the front row are Tom Stafford, John Glenn, Gene Cernan, Dave Scott, Al Worden and Jack Swigert."

The inertial attitude of the vehicle was such that at its closest point of approach to the Moon the SPS engine would be facing the direction of motion, to serve as a brake. The CSM had redundant power buses, but in preparation for the burn these were 'tied' together to ensure that if one power supply were to fail this would not disrupt the operation of the systems at a critical time.

"I've turned the S-Band volume down to get rid of that background noise," Collins announced, "so don't forget that we have to turn it back up on the other side!"

With 2 minutes to go, they emerged from the Moon's shadow. As they had been flying 'on instruments', Collins looked out of the window to visually confirm that they were in the correct attitude. "Yes, the Moon's there, in all its splendour."

As they raced across the terminator, the deeply shadowed terrain appeared to be extremely rough. "Man," exclaimed Aldrin, "look at it!"

"Don't look at it!" said Armstrong, drawing their attention back inside. "Here we come up to ignition."

At 5 seconds to go, the computer flashed '99' in the Verb display of the DSKY, to ask the crew whether they wished to go ahead with the burn as specified. "99", noted Collins.

"Proceed," commanded Armstrong.

Collins hit the PROCEED key to tell the computer to execute the manoeuvre. "Stand by for ignition." Armstrong's heart rate was 106 beats per minute, Collins's was 66, and Aldrin's was 70.

Two RCS thrusters fired briefly to settle the propellants in the main tanks, and then at the appointed time the computer ignited the SPS engine.

"Burning!" confirmed Armstrong.

"What's our chamber pressure?" Collins asked.

The gauge was below the left FDAI. "It's good, 95," confirmed Armstrong.

"The PUGS is oscillating around," noted Aldrin. The Propellant Utilisation Gauging System measured how the engine was drawing fuel and oxidiser. Aldrin was to use a knob to maintain the correct combustion mixture, but this was tricky due to the lag in response.

"Okay, we're steering," noted Collins. "The gimbals are a little busier than I'd have expected, but everything's looking good." The engine gave a load equivalent to one-fifth gravity. "The *g* feels sort of pleasant."

"Tank pressures are good," Aldrin confirmed.

"The chamber pressure is building up a little bit; 96 now," Armstrong noted as he monitored the gauge.

"That's a little more chamber pressure than they were predicting," pointed out Collins, thinking of the miscalibration.

"Chamber pressure is continuing to rise," added Armstrong a moment later, "it's up to about 98 psi."

"We're wandering off a little bit in roll, but that's to be expected," observed Collins. "It's coming back."

The high chamber pressure would reduce the duration of the burn, but that was not the critical issue; what mattered was the change in velocity, and the computer would shut down the engine when the required 2,917.3-foot-per-second change in velocity had been attained. "Cutoff is going to be about 3 seconds early," warned Aldrin.

"Nominal cutoff is at 6 + 02," Collins noted, "so expect it around 6 minutes even, huh?"

"I'm predicting 5 + 58, 4 seconds early," said Armstrong. "Maybe 5 seconds."

"She's steering like a champ," Collins said. "The rates are wandering, but in all three axes they're plus or minus 0.1 degree."

"Five seconds early, at 5 + 57," Armstrong updated. "The chamber pressure is 100 psi even."

Collins counted down the seconds to the predicted cutoff.

"Shutdown!" confirmed Armstrong.

As they ran through the post-burn checklist, Collins prompted the computer to display the discrepancies between the desired and achieved velocity as measured in the three-coordinate system. "Minus 1, minus 1, plus 1; Jesus!" The 'residuals' were only 0.1 foot per second; the burn was essentially perfect. "I take back any bad things I ever said about MIT – which, of course, I never have."

Aldrin glanced out at the Moon. They were now well past the terminator and the terrain was well lit. "I have to vote with the Apollo 10 crew that the surface is brown."

"It sure is," agreed Collins.

"It looks tan to me," said Armstrong.

"But when I first saw it, at the other Sun angle it looked grey," Aldrin noted, referring to just prior to the burn. "It got more brown with increasing Sun angle."

Armstrong again drew his colleagues' attention inside, "Alright, now we've got some things to do."

As they continued through the checklist, Collins said, "Well, I don't know if we're 60 miles or not, but at least we haven't hit that mother."

The computer displayed the parameters of their orbit. "Look at that!" Aldrin exclaimed, "169.6 by 60.9."

"Beautiful, beautiful, beautiful, beautiful!" enthused Collins. "Write it down just for the hell of it: 170 by 60, like gangbusters."

"We only missed [apolune] by a couple of tenths of a mile," Aldrin added in amazement.

It is not possible to orbit Earth at an altitude of 60 nautical miles, as this would be subject to the drag of the upper atmosphere. However, the Moon is airless. It is also smaller and less massive and, since its gravity is weaker, a spacecraft in orbit is not required to travel so fast. On the other hand, because the orbit is smaller, the period – at about 2 hours in this case – is only marginally longer than for a low orbit around Earth.

"Hello, Moon," Collins greeted. "How's the old back side?"

"Now," said Armstrong, "the flight plan says we roll 180 degrees and pitch down 70 degrees."

"What are we pitching down for?" Collins asked. Then he laughed. "I don't know what we're doing."

Armstrong enlightened his CMP, "We're going to roll over and pitch down so that we can look out the front windows down at the Moon!"

"Oh, yes, okay," Collins acknowledged.

As the spacecraft slowly executed the 180-degree roll, Collins asked, "Can we see the Earth on the horizon from here?"

"We should be able to," Armstrong replied.

Collins decided they should take a picture of their first Earthrise. "Big lens or small one?"

"For the Earth coming up, we want the 250-millimetre," decided Aldrin, as he started to assemble the Hasselblad. "Infinity, at f/11 and 1/250th, huh?"

"Is it loaded with black-and-white, or colour?" asked Collins.

"Colour."

"Alrighty!" said Collins, satisfied.

"We ought to wash this window," Aldrin said. "Anybody got a Kleenex?"

Both Armstrong and Collins offered towels.

"Well, one more SPS burn," Collins mused, thinking of completing the lunar orbit insertion sequence.

"Two more!" corrected Aldrin, remembering that the engine would also have to make the transearth injection burn if they were to get home.

Meanwhile, at home

Jan Armstrong had Barbara Young over for lunch. Having been through lunar orbit insertion on Apollo 10, Barbara described how she had had an anxious wait when her husband's flight had reported back fully two minutes later than expected, raising the prospect of the burn having slowed that vehicle so much that it would crash. However, Jan was more concerned that AOS might be *early*, indicating that the burn had not occurred and the spacecraft was heading back to Earth on the free-return trajectory, because she knew that to lose the opportunity to attempt to land would be devastating for the crew's morale. "Don't you dare come around," she muttered as the no-burn time approached. "Don't you dare!" When that moment passed with no signal, she delightedly exclaimed "Yippee!"

Back in space

"Look at those craters in a row. Something really peppered that one," Collins said, drawing attention to the lunar surface. "There's a lot less variation in colour than I would have thought, you know, looking down?"

"But you'd say it's brownish?" Aldrin asked.

"Sure."

"Oh, golly, there's a huge, magnificent crater over here," exclaimed Aldrin. "I wish I had the other lens on. God, that's a big beauty. You should look at that guy, Neil."

"I see him," Armstrong said.

"Well," said Collins, "there's really no doubt that the Moon is a little smaller than the Earth; look at that curvature."

Aldrin suggested that as they still had about 10 minutes to AOS, they ought to switch the Hasselblad to a 'wider' lens in order to shoot the lunar landscape.

"Just don't miss that first one," urged Collins, eager that they should record their *first* Earthrise.

The landscape passing below was fascinating.

"What a spectacular view!" remarked Armstrong.

"There's a hole down here you just wouldn't believe," Collins pointed out. "And there's the biggest one yet. God, it's huge! It's enormous! It's so damned big I can't even get it in the window! You want to look at that? That's the biggest one you ever seen in your life. Neil? God, look at this central mountain peak."

Meanwhile in Houston, Jack Riley, continuing his public commentary, noted, "It's very quiet here in Mission Control. Most of the controllers are seated at their consoles, several are standing up. We're now 7 minutes from the acquisition time for the nominal burn. If Apollo 11 attained only a partial burn, we could receive a signal at any time."

"Isn't that a huge one?" Collins exclaimed, indicating another crater as they continued around the Moon's far side. "It's fantastic! Oh, boy, you could spend a lifetime just geologising that one crater alone!" After pausing to reflect, he added, "Although that's not how I would like to spend *my* lifetime!"

"There's a big mother over here, too," Aldrin pointed out.

"Come on now, Buzz," Collins chastised, "don't refer to them as 'big mothers' – give them some scientific names."

"It sure looks like a lot of them have slumped down," noted Aldrin, referring to the terraced rims of the craters, which had indeed slumped into the pit.

"A slumping big mother!" exclaimed Collins. "Well, you see those every once in a while."

"Most of them are slumping," continued Aldrin, ignoring Collins. "The bigger they are, the more they slump."

"We're at 180 degrees," announced Armstrong as he terminated the roll, "and now we start a slow pitch down about 70 degrees."

"We've got 4 minutes to get pitched down before AOS," pointed out Aldrin. "We'll never make it."

"Goddamn, a geologist up here would just go crazy," suggested Collins. "We shouldn't take any more pictures on this roll until we get Earth."

"We might make it in time," said Armstrong, referring to the timing of their pitch manoeuvre.

"There it is," exclaimed Aldrin. "It's coming up!"

"What is?" Collins asked.

"The Earth!" explained Aldrin. "See it?"

"Yes. Beautiful."

"It's halfway up already," Aldrin noted. He snapped a picture, but not having changed back to the 250-millimetre lens it was not the close up Collins desired.

"We ought to have AOS now," Armstrong pointed out.

The antenna in Madrid acquired the carrier signal right on time, indicating a good burn.

"AOS!" Riley announced to the public.

"Apollo 11, this is Houston. Do you read?" prompted McCandless. Although Armstrong replied with a full burn report, the high-gain antenna had yet to lock on and the transmission by omnidirectional antenna was so noisy that he was mostly unintelligible. "Could you repeat your burn status report?" McCandless requested.

"It was perfect!" Armstrong replied simply.

Once the high-gain link had been established, the flight controllers examined the telemetry. Going 'over the hill', the combined mass of the two spacecraft was 96,012 pounds. The LOI-1 burn had consumed 24,008 pounds of propellant.

Meanwhile, at home

Joan Aldrin, who was at the hair-dresser, was listening to the radio coverage of the mission. In her house, Rusty Schweickart was explaining to the assembled crowd that the manoeuvre had gone to plan.

SIGHTSEEING

As Apollo 11 started across the near side of the Moon, essentially in the equatorial plane and going east to west,[2] Armstrong called, "Apollo 11 is getting its first view of the landing approach. We're going over the Taruntius crater." This crater, some 30 nautical miles in diameter, was in the northwestern part of the Sea of Fertility. It had a flat floor and a complex of several low peaks at its centre. "The pictures and maps brought back by Apollos 8 and 10 have given us a very good preview of what to look at here – it looks very much like the pictures, but like the difference between watching a real football game and watching it on television." The clarity of viewing by eye-ball far exceeded that of the pictures. "There's no substitute for actually being here."

"We're going over the Messier series of craters now, looking vertically down on them," announced Aldrin, "and we can see good-sized blocks in the bottom of

[2] Prior to the dawning of the space age, astronomers had defined lunar longitudes in terms of their view of the Moon in the terrestrial sky, with the leading limb being east. However, in 1961 the International Astronomical Union had redefined the system to place east in the direction of sunrise as seen from the lunar surface, which reversed the old scheme.

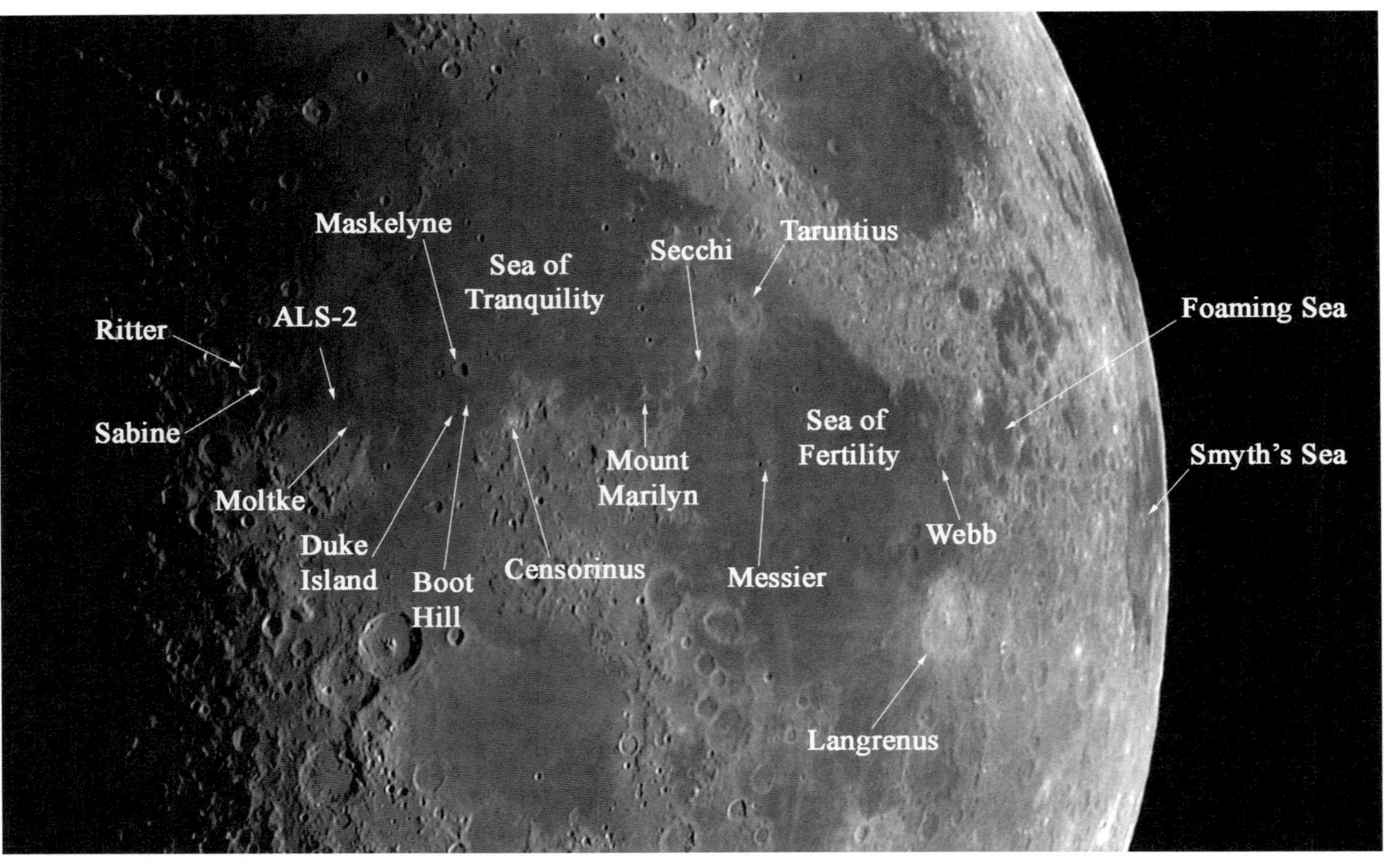

The features noted by the Apollo 11 crew during their first telecast in lunar orbit.

Messier-A." This was a pair of craters, each about 6 nautical miles across. They had originally been known as Messier and Pickering, after French and American astronomers, but Pickering had been renamed Messier-A. The craters were notable for having produced a pair of divergent bright streaks that ran westward across the dark plain of the Sea of Fertility for a distance of about 60 nautical miles.

"And there's Secchi," added Armstrong. Named after an Italian astronomer, it was 14 nautical miles in diameter.

Having crossed the Sea of Fertility, they started across the Sea of Tranquility, where they were to land. "We're going over Mount Marilyn at the present time," said Aldrin. "That's the ignition point."

"Jim's smiling," pointed out McCandless. Jim Lovell had named the peak after his wife.

Tom Stafford, who had reconnoitred this route on Apollo 10, explained to Jan Armstrong how the astronauts had assigned names they would readily remember to various features that they intended to use as landmarks. However, such names were unofficial. In 1961 the International Astronomical Union had specified how features were to be named: the flat plains would be named in Latin after states of mind, using the traditional (if inaccurate) terms *oceanus* (ocean), *mare* (sea), *sinus* (bay), *lacus* (lake) and *palus* (marsh); mountain ranges were to be named after their terrestrial counterparts; and craters were to be named after deceased scientists. In 1967 the IAU had met to discuss the assignment of specific names, but decided to defer naming until the next meeting in 1970. Meanwhile, of course, the crews of Apollo 8 and Apollo 10 had named some features. As the Moon was progressively transformed from an object studied by astronomers using telescopes into a world that was being explored by astronauts, it was inevitable the astronomers would lose 'ownership' over it.

As soon as the spacecraft appeared around the limb, the Manned Space Flight Network had begun to track it, to determine its orbit independently of the onboard navigation. "Our preliminary tracking data for the first few minutes shows you in a 61.6 by 169.5 orbit," McCandless announced. At this time, the spacecraft was at an altitude of 127 nautical miles, and climbing towards apolune. The period of the elliptical orbit was 2 hours 8 minutes 37 seconds.

Armstrong called out landmarks on the line of approach. "We're going over Maskelyne, Boot Hill, Duke Island, Maskelyne-W – our yaw-around checkpoint – and Sidewinder." Boot Hill, Duke Island and Sidewinder were unofficial names. The crater Maskelyne, named after an Astronomer Royal, was 13 nautical miles in diameter. Maskelyne-W followed the IAU convention that small craters close to a 'named' crater were identified by a letter postfix. The astronauts had dubbed it the Wash Basin. "Now we're coming up to the terminator. The landing site is well into the dark." By design, they had arrived prior to sunrise at the landing site. The landing was 26 hours away, and with the Sun rising at 12 degrees per day, it would be some 10 degrees elevation at the time of landing. On the question of the colour of the surface, Armstrong noted that it had appeared tan in the vicinity of the subsolar point, where the Sun shone straight down, and had then faded to grey until, at the terminator, it was ashen.

While the spacecraft had been behind the Moon, its telemetry was taped and, on AOS, was downloaded to Earth. The flight controller responsible for the CSM promptly studied the performance of the SPS during the burn, and noticed that the tank pressure for the nitrogen used to drive the propellant feed valves of Bank-B was anomalously low, although it was holding steady now. Bank-A appeared nominal. McCandless sought clarification. "When you have a free moment, could you give us an onboard readout of nitrogen tank Bravo."

"We're showing the pressure in tank Bravo to be 1,960 psi, something like that," replied Aldrin, "and Alfa is about 2,250 psi." These matched the telemetry, which showed corresponding values of 1,946 and 2,249. Now able to eliminate a telemetry problem, the flight controller delved deeper into the record to further characterise the problem, with a sense of urgency because the SPS engine would soon be called upon to perform the second burn of the lunar orbit insertion sequence.

As the spacecraft passed over the Ocean of Storms on the darkened part of the near side, McCandless made an unscheduled request. "We've got an observation you might make – if you've time. There's been a lunar 'transient event' reported in the vicinity of Aristarchus." An astronomer had suggested that the astronauts take a look at the 22-nautical-mile-wide crater Aristarchus which, although far to the north of the ground track, should be on the near side of the horizon at their current altitude of 167 nautical miles. For many years, astronomers had been actively watching for transient lunar phenomena such as glows in the night. Of course, the sightings were disputed. The best evidence was by the Russian astronomer Nikolai A. Kozyrev of the Crimean Astrophysical Observatory on 3 November 1958, when he secured a spectrogram of a 'red glow' that persisted for almost an hour near the central peak of the large crater Alphonsus, but even this was disputed. The Ranger 9 probe was sent diving into this crater in 1965 to investigate the possibility that the peak was a volcano. In fact, the origin of lunar craters was disputed: the traditional theory was they were volcanic calderas, but there was mounting evidence that they were made by large impacts.

"With Earthshine, the visibility is pretty fair," Aldrin pointed out.

"Take a look, and see if you see anything worth noting up there," McCandless said.

"You might give us the time that we'll cross 45°W," called Aldrin, "and then we'll know when to start searching for Aristarchus."

McCandless had the times at hand, "Aristarchus should become visible over your horizon at 077:04, and the point of closest approach will be at 077:12."

As they passed due south of the crater, Collins reported, "I'm looking north up toward Aristarchus now, and there's an area that is considerably more illuminated than the surrounding area."

"It does seem to be reflecting some of the Earthshine," Aldrin said. In fact, as Aristarchus has a high albedo (it is one of the brightest features on the near side) it was visible to telescopic observers. "There is one wall of the crater that seems to be more illuminated and – if we are lining up with Earth correctly – that does seem to place it near to the 'zero phase'." He suspected that the 'bright patch' was merely selective reflection of Earthshine.

"Is that the inner or the outer wall?" McCandless asked. "And can you discern any difference in colour of the illumination."

"That's an inner wall of the crater," Aldrin confirmed. "There doesn't appear to be any colour."

"Have you used the monocular on this?" McCandless persisted.

"We'll give it a try if we have the opportunity next time around," Armstrong promised. "We're in the middle of lunch."

"We've about 6 minutes remaining until LOS," McCandless called 15 minutes later, "and to enable us to configure our ground lines, we'd like to know if you're still planning to have the television up at the start of the next pass."

"It never was our plan to do so," noted Armstrong, recalling a debate during mission planning, "but it's in the flight plan and so I guess we'll do it." A telecast was to begin as soon as communication was re-established following the far-side passage.

Passing behind the Moon for the second time, the astronauts connected up the television camera. "Which window are we going to use," Aldrin asked, "so that I can figure out how to put the monitor on?"

"I suppose the best one would be the centre window," Armstrong said. "Don't you think?"

"Probably," Aldrin said.

Armstrong, being practical, decided, "Let's get into attitude, and see what we think."

"We're not going to have much of a television show unless we get high-gain," pointed out Aldrin. Once they were in attitude, he steered the antenna to point at where Earth should rise over the lunar horizon.

"We've had AOS by Goldstone," announced Riley publicly. "Television is now on."

"We have a good clear television picture," reported McCandless. "We can see the horizon against the blackness of space, and without getting into the question of greys and browns, it looks, at least on our monitor, sort of a brownish-grey."

"That's a good, reasonable way of describing it," agreed Aldrin. "I'd say we're about 95°E, coming up on Smyth's Sea."

"For your information, your altitude is about 92 nautical miles," McCandless advised.

"We'll try and pick up some of the landmarks that we'll be seeing during our approach to the powered descent," Aldrin explained.

"Smyth's Sea doesn't look much like a sea," observed Collins. "The area that is devoid of craters, of which there's not very much, is sort of hilly looking."

"We're now at about 83°E, which is equivalent to 13 minutes before ignition," said Armstrong.

"Of course, you'll be considerably lower at the initiation of powered descent," McCandless pointed out.

"We're coming up on the Foaming Sea," said Collins.

The main wall display in the Mission Operations Control Room showed a map of the spacecraft's ground track across the Moon, with a moving symbol showing its progress. "We show you coming up on landmark Alfa 1 shortly," McCandless noted.

"Alfa 1 isn't large," Collins pointed out, "but it's extremely bright." The small bright-rayed crater designated A-1/11 was one of four craters in a featureless part of the small dark plain known as the Foaming Sea. If provided sextant sightings on landmarks of known positions, program P22 would calculate the parameters of the spacecraft's orbit and, if so instructed, update the state vector. Alternatively, with knowledge of the orbit, it could process the sightings to calculate the positions and elevations of the terrain. Collins was to track this crater as practice for supporting the subsequent powered descent by Eagle.

"We show you over the Sea of Fertility now," McCandless prompted.

"The crater that's in the centre of the screen now is Webb," explained Aldrin. "We will be looking straight down into it about 6 minutes before the powered descent." This crater, named after a British astronomer, was 11 nautical miles in diameter and was located on the eastern part of the Sea of Fertility. Aldrin moved the camera to a side window to show an oblique view to the south of their ground track.

"We're getting a beautiful picture of Langrenus, with its rather conspicuous central peak," noted McCandless.

"The Sea of Fertility doesn't look very fertile to me," mused Collins. "I don't know who named it."

Armstrong speculated, "It may have been named by the gentleman this crater was named after – Langrenus – who was a cartographer to the King of Spain and made one of the early reasonably accurate maps of the Moon." On Earth, his wife at home exclaimed, "So that's what he was doing with the *World Book* in his study!" In 1645 Michel Florent Van Langren (Langrenus in Latin) issued the first map of the Moon to include names, although his successors rejected most of his names. The crater later named in his honour was 74 nautical miles across. In fact, the Sea of Fertility was named by the Jesuit priest Giovanni Battista Riccioli, who, working with Francesco Grimaldi in Italy, published a map of the Moon in 1651. They had their own craters near the western limb.

Aldrin changed window again in order to view straight down, and announced, "Crater Secchi."

"We're getting a good view of the track leading into the landing site now," said McCandless.

"This is very close to the ignition point for the powered descent," Aldrin noted. "We're passing Mount Marilyn, that triangular-shaped mountain in the centre of the screen at the present time, with crater Secchi-Theta on its far northern edge." And then, as another crater came into view, "The bright, sharp-rimmed crater at the right edge of the screen is Censorinus-T. We're now at the 1-minute point in the powered descent." Continuing west, they passed from the Sea of Fertility onto the Sea of Tranquility.

"For your information, your altitude is 148 nautical miles," said McCandless.

"I'm unable to determine altitude at all by looking out the window," Collins remarked.

"I bet you could tell if you were down at 50,000 feet," quipped McCandless.

"There's a good picture of Boot Hill," said Aldrin. "That'll be 3 minutes 15 seconds into the descent." Then, "That's Duke Island to the left. The biggest of the craters near the centre of the picture right now is Maskelyne-W. It'll be a position check in the descent at about 3 minutes 39 seconds; it'll be our downrange position check and crossrange position check prior to the yaw-around to acquire the landing radar. Past this point, we'll be unable to see the surface until very near the landing area." Nearing the terminator, the illumination highlighted the shallow undulations on the Sea of Tranquility, in particular a pair of sinuous rilles whose snake-like appearance had prompted their names of Sidewinder and Diamondback. "The landing point is just barely in the darkness." The crater Moltke was named after the nineteenth-century Prussian strategist Count H.K.B. von Moltke, who arranged for the publication in 1874 of a map of the Moon prepared by J.F.J. Schmidt. The crater was 3.5 nautical miles in diameter, and situated about 28 nautical miles southeast of where Apollo 11 hoped to land. The eastern crest of its raised rim was catching the Sun's rays, but the rest of it was still in darkness.

Collins, who was doing the 'flying', had observed that after he set the docked vehicles into a given attitude, the main axis tended to drift (despite counteracting thruster firings) towards vertical with the LM on the bottom, and he thought this instability might be a gravity-gradient effect produced by the mascons. "It looks like that LM just wants to head down towards the surface."

"That's what the LM was built for!" McCandless retorted.

Now in darkness, Aldrin switched off the television, and communication with the spacecraft lapsed as the crew prepared for the second manoeuvre of the lunar orbit insertion sequence, which was to occur on the far side of the Moon at the end of the current revolution.

Meanwhile, at home

It was an excellent telecast, lasting over 30 minutes. Joan Aldrin had returned from the hairdresser in time to watch it. She had been accompanied by Jeannie Bassett who, driving, had tried in vain to evade the photographers. As the lunar landscape passed by, Joan lost interest. When they crossed the terminator into darkness, she said, "Well, now I just have to get through the next 24 hours." Lurton Scott had taken the Collins children to the cinema while Pat Collins and Clare Schweickart reviewed newspaper coverage of the mission; they broke off to watch the telecast. After watching, Jan Armstrong spread a large-scale map on the floor and reviewed the features on the approach route and in the immediate vicinity of the landing site.

ORBIT REFINEMENT

Collins terminated the orbital rate and adopted an inertial attitude to undertake the P52 platform alignment in preparation for the forthcoming LOI-2 manoeuvre. As a contingency against a total loss of communications, McCandless read up the data for two transearth injection manoeuvres: one in case the LOI-2 burn failed to occur, and the other in case it succeeded and contact was then lost. In both eventualities, the LM would be ditched prior to leaving orbit.

Meanwhile, it had been found that the anomalous decrease in the pressure in the nitrogen tank of Bank-B during the LOI-1 burn had occurred only while the system was active, which prompted the speculation that it was merely a thermal effect that made the orifice of the solenoid valve open wider on Bank-B than on Bank-A, thereby increasing the nitrogen flow rate (in other words, there was no external leak). The engine could be operated in single-bank mode; the dual-bank mode was to provide redundancy against one bank failing and shutting down the SPS. As the pressure of the nitrogen tank in Bank-B was now stable and was well above the 'red line' value of 400 psi (below which it would be incapable of holding the propellant feed valves open), it was decided to conserve this tank in order to retain redundancy against the total loss of Bank-A. McCandless relayed the decision to make the LOI-2 burn in single-bank mode using Bank-A, then as they approached the limb he gave the formal go-ahead to attempt the burn. The attitude of the spacecraft placed them 'heads down' in order to point the sextant towards space, and once they were on the far side Collins took a star sighting on Denebola to confirm their attitude.

When Apollo 8 and Apollo 10 had executed LOI-2, they had entered more or less circular orbits at 60 nautical miles. However, when the time came to perform the transearth injection manoeuvre, the gravitational perturbations of the mascons had transformed their paths into ellipses ranging between 54 and 66 nautical miles. Although this did not compromise Apollo 8's objectives, if a landing had been tried on Apollo 10, the ground track would have been displaced several miles from that intended. In the case of Apollo 11, however, it was essential for Eagle to fly on course and, for the rendezvous, for Columbia to be waiting in as circular an orbit as possible. On the previous missions, the longitude of perilune induced by the perturbations had been about 100°E. The flight dynamics team had calculated that if the post-LOI-2 orbit were to have a given ellipticity and a perilune at 85°W, the mascons would tend to circularise the orbit. The option of having Columbia fire its engine to recircularise its orbit shortly prior to rendezvous was dismissed as a waste of propellant. Furthermore, it was arranged that as the Moon slowly rotated on its axis, the plane of the spacecraft's orbit would drift such that when Eagle made its descent its ground track would intersect the landing site. At the time of the LOI-2 manoeuvre the combined mass of the vehicles was 71,622 pounds. The 17-second burn was to start at 080:11:36, and achieve a retrograde delta-V of 159.2 feet per second for a spacecraft velocity of 5,364 feet per second in order to revise the orbit to 53.7 by 65.7 nautical miles.

As they waited, they entered sunlight, and Collins noticed lots of little bright spots on the LM reflecting the light. "The poor old LM's contaminated – it's got urine particles all over it! By the way the light's shining, they look yellow. You know, I guess everything else has boiled off and it's left a little solid deposit."

"Wait until the 'forward contamination' people hear about that!" said Aldrin. "There will be no more urine dumps on the way to the Moon; those fellows will have to store it all in a nice little bag."

Ignition was precisely on time. Collins was timing it using a stopwatch, and if the engine did not shut down on time they would intervene. Even a 2-second over run would slow them sufficiently to cause their trajectory to dip dangerously close to the surface during the near-side pass, but cutoff was as planned. Armstrong asked the computer for the new orbital parameters. "66.1 by 54.4 – you can't beat that!"

"That's about as close as you're going to get," Collins agreed.

With the spacecraft in a 2-hour orbit, Mission Control would receive telemetry for about 75 minutes. This was more than 50 per cent of the orbital period because, at an altitude of about 60 nautical miles, the vehicle was in line of sight for several minutes beyond the 90-degree angle as measured from the centre of the lunar disk.

As before, Armstrong initiated a 180-degree roll, as a preliminary to resuming a pitched-down attitude.

"While this thing's rolling over, I'm going to take a pee," Collins announced. On venting the urine to space, he pondered how long it would take before its orbit decayed and it struck the surface – more contamination of the lunar environment! As Armstrong began the pitch manoeuvre, Collins noted that they had missed out on a long-lens picture of Earthrise the first time around, and had been too busy the second time. "Gee, it's too bad that we can't stop right here and observe the Earth come up."

"We probably can do it, if we stop it right here," Armstrong noted. "That is, if you want to spend the gas."

"That's the only trouble," Collins mused, "the doggone gas."

"Why don't we stop it?" prompted Aldrin.

"Okay!" said Collins.

Armstrong decided to adopt an inertial attitude that would put Earthrise in one of the windows and also enable the high-gain antenna to lock on. In commanding the manoeuvre, he selected the wrong direction. "Oh, son of a gun!" he chuckled. "We are going backwards. Oh, well."

"Dummkopf!" said Aldrin.

"Neil, pitch *down*," encouraged Collins.

"Prior planning prevents poor performance," recited Aldrin.

"Is that right, Buzz?" asked Armstrong.

"Where'd you ever hear that one, Buzz?" Collins demanded.

"I can't think," said Aldrin.

A few minutes later, Collins exclaimed, "Here's the Earth. Hey, I've got the view over here."

John McLeaish, who had taken over as the Public Affairs Officer, informed the waiting world that both Goldstone and Hawaii had acquired Apollo 11 on time, an indication that the burn had gone to plan. During the far-side passage, the shift had changed in Mission Control. Although Milton Windler's Maroon Team had taken the White Team's slot in the daily cycle to give Kranz's controllers a 32-hour rest prior to tackling the powered descent, Charlie Duke had opted to work the coming shift. Armstrong reported that the LOI-2 burn had been nominal. It was all strictly business. For most of this near-side pass there was little communication with the spacecraft. "This pass is fairly quiet," noted McLeaish almost apologetically. "No doubt the crew is occupied with preparations to enter the LM, which we expect to occur over the far side of the Moon." Aldrin was to spend 2.5 hours transferring items into the LM, and methodically configuring the switches in preparation for the next day's operations.

About 50 minutes into the pass, Duke prompted for a progress report. "We're wondering if you've started into the LM yet?"

"We have the CSM hatch out, the drogue and probe removed and stowed, and are just about ready to open the LM hatch now," replied Armstrong. And then, a few minutes later, "Okay, Charlie. We're in the LM."

McLeaish observed, "They appear to be a little ahead on their time line."

After making the post-manoeuvre P52 platform check, Collins also decided to get ahead by adopting the attitude for the P22 landmark tracking that he was to do on revolution 4 of Alfa 1, the small bright crater on the Foaming Sea, with the objective of measuring its elevation relative to the intended landing site.

During the far-side pass, Armstrong shaved using cream and a razor.

"I see Earth," said Aldrin, as the spacecraft rounded the limb on revolution 4, "but it's a lousy picture."

Goldstone and Hawaii both acquired Apollo 11, and Duke put in a call. Collins told him that he was set up for the P22 landmark tracking. "Ho-hum, ho-hum. I only got set up for this thing about an hour early," he chastised himself, provoking laughter on board. As they flew overhead, Collins made five sightings of Alfa 1 at timed intervals. On completion, he quipped to Armstrong, "Well, that's one P22 out of the way. Ho, ho, ho!" When the sightings were processed, it was calculated that the landmark was 500 feet above the landing site – knowledge that would assist Armstrong and Aldrin to monitor their approach to the point at which they were to initiate the powered descent.

While in the landmark tracking attitude, the spacecraft had been unable to point its high-gain antenna at Earth, and the downlink using one of the omnidirectional antennas was noisy. However, this had no impact because, as McLeaish observed, "So far on this pass we've had just one contact with Apollo 11, and that was Mike Collins at AOS."

Duke called, "You can proceed to sleep-attitude now."

"Let's hold this attitude a bit," Aldrin prompted Collins, "I want to look at the PDI approach. Man, this is really something. To see our approach into the landing site, you've got to watch it through *the LM's* window."

"Houston," Collins replied, "we're holding inertial for a little while to study the approach to the landing zone." On looking out for himself, he said to Armstrong, "There go Sidewinder and Diamondback – God, if you ever saw check points in your life, those are it."

"But we don't get to see them," Armstrong noted.

"You don't?"

"No, we're yawed face-up."

Since their previous inspection, the terminator line had migrated westward sufficiently to reveal their landing site. "I think I can see it," Aldrin called with delight. "Yes, I can! I've got the whole landing site." He depressed his Push-to-Talk, "Houston, this is Eagle. I can see the entire landing area." While this was the first use of the call sign, the communication was via his umbilical to Columbia's system.

"Roger, Buzz," replied Duke matter-of-factly.

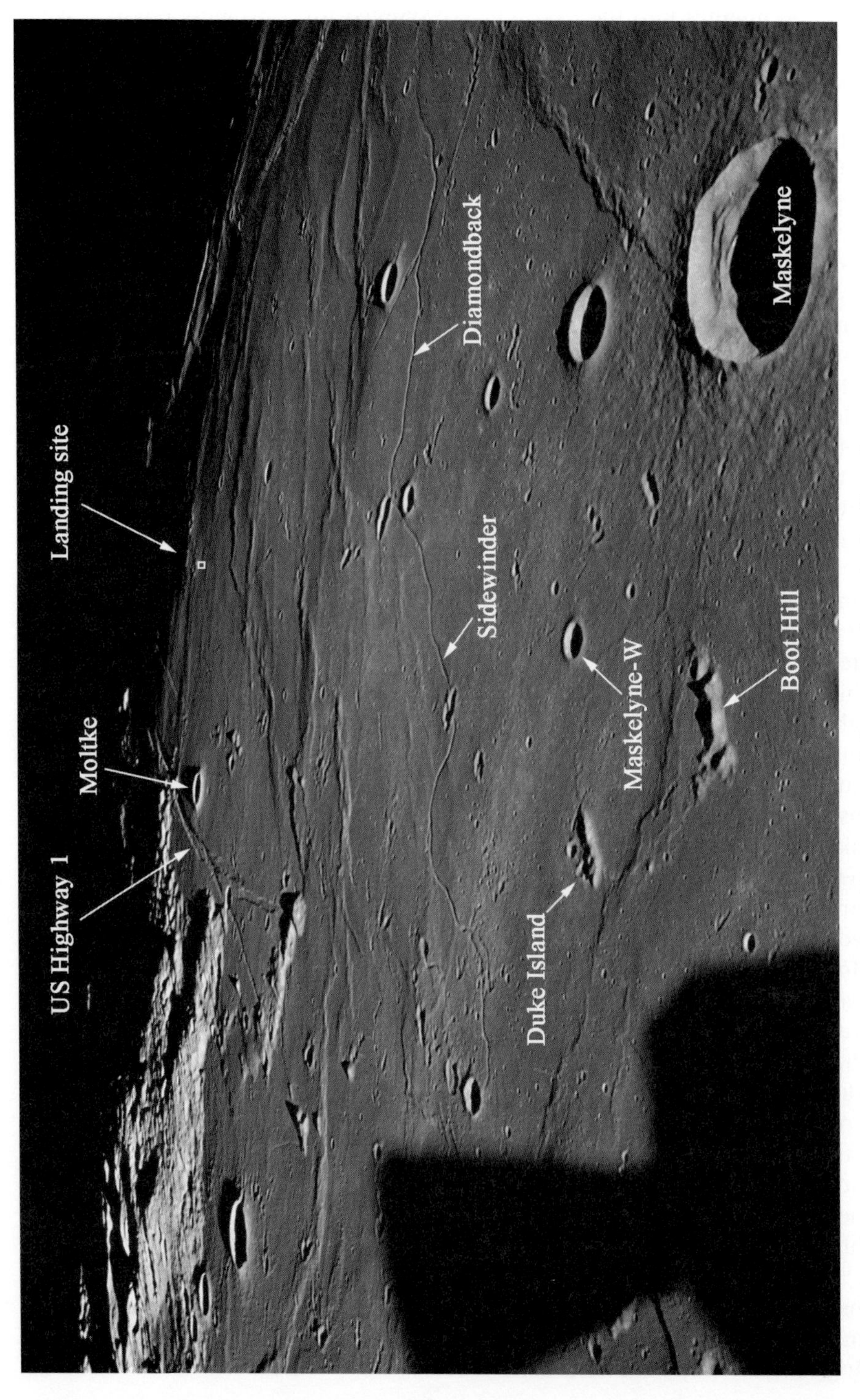

An oblique view of the approach to the landing site.

With the Sun barely above the horizon at the landing site, the lengthy shadows were highlighting the topography sufficiently to make the mildest of surface relief appear very rough. "Boy, that sure is eerie looking," Aldrin mused.

"Isn't that something?" said Armstrong, viewing from the command module. This was a significant moment in the mission.

"I missed taking a picture of it!" exclaimed Aldrin. It did not matter, he would be able to take one later.

"Shall we manoeuvre to the sleep attitude?" Collins asked.

"All right," Armstrong agreed.

Although the LM's power margin was narrow, the flight plan included a test of its VHF and S-Band radio systems. "Houston, Apollo 11 – Eagle – over," called Aldrin directly.

"Roger, Eagle," replied Duke.

"I'll go ahead with the sequence camera checkout," said Aldrin. "I'm still on low voltage taps, and I assume there's no problem doing that."

"That's affirmative," Duke confirmed.

Twenty minutes later, having tested the Maurer 16-millimetre camera, Aldrin announced, "Eagle is powering down. Out."

Just after Apollo 11 passed 'over the hill' Aldrin rejoined his colleagues, and Collins suggested that they need not reinstall the probe and drogue overnight.

"It's okay with me," agreed Armstrong.

"How's that going to affect sleeping?" Aldrin wondered.

"I'd rather sleep with the probe and drogue than have to dick about with them in the morning," Collins insisted. As long as they could squeeze their legs around the hardware stowed under the couches, there ought to be room to sleep. With the decision made, he installed the command module hatch over the tunnel.

"It's amazing how quickly you adapt," said Collins. "It doesn't seem weird at all to me to look out there and see the Moon going by, you know?" This prompted general laughter.

While Apollo 11 was behind the Moon, Owen Garriott took over as CapCom from Duke for the remainder of the Maroon Team's shift. On the near-side pass, Collins undertook miscellaneous chores, then prepared their suppers. Over the far side once again, Aldrin, who had stood 'watch' the previous night, offered to do so again, "Why don't you guys sleep underneath tonight? I'll sleep top-deck."

"Unless you'd rather sleep up top, Buzz," Collins said, "but you guys should get a good night's sleep before going in that damn LM. Which would you prefer? Is that probe and drogue going to be in your way over there?"

Aldrin took a look. "No, I don't think so."

It was decided to follow the flight plan, and have Armstrong and Aldrin sleep in the hammocks. "Well," said Collins, "I think today went pretty well. If tomorrow and the next day are like today, we'll be safe."

"We're ready to go to bed," announced Garriott when the spacecraft appeared around the trailing limb on revolution 6.

"We're about to join you," Collins replied.

With that, communications ceased. However, prior to retiring, Armstrong and

Aldrin, seeking to 'get ahead', prepared the clothing and equipment they would need the next day. Collins had the satisfaction of seeing his crewmen fall asleep before he himself settled down in the left couch.

In Houston, Glynn Lunney's Black Team took over for the 'graveyard' shift, and his flight dynamics team analysed the tracking by the Manned Space Flight Network in order to verify that the spacecraft's orbit was evolving in the manner designed to counteract the mascon perturbations. On the evening of Saturday, 19 July, Gene Kranz attended a Mass "to pray for wise judgement and courage, and pray also for my team and the crew". The astronauts' wives endeavoured to find solitude in which to contemplate what Armstrong and Aldrin were to attempt the next day.

FLIGHT DAY 5

Early on the morning of Sunday, 20 July, Ron Evans made the wake-up call.

"Good morning," replied Collins half a minute later. "You guys sure do start early."

"It looks like you were really sawing them away." Evans said, having noted the telemetry indicating that all three astronauts had been sleeping soundly.

"You're right," Collins agreed. "How are all the CSM systems looking?"

"It looks like the command module's in good shape. The Black Team's been watching it real closely for you."

"We appreciate that, because I sure haven't."

Moments later, the spacecraft passed 'over the hill'. While on the far side, the crew tidied up and prepared the breakfast. On their reappearance on revolution 10, Evans, making the most of his opportunity to converse, announced, "The Black Bugle just arrived with some morning news briefs, if you're ready."

"Go ahead," Armstrong replied.

"Today church services around the globe will be mentioning Apollo 11 in their prayers. President Nixon's worship service at the White House is also dedicated to the mission, and fellow astronaut Frank Borman is still in there pitching – he will read the passage from *Genesis* that was read out on Apollo 8 last Christmas. The Cabinet and members of Congress, with emphasis on the Senate and House space committees, have been invited, together with a number of other guests. Buzz, your son, Andy, got a tour of the Manned Spacecraft Center yesterday which included the Lunar Receiving Laboratory; he was accompanied by your uncle, Bob Moon."

"Thank you," said Aldrin.

"Among the headlines about Apollo this morning," Evans continued, "there is one asking that you watch for a lovely girl with a big rabbit. An ancient legend says a beautiful Chinese girl called Chang-o has been living there for 4,000 years. It seems she was banished to the Moon because she stole the pill of immortality from her husband. You might also look for her companion, a large Chinese rabbit, who is easy to spot since he is always standing on his hind feet in the shade of a cinnamon tree; the name of the rabbit is not reported."

The astronauts promised that they would "keep a close eye out for the bunny girl".

Evans went on, "You residents of the spacecraft Columbia may be interested in knowing that today is Independence Day in the country of Colombia. Gloria Diaz of the Philippines was crowned Miss Universe last night, beating sixty other girls for the global beauty title. Miss Diaz is 18, has black hair and eyes, and measures thirty-four-and-a-half, twenty-three, thirty-four-and-a-half. The first runner up was Miss Australia, then Miss Israel and Miss Japan. When you are on your way back, Tuesday night, the American and National League All Stars will be playing ball in Washington. Mel Stottlemyre of the Yankees is expected to be the American League's first pitcher. No one's predicting who'll be first pitcher for the National League yet; they have nine on the roster." And then he rounded off with a funny: "Although research has certainly paid off in the space program, research doesn't always pay off, it appears. Woodstream Corporation, the parent company of the Animal Trap Company of America that has made more than a billion wooden spring mousetraps, reported that it built a better mousetrap but the world didn't beat a path to its door. As a matter of fact, it had to go back to the old-fashioned kind. They said, 'We should have spent more time researching housewives, and less time researching mice'. And with that the Black Bugle is completed for this morning."

"Thank you, very much," acknowledged Armstrong.

A few minutes later, the spacecraft passed around the far side again.

Meanwhile, at home

On his arrival in Mission Control, Kranz was astonished to find Dick Koos absent; Koos had rolled his new Triumph TR3 driving in, but was uninjured and arrived in time for the powered descent. On reviewing Lunney's console log, Kranz was pleased to discover that he had not inherited any problems – the spacecraft was in excellent condition. Chris Kraft arrived, patted Kranz on the shoulder and wished him "good luck", then took his seat on Management Row. When Kranz was made a flight director early in the Gemini program, his wife Marta had begun the tradition of making him a waistcoat specifically for each mission. For Apollo 11 she had made one of white brocade inlaid with very fine silver thread. At 095:41 Kranz took over the flight director's console, and Lunney went to brief the press. During the far-side pass, the other members of the White Team settled in for what was to be a momentous shift – the landing was about 7 hours off. Man could land on the Moon for the *first time* only once. As the shift began, this task had not yet been attempted. Soon it would be. Once achieved, the moment of its attainment would become part of the historical record. On looking around into the viewing gallery, Kranz noticed Bill Tindall, and waved him down to sit alongside him at his console. Kranz would later write of Tindall, "he was the guy who put all the pieces together, and all we did was execute them."[3]

At 9.30 am Joan Aldrin, her children and Robert and Audrey Moon, attended

[3] At the post-flight party, the flight controllers voted Bill Tindall an honorary flight director, with the team colour grey.

Webster Presbyterian Church, where her husband served as an elder. The church was packed, with folding chairs in place to accommodate the extra worshippers. As in Mission Control, the mood was tense. The Reverend Dean Woodruff began his sermon: "Today we witness the epitome of the creative ability of Man. And we, here in this place, are not only witnesses but also unique participants." Everyone knew that by the day's end Armstrong and Aldrin might well be dead. Pat Collins, her children and sister Ellie Golden, went to morning Mass at St Paul's Roman Catholic Church. Jan Armstrong remained at home and impatiently watched the clock. At noon some of the churchwomen delivered a cold luncheon to the Aldrin home, together with a cake that had been frosted with the Stars and Stripes and the words 'We came in peace for all mankind'. Woodruff arrived later, and remained for the powered descent.

PREPARING EAGLE

After breakfast Aldrin went into the lower equipment bay, removed his constant-wear garment, reinstalled his urine-collection and fecal-containment utilities and put on his liquid-cooled garment, the fishnet fabric of which had a network of narrow flexible plastic tubes sewn into it through which cold water would be pumped to manage the heat generated by the exertion of the moonwalk. It had to be donned now, since they were to remain suited while in the LM. Aldrin then went into the LM, vacating the lower equipment bay to Armstrong, who suited up with Collins assisting with the zippers and checking fixtures – a process that took 30 minutes. When Armstrong went into the LM, Aldrin returned to suit up. Collins also suited up as a precaution against inadvertent decompression during undocking, or the need for an early abort in which Eagle's crew would conduct an external transfer using the side hatch.[4]

At AOS on revolution 11, Aldrin was well into powering up Eagle's systems. When Duke requested a status report, it became evident that Aldrin was running about 30 minutes ahead of the flight plan. When the steerable high-gain antenna mounted on a boom on the right-hand side of the roof was pointed to Earth, the LM flight controllers received their first significant telemetry of the mission.

The two crews worked independently in their preparations, but certain events required coordination. One item was setting Eagle's clock, which was to be done by synchronising it with its counterpart in Columbia.

"I have 097:03:30 set in," called Armstrong.

Collins counted down, "15 seconds to go. 10, 5, 4, 3, 2, 1. MARK."

"Got it," confirmed Armstrong.

Another task, some time later, was to coarsely align the platform of Eagle's

[4] Only once Eagle was safely on the surface would Collins remove his suit, and he would don it again shortly prior to liftoff.

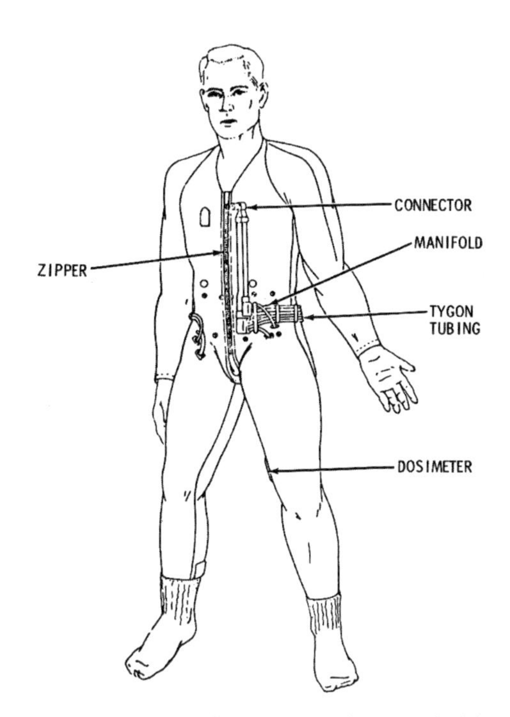

The Liquid Cooled Garment to be worn inside the Extravehicular Mobility Unit.

inertial navigation system. In essence, two humans acted as an interface between two machines.

"I'm ready to start on a docked IMU coarse-align," Armstrong said. "When you're ready, go to Attitude Hold with Minimum Deadband." Once Collins had confirmed that he was holding the docked vehicles steady, Armstrong said, "I need your Noun 20."

"I've got Verb 06, Noun 20. Give me a Mark on it," Collins replied.

"MARK!" called Armstrong.

"Register 1, plus 11202, plus 20741, plus 00211," said Collins, reading the display on his DSKY. The numbers represented the CSM's attitude with respect to the current REFSMMAT.

Armstrong gave a read-back for confirmation, "11202, 20741, 00211."

"That's correct," Collins agreed.

After performing an arithmetical transformation to allow for the fact that the designers of the vehicles had specified their Cartesian axes differently, Armstrong keyed the attitude into his DSKY.

The alignment showed up in the telemetry, and Duke reported, "That coarse align looks good to us."

"Okay, Mike," said Armstrong. "Your Attitude Hold is no longer required."

If necessary, Eagle would later make star sightings for its equivalent of a P52 to refine the alignment.

Once he had installed the probe and drogue assemblies in the tunnel and fully extended the probe, Collins called to Eagle, "The capture latches are engaged in the drogue. Would you like to check them from your side?"

"Stand by," Armstrong replied. He looked up through the open hatch to verify that the tip of the probe projected through the hole in the centre of the drogue and the three small latches, each no larger than a finger nail, were engaged. "Mike, the capture latches look good." Armstrong then closed the upper hatch to make Eagle air-tight.

At this point, Apollo 11 passed around the far side of the Moon. They were to continue with the preparations during revolution 12, and undock prior to AOS on revolution 13.

Collins slowly retracted the probe until the latches established a firm grip of the drogue. The next task was to release the main latches in the docking units. To guard against the possibility of depressurisation, he donned his helmet and gloves.

Five minutes later, Aldrin called, "Mike, let us know how you're coming up there, now and then."

"I'm doing just fine," Collins replied. He was physically priming the latches, imparting the stored energy they would need to re-engage on the redocking. "I've cocked eight latches, and everything is going nominally." And then a minute later, "All 12 docking latches are cocked."

"Okay," acknowledged Aldrin.

"I'm ready to button up the hatch," Collins announced.

Although the vehicles were now held together only by the capture latches, these were able to maintain the hermetic seal in the tunnel because the interface had been compressed by the hard docking. "Mike, have you got to the Tunnel Vent step yet?" Aldrin asked.

"I'm just coming to that," replied Collins. "What can I do for you?"

"Well, we're waiting on you," Aldrin noted. Although ahead of schedule, the LM crew had to wait for Collins to vent the tunnel before they could proceed.

Two minutes later, Collins reported, "I'm ready to go to LM Tunnel Vent." He opened the valve to space. The process was expected to take about 8 minutes.

"How're you doing, Mike?" Armstrong asked several minutes after that time had elapsed.

"Stand by, and I'll give you the delta-P reading," replied Collins. He reset the valve to enable a nearby gauge to measure the difference in pressure between the tunnel and the command module, which was at about 5 psi. "It's 3.0 psi." There was still a significant amount of air remaining in the tunnel.

In Eagle, Armstrong and Aldrin donned their helmets and gloves to check the hermetic integrity of their suits.

Meanwhile, Collins started to manoeuvre into the attitude required for the next P22 landmark tracking, which would be Site 130-prime, a crater inside Crater 130 in the Foaming Sea. Because this had been selected as a reference by John Young on Apollo 10 for the reason that it was both readily identified and small enough to be accurately marked using the sextant, it was also referred to as John Young's crater. The sightings were to be used to update Houston's knowledge of the orbit, and where the spacecraft was in that orbit, in order to calculate the precise time at which to initiate Eagle's powered descent.

When the manoeuvre was finished, Armstrong called Collins, "We're going to put our gear down."

"Master Arm," said Aldrin on intercom, reading the checklist. "Landing Gear Deploy, Fire."

"Here we go, Mike," Armstrong warned before detonating the pyrotechnics to release the spring-loaded legs.

"Bam, it's out. There ain't no doubt about that," Aldrin mused. "Master Arm, Off."

"The gear went down okay, Mike," Armstrong called. There were redundant circuits, but the primary had successfully fired the pyrotechnics to deploy the legs. The 67-inch-long probes, whose tips had been latched against the inner parts of the legs, hinged on the undersides of the lateral and rear foot pads to project 'straight down'. For Apollo 5 in February 1968, a Saturn IB had launched LM-1 absent its legs for an unmanned test. LM-3 had demonstrated the deployment of the legs on Apollo 9 in March 1969. At that time the design had included a probe on each of the pads but, at Armstrong's request, it had been decided to delete the probe from the forward leg lest it be bent on touchdown in such a manner as to cause him to slip (or worse, puncture his suit) as he jumped backwards down off the ladder.

As Apollo 11 appeared around the trailing limb on revolution 12, Duke made them aware that communications had been restored. "Apollo 11, Houston. We're standing by." There was a lot of static on the downlinks. With no response, in all likelihood owing to the fact that he had not directed his call to a specific vehicle, Duke persisted. "Columbia, Houston. Do you read?"

"Loud and clear," Collins acknowledged.

"Eagle, Houston. Do you read?" No response.

"Eagle, do you read Columbia?" asked Collins.

"Yes," acknowledged Aldrin. "I'm working on the high-gain right now." He slewed the steerable dish as per the flight plan, but could not establish contact with Earth. "Are you in the right attitude, Mike?"

"That's affirm."

"Columbia, Houston," called Duke.

"Houston, Columbia. You're loud and clear."

"Eagle, Houston. Will you verify you are on the forward omni?" No response. "Columbia, Houston. We have no voice with Eagle. Would you please verify that Eagle is on the forward omni."

"Buzz," Collins called. "Are you on the forward omni?" When there was no response, he repeated the call.

"Roger. I am," confirmed Aldrin.

"Houston, Columbia. Eagle is on the forward omni."

Duke tried again, "Eagle, Houston."

"Roger, I've got you now," acknowledged Aldrin. "I fed in those angles for the S-Band, and couldn't get a lock-on. It appears as though the antenna would have to be looking through the LM in order to reach the Earth."

Because the docked vehicles were oriented to facilitate P22 landmark tracking shortly after flying around the limb, it was difficult for the boom-mounted S-Band antenna cluster on Columbia to point at Earth, and the body of the LM blocked the line of sight of its steerable dish. In this attitude the vehicles would have to rely on their respective omnidirectional antennas.

"Eagle, Houston. Could you give us an idea where you are in the activation?"

"We're just sitting around waiting for something to do," Aldrin replied. "We need a state vector and a REFSMMAT before we can proceed to the AGS calibration, and we need you to watch our digital autopilot data load, the gimbal drive check and the throttle test."

Although Armstrong and Aldrin were well ahead in their LM activation, they were again obliged to wait until Houston was able to upload information and monitor their telemetry, which could not be done until they could use their high-gain antenna, which in turn meant waiting until Collins had performed his landmark tracking. While getting ahead created a margin against encountering a problem that might slip them behind schedule, the need to do certain tasks at given times meant that being ahead early on did not in itself enable the process to be completed ahead of the flight plan.

"It'll be about another 10 minutes or so before we get the P22 and manoeuvre to an attitude for the high-gain," Duke pointed out.

Armstrong and Aldrin proceeded with those items that could be done using the low data-rate provided by an omnidirectional antenna.

"We're ready to pressurise the RCS," Aldrin announced.

"You can go ahead with RCS pressurisation," Duke agreed, "but we'd like to hold off on the RCS hot-fire checks until we get the high bit-rate."

"Eagle, Columbia. My P22 is complete," Collins reported. He manoeuvred to let the high-gain antennas on both of the vehicles see Earth, and communications markedly improved. With high data-rates on both its uplink and downlink, Eagle was able to complete data uploading and checkout.

"Houston, Eagle," Aldrin called. "Both RCS helium pressures are 2,900 psi."

"Let me know when you come to your RCS hot-fire checks," said Collins, "so I can disable my roll thrusters."

Fifteen seconds later, Aldrin announced that they were ready. "Columbia," he called, "We'd like Attitude Hold with Wide Deadband."

"You got it," Collins replied. A wide deadband on the Attitude Hold would allow the testing of Eagle's thrusters to disturb the attitude of the docked vehicles without prompting Columbia's control system to waste propellant in attempting to intervene. "My roll is disabled. Give me a call as soon as your hot-fire is complete, please."

"Houston, Eagle," Aldrin called several minutes later. "The RCS hot-fire test is complete. How did you observe it?"

"It looked super to us," Duke confirmed.

"I've got my roll jets back on now," Collins announced.

At this point, Kranz polled his flight controllers, and Duke relayed the result, "Apollo 11 Houston. You're Go for undocking."

"Understand," replied Aldrin.

UNDOCKING

After the P22 landmark tracking, Collins had initiated a manoeuvre to orient the docked vehicles to enable Eagle to calibrate its abort guidance system (AGS). As he waited for this to finish, he noted that there was an advantage to being behind the Moon, as they then were, "It's nice and quiet over here, isn't it?"

"You bet," agreed Aldrin.

On finishing the manoeuvre, Collins nulled out the rates and then went 'free' while Eagle performed the calibration. After five minutes Collins asked, "How's the Czar over there? He's so quiet."

"I'm punching buttons," replied Armstrong, referring to the DSKY activity.

A few minutes later, Collins, having been thinking ahead, urged, "You cats take it easy on the lunar surface. If I hear you huffing and puffing, I'm going to start bitching at you."

"Okay, Mike," Aldrin promised.

During the calibration, the only tasks that could be performed were those that would not induce vibrations. One permissible operation was to open the helium valves to pressurise the DPS propellant tanks. Once the calibration was finished, Collins called, "I'm going to manoeuvre to the undocking attitude." He aligned the docked vehicles 'vertically', with the CSM beneath. "How about 100 hours and 12 minutes as an undocking time? Does that suit your fancy?"

"That'll be fine," agreed Armstrong.

"Are you guys all set?" Collins asked, as the clock ticked down.

"We're all set when you are, Mike," Armstrong confirmed.

"15 seconds," called Collins. He released the capture latches. If these were to fail to release, the design allowed for a suited crewman to manually release them, either by Collins pulling a handle or by a LM crewman pushing a button on the tip of the probe; in either case, the cabin would require to be depressurised and the appropriate hatch opened to gain access to the mechanism. The latches, however, did release. As the residual air in the tunnel escaped, it made the vehicles slowly drift apart.

Since it was desired that the LM remain in the orbit resulting from the LOI-2 manoeuvre, the parameters of which had been precisely defined by Manned Space Flight Network tracking, the same state vector had been loaded into both vehicles and as soon as Eagle was free Armstrong cancelled the 0.4-foot-per-second rate of separation that the PGNS indicated had resulted from the undocking.[5]

"I've killed my rates, Mike," Armstrong announced, "so you drift on out to the distance you like and then stop."

When Collins was about 65 feet away, he halted to station-keep.

Meanwhile, in the Aldrin home, son Andrew wondered aloud why NASA had not installed a communications satellite to relay while the spacecraft was behind the Moon.[6] In the Collins home, Joe and Mary Engle were looking after the children. Pat Collins, flight plan on her lap, was eager for AOS to find out if the undocking had occurred. Jan Armstrong was also at home with her flight plan, thinking the same thoughts.

"Eagle. Houston. We're standing by," called Duke as the vehicles appeared on revolution 13.

"The Eagle has wings," replied Armstrong.

Collins had installed his 16-millimetre Maurer in window 4 to document this part of the mission. On the flight plan, the television camera was to have been set up

[5] As was realised later, however, although the impulse from the tunnel venting was cancelled, this manoeuvre, and others made while 'displaying' Eagle to Collins, imparted slight residuals which, when propagated forward in time through the DOI manoeuvre, nudged Eagle's trajectory slightly 'off' at the PDI point.

[6] Some at NASA would later suggest doing precisely this for later missions.

The Moon illuminated essentially as it was at the time of Eagle's descent. (Courtesy W.D. Woods)

alongside it to provide 'live' television, but about 57 hours into the translunar coast Houston had cancelled this telecast owing to the lack of an available channel on a geostationary satellite to relay the transmission from the Madrid receiving station to Houston for conversion. In any case, as Collins had said shortly prior to LOS, he was too busy to set up the television system. The loss of 'live' views of Eagle in flight was a disappointment to the national television networks, which had hoped to use it to introduce their uninterrupted coverage of the next phase of the mission.

Several minutes into the near-side pass, Armstrong yawed Eagle around and pitched it up in order to place it 'side on' to Columbia, and then slowly yawed it through 360 degrees to enable Collins to visually confirm that the legs had fully deployed and the probes were in position. An unusual sound late in the translunar coast had prompted Armstrong to speculate that the hinged panel on the right side of the vehicle had prematurely deployed, but Collins confirmed that this was in its stowed configuration. "You've got a fine-looking flying machine," he assured.

"See you later," promised Armstrong, as the separation manoeuvre loomed.

"You guys take care," said Collins. With his spacecraft oriented apex-up, he fired his forward-facing RCS for 8 seconds to impart a downward radial thrust in order to withdraw at 2.5 feet per second, during which time the rendezvous radar mounted on Eagle's 'forehead' tracked Columbia as a test of the radar's ability to lock onto the transponder on the other vehicle, and Columbia tested its VHF ranging apparatus; these tests being designed to verify the rendezvous systems prior to Eagle entering the descent orbit. The separation burn occurred about 10 degrees east of the landing site, and placed Columbia into an equi-period orbit with its perilune 90 degrees later, and some 5 nautical miles lower.

Meanwhile, at home

Astronauts cycled back and forth between the Armstrong and Aldrin homes, as indeed did their wives, although rarely together because their efforts were divided between the families – it was a routine that Jan and Joan understood, as they had done the same thing themselves, and there was no need to play host because it was a self-organising process.

Colour Section

On 14 April 1969 Neil Armstrong, Buzz Aldrin and Mike Collins donned their training suits to have their Apollo 11 portrait taken in front of a 5-foot-diameter picture of the Moon.

As Apollo 11 lifts off, the lower arms of the tower swing away.

Apollo 11 clears the tower.

A view from Apollo 11 while in 'parking orbit' around Earth.

After LOI-1, the Apollo 11 crew were eager to witness Earth's reappearance.

Following undocking, Collins inspected Eagle's landing gear.

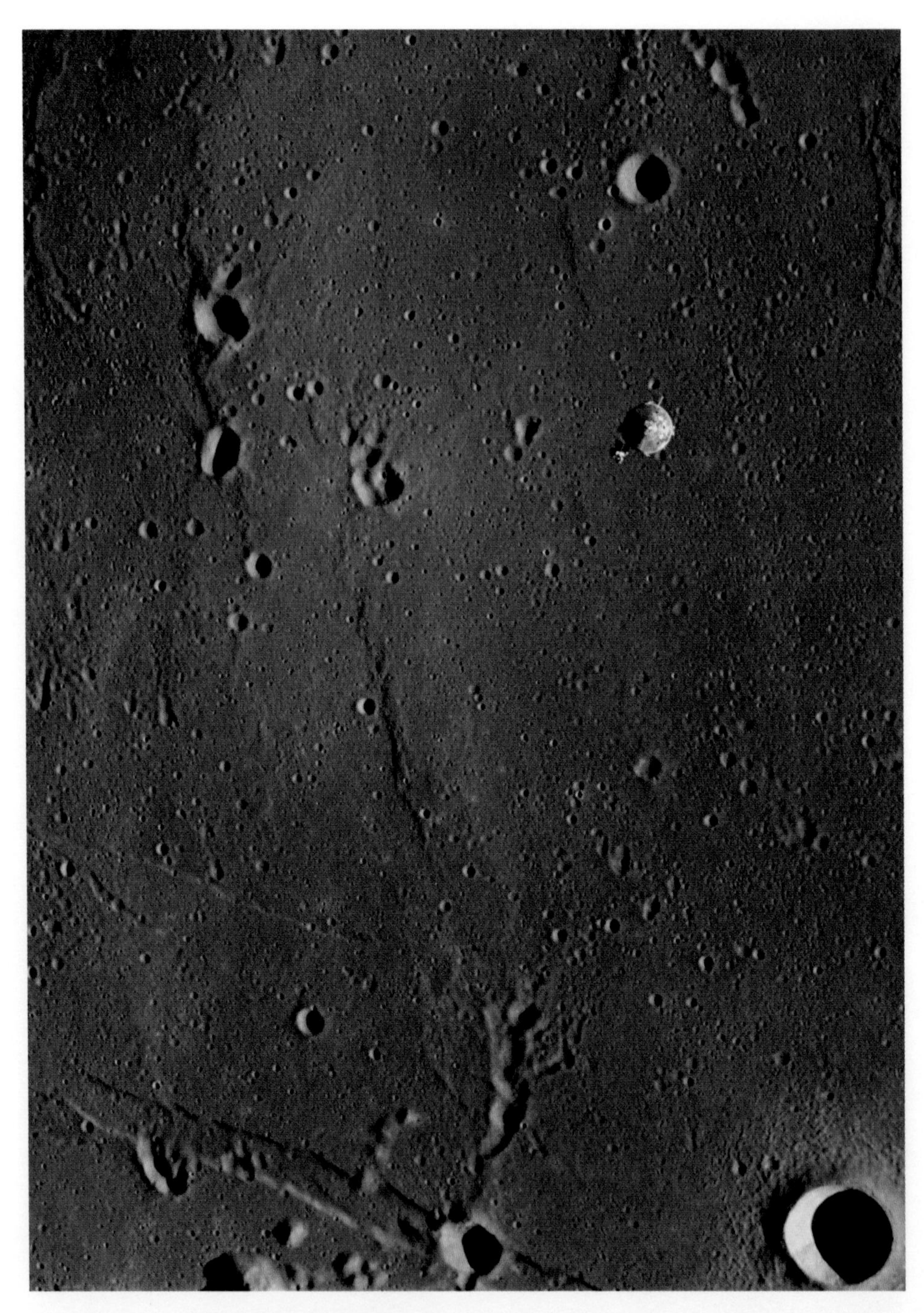

Following undocking, Eagle viewed Columbia against the backdrop of the landing site on the Sea of Tranquility (see annotation elsewhere).

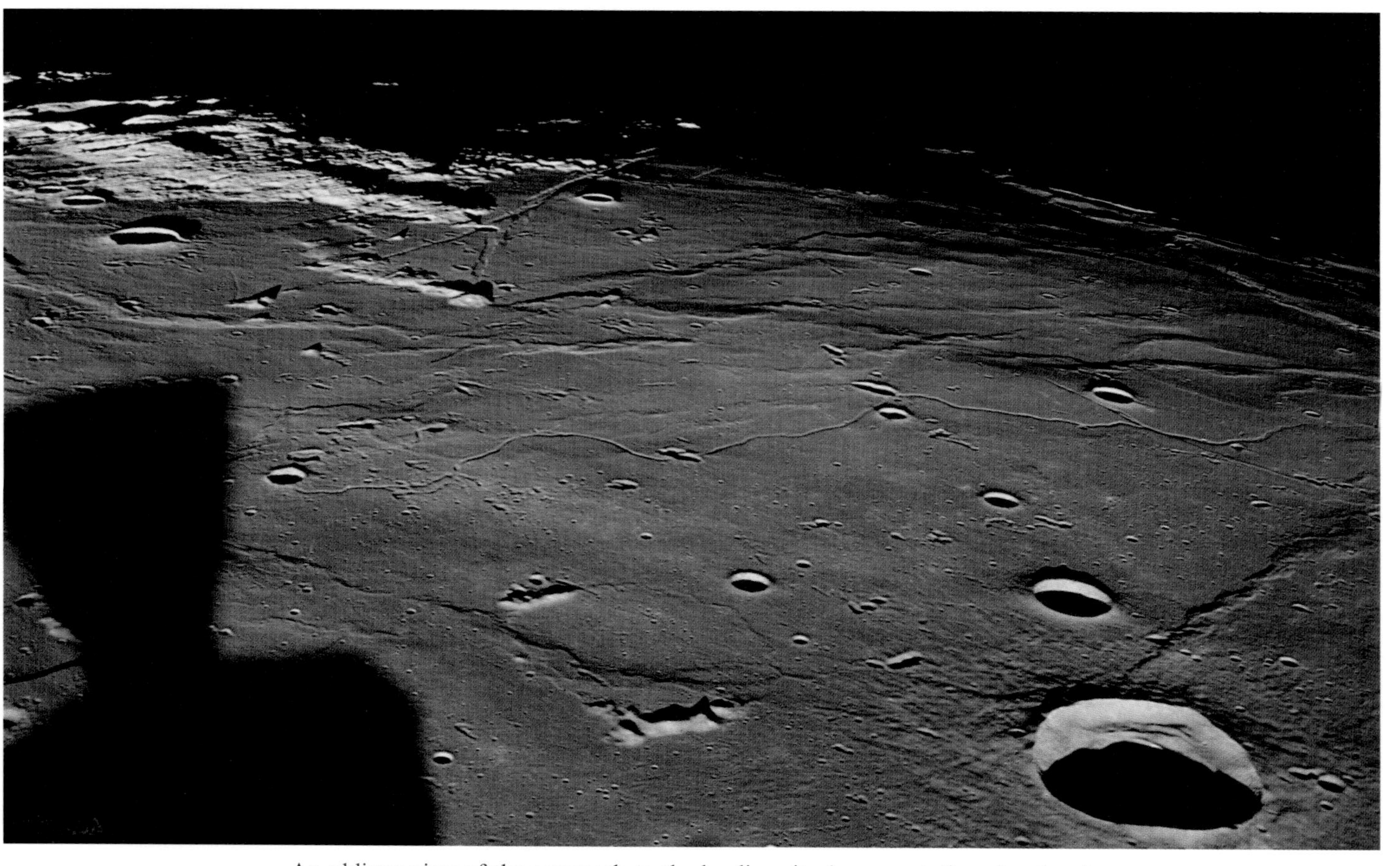

An oblique view of the approach to the landing site (see annotation elsewhere).

Frames from the 16-millimetre camera showing Neil Armstrong collecting the contingency sample alongside Eagle, setting up the television camera, and, with Buzz Aldrin, erecting the Stars and Stripes.

Buzz Aldrin on Eagle's foot pad.

The commemorative plaque on Eagle's forward leg.

Buzz Aldrin stands alongside the SWC. The rim of the crater that Eagle passed over immediately prior to landing forms the horizon, marred by the glare of the Sun.

Buzz Aldrin with the Stars and Stripes.

Part of a panoramic sequence taken by Buzz Aldrin looking north across Eagle's shadow, showing the television tripod, the Stars and Stripes and Neil Armstrong working at the MESA.

Buzz Aldrin inspects the pad and probe on the north side of the LM.

An impromptu (but iconic) picture of Buzz Aldrin.

A view of Eagle and the SWC taken by Buzz Aldrin while taking a panoramic sequence from a position north of the vehicle.

The equipment section at the rear of Eagle's ascent stage with Earth above.

Buzz Aldrin prepares to extract the PSE from the SEQ bay at the rear of Eagle's descent stage.

Having left the ALSCC where he took the previous picture, Neil Armstrong moved further out to take a panoramic sequence, catching Buzz Aldrin placing the PSE on the ground. The LRRR is still in the SEQ bay. Notice the 'washed out' landscape down-Sun, due to backscattered sunlight and the fact that shadows are masked by the objects that cast them.

Buzz Aldrin carries the PSE (left) and LRRR (right) to the deployment site.

Having deposited the LRRR, Buzz Aldrin moves further out with the PSE.

Having deployed the LRRR, Neil Armstrong photographed it for the scientists.

Neil Armstrong photographed Buzz Aldrin in the process of deploying the PSE.

With the two EASEP instruments deployed, Neil Armstrong took this 'locator' shot to document their positions relative to Eagle.

Buzz Aldrin working on the first 'core' sample.

The view from Aldrin's window after the moonwalk.

Safely back in Eagle after the moonwalk, Neil Armstrong looks very contented.

As Eagle completed its rendezvous with Columbia, Mike Collins took this picture with Earth in the background.

With the three BIG-clad astronauts safely in a raft, Clancey Hatleberg tends to Columbia's hatch.

Richard Nixon greets the Apollo 11 crew through the window of the MQF.

7

Lunar landing

DESCENT ORBIT INSERTION

During Eagle's near-side pass on revolution 13, Gene Kranz polled his team on the descent orbit insertion (DOI) manoeuvre. After studying their telemetry, the flight controllers pronounced Eagle to be in excellent condition, and 10 minutes before it went 'over the hill' Duke relayed the decision to proceed.

Kranz recalled of this loss of signal, "the adrenaline, no matter how you tried to hide it, was really starting to pump". He instructed his team to "take five", thereby prompting an exodus to the toilets. "You're standing in line, and for once there isn't the normal banter – no joking. The preoccupation is the first thing that hits you. Today is different." This was no simulation! "This is the day we're either going to land, abort or crash – the only three alternatives." While Apollo 11 was behind the Moon, the flight control team was rearranged to handle Columbia and Eagle in parallel, using independent communications links and telemetry – although the same CapCom. Spencer Gardner, monitoring the flight plan, ensured that this transition was done in an orderly manner.

The glassed-in viewing room of the Mission Operations Control Room had 74 seats, but it was a case of turn up early or stand in the aisles. As Douglas K. Ward, the Public Affairs Officer for the White Team, observed: "I believe in the viewing room we probably have one of the largest assemblages of space officials ever seen in one place." They included Thomas O. Paine, NASA Administrator; Abraham Silverstein, Director of NASA's Lewis Research Center; James C. Elms, Director of the Electronics Research Center in Massachusetts; Kurt H. Debus, Director of the Kennedy Space Center; Rocco A. Petrone, Director of Launch Operations; Edgar M. Cortright, Director of the Langley Research Center; Wernher von Braun, Director of the Marshall Space Center; Eberhard F.M. Rees, von Braun's deputy; John C. Houbolt, who championed the use of the lunar orbit rendezvous mission mode; and Charles Stark Draper, Director of the Instrumentation Laboratory at the Massachusetts Institute of Technology. A large number of astronauts including Tom Stafford, Gene Cernan and Jim McDivitt were in the viewing room, together with former astronaut John Glenn. On Management Row of the main floor, were Robert R. Gilruth, Director of the Manned Spacecraft Center; Samuel C. Phillips, Apollo

Program Director; Robert C. Seamans, formerly Deputy Administrator of NASA, now Secretary of the Air Force; Christopher C. Kraft, Director of Flight Operations; and George M. Low, Apollo Spacecraft Program Manager.

The control room was considerably smaller than it appeared on television and, being in constant use, reeked of smoke, pizza, stale sandwiches and coffee burned onto hot plates. It comprised four rows of consoles, rising successively to the rear. Each console had one or two display units which, while there was telemetry from the spacecraft, presented alpha-numerical data on a television screen that refreshed once per second. All of the flight controllers were male. Each wore a headset with an earpiece and microphone plugged into his console, and needed to monitor several intercom loops simultaneously, with his principal loop at a higher volume.

Irrespective of their seniority in NASA, at this point the VIPs had the status of observers; the same applied to Management Row. The flight director had *absolute* authority and, with it, total responsibility. He, in turn, relied on his team of flight controllers. The average age of the White Team for this mission was 26 years; Kranz was 35 years of age and Kraft was 45. Although America's manned space program was barely a decade old, the people on the control room floor this day, 20 July 1969, represented three generations in the flight control business.

As chief of the flight dynamics branch, Jerry C. Bostick supervised the team in the 'Trench', as the front row was informally known. Knowing Kranz to be 'weak' on trajectories, Bostick assigned Jay H. Greene as his flight dynamics officer (call sign 'FIDO'). A pipe-smoker with a heavy Brooklyn accent, Greene was thoroughly knowledgeable on trajectory issues. During the powered descent his job would be to monitor the Manned Space Flight Network's tracking of Eagle. To his left was the retrofire officer (call sign 'Retro'), Charles F. 'Chuck' Deiterich, who wore a large moustache and had a Texas drawl. To Greene's right sat Steven G. Bales (call sign 'Guidance'), whose job was to monitor Eagle's guidance systems, particularly the computer and landing radar. As one of the first generation of computer graduates hired by the agency straight out of college, Bales had more in common with the 'systems' people in the row behind. He wore dark-rimmed glasses, spoke rapidly, and the inflection in his voice was an indicator to his state of mind. Today was his 27th birthday. Granville E. Paules was to assist Bales to monitor Eagle's systems. Tall, blond and taciturn, Paules had the unusual habit of turning to face his interlocutor on the intercom; like Bales, he was young and systems minded.[1]

The flight surgeon for this shift, John F. Zieglschmid, had a console on the far left of the second row. To his right sat Charlie Duke, a South Carolinan with an associated drawl. Duke became an astronaut in 1966 but had yet to be assigned a specific mission. Kranz considered him to be unique among the astronauts, in that he would have made an excellent flight director. His assignment to the White Team was no fluke. "To be a good CapCom," Duke would later explain, "you've got to know the procedures used by the crew, the software, and the flow of information to the

[1] The character sketches of the White Team are courtesy of Gene Kranz.

The White Team formed for the Apollo 11 lunar landing. Left to right (front row) Don Puddy, Bob Carlton, Gene Kranz, Charlie Duke, John Zieglschmid, Captain George Ojalehto, Spencer Gardner, Frank Edelin, Arnie Aldrich and Buck Willoughby; (back row) John Aaron, Dick Brown, Chuck Lewis, Larry Armstrong, Bill Blair, Ed Fendell, Jim Hannigan, Jerry Bostick, Jay Greene, Gran Paules, Steve Bales, Chuck Deiterich and Doug Wilson. (Courtesy of Gene Kranz)

crew.'' As his astronaut specialism, Duke had monitored the development of the LM propulsion systems. On Apollo 10, Tom Stafford had asked Duke to be prime CapCom, and Duke had helped to develop the time line for checking out the LM. Armstrong had asked Slayton to have Duke do the same for Apollo 11, and then follow through with the powered descent. Duke joked that he had been given the job for the simple reason that he was the astronaut most knowledgeable of the LM who *wasn't* already on a crew! The systems controllers occupied a set of consoles on the right of the second row. Looking after Columbia were John W. Aaron, the electrical, environmental and communications officer (call sign 'EECOM') and Briggs W. 'Buck' Willoughby, the guidance, navigation and control officer (call sign 'GNC'). The equivalent of EECOM for Eagle was Donald R. Puddy (call sign 'TELCOM'). A tall, intense Oklahoman who never wasted a word, Puddy joined NASA from college. On the far right was Eagle's equivalent of GNC, Robert Carlton (call sign 'Control'). Dubbed the 'Silver Fox' owing to his prematurely greying hair, Carlton was dry, deliberate and unperturbable. To Kranz's left on the third row was Charles R. 'Chuck' Lewis, a college graduate who had initially been a remote-site CapCom for the Gemini flights. As assistant flight director (call sign 'AFD') Lewis was, in effect, Kranz's 'wing man', monitoring, and ready to deal with issues that occurred while Kranz was busy. At Lewis's left, Edward I. Fendell (call sign 'INCO') was working to consolidate instrumentation and communications into a single console, although at the time of Apollo 11 some of these tasks were still with EECOM and TELCOM. The 'core' of the team for the powered descent was Bales, Paules, Greene and Deiterich in the front row, Puddy, Carlton and Duke in the second row, and Kranz in the third row.

Once the flight controllers had resumed their consoles, Kranz directed them to switch to the assistant flight director's intercom loop, which was not available to the viewing gallery, in order to speak to them privately. Unfortunately, his speech was not recorded. On one occasion Kranz recalled, "I had to tell these kids how proud I was of the work that they'd done; that on this day, from the time that they were born, they were destined to be here and destined to do this job; they were the best team ever assembled and today, without a doubt, they were going to write the history books by being the team that took an American to the Moon." On another occasion he recalled saying, "Okay, all flight controllers, listen up. Today is our day, and the hopes and dreams of the entire world are with us. This is our time and our place, and we will remember this day and what we will do here always. In the next hour we will do something that has never been done before – we will land an American on the Moon! The risks are high, but that is the nature of our work. We have worked long hours and had some tough times, but we've mastered our work. Now we are going to make that pay off. You are a hell of a good team; one that I feel privileged to lead. Whatever happens, I'll stand behind every call you make. Good luck, and God bless us today!"

As a management interface between the flight controllers and the industrial community, Thomas J. Kelly, who led the Grumman team that developed Eagle, Dale D. Myers, his counterpart at North American Rockwell for Columbia, and senior executives of other hardware suppliers, sat at consoles in the Spacecraft Analysis Room along the corridor from the Mission Operations Control Room. Each company had engineers in other support rooms at the Manned Spacecraft Center, as well as at their own plants, and their key engineers were 'on call' to receive telephone messages wherever they might be.

With acquisition of signal imminent, Kranz ordered the doors of the MOCR locked, which was standard procedure for critical phases of a mission. He also ordered "battle short", in which the main circuit breakers for Building 30 were physically jammed in position to prevent an inadvertently tripped breaker from cutting the power. In fact, Building 30 comprised a pair of interlinked structures, one of conventional design housing offices and the other, a windowless concrete box, housing the control facilities. As Kranz put it, if a fire were to break out in one of the offices he would rather let it burn than have the automatic protective systems cut the power.[2] The display on the front wall was set to compare Eagle's onboard guidance

[2] The ground level of Mission Control held the Real-Time Computer Complex, and each of the two upper levels held a Mission Operations Control Room. Apollo 11 was managed from the top level.

Charles Duke (nearest), Jim Lovell and Fred Haise during Eagle's descent.

systems with tracking provided by the Manned Space Flight Network. Although headphone sockets were at a premium, Deke Slayton was with Duke at the CapCom console. Jim Lovell and Fred Haise of the backup crew were nearby, as were Pete Conrad, who was to command the next mission, and Dave Scott, his backup. The simulation room was packed with people who were unable to find a socket on the main floor. As Kranz later expressed it, being in the Control Room during a flight was second best only to being in the spacecraft. Mission Control's photographer, Andrew Patnesky, was snapping away from a position just behind the CapCom's console, from where he had a good view across the room. In addition, artist Bob McCall was unobtrusively sketching with pencil on paper.

Observing the growing sense of apprehension, George Low said to Chris Kraft, "I've never seen things so tense around here."

At undocking, Columbia's mass was 36,651 pounds, and Eagle's was 33,627 pounds. In terms of dry mass, the ascent stage, at 4,804 pounds, was 321 pounds heavier than the descent stage. The descent propulsion system had 18,000 pounds of propellants. The ascent stage had 5,214 pounds of propellant for its main engine and an initial load of 604 pounds of propellant for the attitude control system.

At 101:36, some 180 degrees after the separation manoeuvre, and about 2 pm in Houston, Eagle fired its thrusters for 7 seconds to slow down and impart a force to settle the propellants in the main tanks. The first 15 seconds of the DOI burn were made with the DPS throttle at 10 per cent of its designed power of 10,500 pounds of thrust, during which time the engine was gimballed to aim the thrust through the centre of mass of the vehicle, then the computer opened the throttle to 40 per cent until it had accomplished the requisite 76.4-foot-per second retrograde manoeuvre – nominally after a total duration of 29.6 seconds. Since there was no sound or vibration from the engine at the 10 per cent level, the instruments gave the only indication that Armstrong and Aldrin had to confirm ignition, but when it throttled up they felt a sagging in their knees. To restrain them while 'standing' during manoeuvres, the soles of the boots of their suits had velcro to engage with strips on the floor, and they secured themselves by spring-loaded harnesses affixed to loops at their waists. If they needed to brace themselves, they had wall bars, and there were armrests for the hand controllers. The descent orbit had its apolune at 57.2 nautical miles, and its perilune at 8.5 nautical miles or, as the crew thought of it, 50,000 feet. Because the manoeuvre was made 7 minutes after going 'over the hill', the perilune would be above the Sea of Tranquility, some 16 degrees (260 nautical miles) east of the landing site, whereupon, if all was well, Eagle would initiate the powered descent. Until that moment Armstrong and Aldrin would be retracing the trail blazed by Apollo 10.

As Eagle lost altitude in the descent orbit, it speeded up and drew ahead. As it continued downhill, its primary guidance and navigation system (PGNS) was told to manoeuvre the vehicle into an attitude calculated to place the Sun in the field of view of the navigational telescope. If the cross hair was centred on the solar disk (which spanned half a degree of arc) this would confirm that the platform had not drifted from the alignment lifted from Columbia as part of the pre-undocking procedure. If it had drifted significantly, the platform would have to be realigned by making star sightings, which would take time. As long as the cross hair was well onto the solar disk Armstrong would accept the alignment; and this proved to be the case. Aldrin then verified that the abort guidance system (AGS), which utilised body-mounted gyroscopes instead of an inertial platform, was in satisfactory agreement with the PGNS. Meanwhile, Collins used his sextant to monitor Eagle as it drew away, to verify that he would be able to make accurate long-range sightings should he have to rescue the LM. At the time of the DOI burn the vehicles were about 1,100 feet apart, and although Eagle drew ahead as it descended towards perilune, Columbia, orbiting higher, was first to acquire a line of sight to Earth. By design, there was redundancy in coverage, because although the Moon would soon set for Madrid it had already risen at Goldstone.

"Columbia, Houston," called Duke. "We're standing by."

"Reading you loud and clear," replied Collins. "How me?"

"Five-by-five, Mike," responded Duke, using radio code for 'loud and clear'. "How did it go?"

"Everything's going just swimmingly. Beautiful."

"Great. We're standing by for Eagle," said Duke.

"He's coming along."

"We expect to lose your high-gain some time during the powered descent."

"You don't much care, do you?" observed Collins.

"No, sir."

Because Collins was tracking Eagle using his sextant, his attitude relative to Earth was changing, and although the boom-mounted high-gain antenna was gimballed, at some stage during the manoeuvre its field of view would be blocked and Columbia would have to switch to omnidirectional antennas, at which time telemetry would be impaired.

Almost 2 minutes after Columbia appeared, Eagle, now down to 18 nautical miles, reported in.

"We're standing by for your burn report," Duke prompted.

"The burn was on time," replied Aldrin.

"We're off to a good start," Kranz told his team. "Play it cool."

At AOS, Joan Aldrin left the couch to stand by the fire place, laid her forearms on the mantelpiece and rested her head on her hands. Clare Schweickart made her husband, Rusty, promise to alert everyone if he heard anything untoward over the squawk box. Rusty sat with Gerry Carr, flight plan open. Aldrin's son, Mike, was upstairs, watching on another television on his own.

Communications with Eagle were poor, with static on the voice downlink and the telemetry becoming intermittent, or 'ratty' in the parlance of the flight controllers. "Columbia, Houston," Duke called. "We've lost all data with Eagle. Please have him reacquire on the high-gain."

"Eagle, this is Columbia. Houston would like you to reacquire on the high-gain. They've lost data." No response. "Eagle, did you copy Columbia?"

Aldrin adjusted the system that was to steer the antenna in order to maintain the strongest signal strength, and the static cleared. "How do you read us now?"

"Five-by," replied Duke.

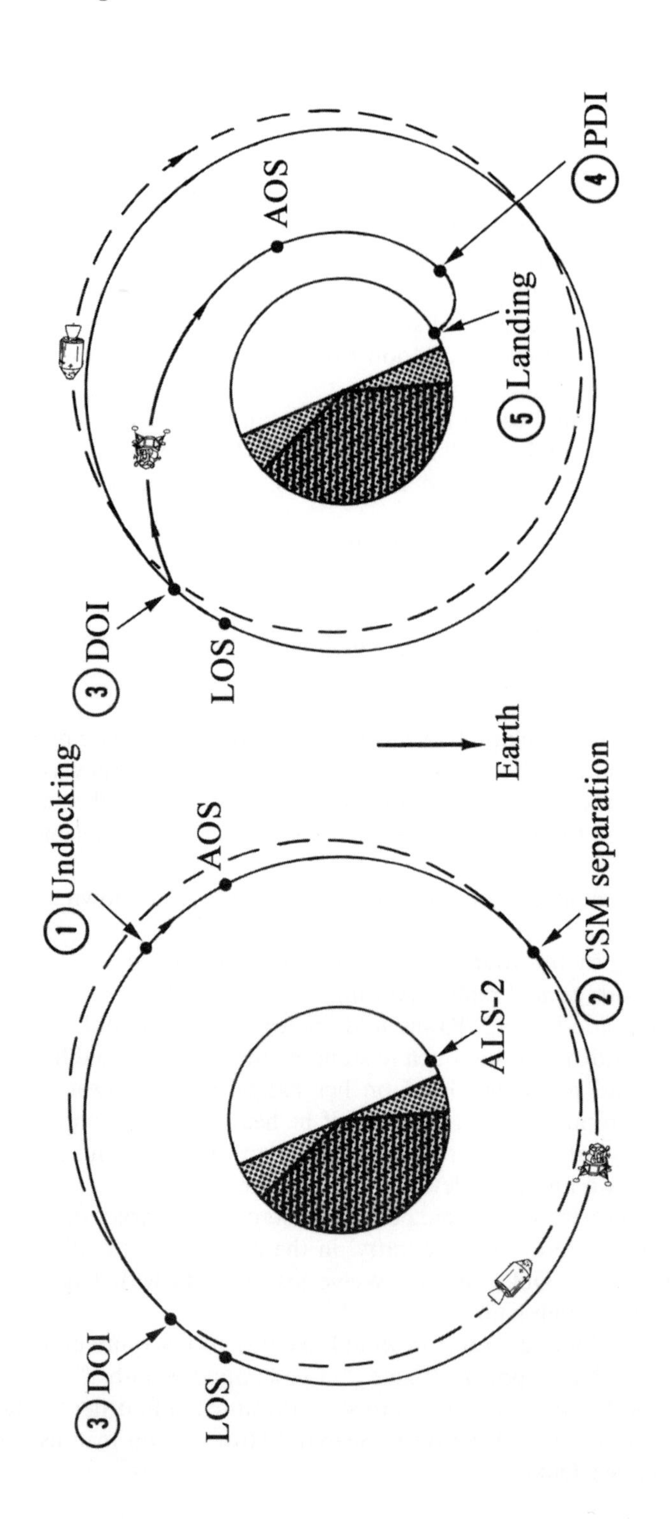

The manoeuvres required to enable Eagle to land on the Moon.

There were 13 minutes remaining to powered descent initiation (PDI).

The crew of Eagle were now flying 'feet first' and 'windows down' to enable Armstrong to check the landmarks surveyed by previous missions. This not only confirmed their ground track, but also provided an estimate of their altitude as a perilune check. That is, by using a stop watch to measure the intervals taken by a series of landmarks to pass between the angular reference marks inscribed on his window, reading their speed from the computer, and referring to a chart prepared by Floyd Bennett, a guidance engineer in Houston, he could estimate their current altitude and, by extrapolating, estimate what it would be at perilune. This manual technique could measure this vital parameter more accurately than Manned Space Flight Network tracking. Meanwhile, Aldrin was using the rendezvous radar to measure the range to Columbia in order to calculate their altitude relative to that spacecraft's orbit. If they were to initiate the descent from too high a point, then Eagle would run out of fuel before it could finish the descent. It was concluded that perilune would occur at an altitude of 51,000 feet and just prior to the PDI point, which was about as good as anyone could have hoped. If the perilune had been too high, Eagle would have had to forgo PDI on this pass and perform a manoeuvre at apolune to trim the perilune in order to try again on the next pass. Collins had been tracking Eagle using his sextant, but in the run up to PDI, with the separation now 100 nautical miles and the LM just a tiny dot moving rapidly against the backdrop of the sunlit lunar landscape, he lost it.

The Go/No-Go decision for PDI was scheduled for 5 minutes after coming around the limb, during which interval the flight controllers were to make a final check of Eagle's systems. However, with the decision imminent, communications deteriorated again. "Columbia, Houston," Duke called. "We've lost Eagle again. Have him try the high-gain."

"Eagle, this is Columbia. Houston has lost you again. They're requesting another try at the high-gain."

The signal improved.

"I don't know what the problem was there," Aldrin said. There was a display on board to indicate how the antenna was steering to maintain the strongest possible signal. "The steerable just started oscillating around in yaw."

"We'll work on it," promised Duke.

The intermittent communications were worrying. In view of the high number of crashes during simulations, Kranz had added a mission rule that there must be adequate telemetry to enable an investigation to determine the reason for a crash. He added another rule that it would be up to the flight director to decide whether this condition was satisfied. As with the voice link, the telemetry was 'dropping out' and the flight controllers' screens were freezing, and without their telemetry the flight controllers could not reach a judgement. Power constraints permitted PDI only on two consecutive perilune passes after DOI. One option (in the absence of other concerns) was to proceed with the descent and re-evaluate after 5 minutes whether to push on or to abort. Kranz faced having to decide whether it would be better to pursue this option and possibly end up aborting, or to 'wave off' in the hope that communications might have improved by the next pass.

Reasoning that with Eagle 'windows down', the steerable antenna on a boom on the right-hand side of the vehicle might have difficulty seeing around the body of the spacecraft to gain a clear line of sight to Earth, Pete Conrad suggested to Duke that Eagle make a slight yaw manoeuvre to improve the geometry.[3,4] Duke passed this on to Kranz, who asked Puddy, who agreed. Duke then called, "Eagle, Houston. We recommend you yaw 10 degrees right in order to help us on the high-gain signal strength." There was no response from Eagle, and the static continued. Kranz gave his team another 30 seconds to inspect the most recent telemetry, then announced, "Okay all flight controllers, Go/No-Go for powered descent. Retro?"

"Go!" replied Deiterich.

"FIDO?"

"Go!" replied Greene.

"Guidance?"

"Go!" replied Bales.

"Control?"

"Go!" replied Carlton.

"TELCOM?"

"Go!" replied Puddy.

"GNC?"

"Go!" replied Willoughby.

"EECOM?"

"Go!" replied Aaron.

"Surgeon?"

"Go!" replied Zieglschmid.

"CapCom we're Go for powered descent," Kranz declared.

"Eagle, Houston," Duke called, "if you read, you're Go for powered descent."

Hearing no acknowledgement, Collins voluntarily relayed. "Eagle, Columbia. They just gave you a Go for powered descent."

Having heard no response, Duke asked Collins to pass on the yaw suggestion. "Eagle, this is Columbia. They recommend that you yaw right 10 degrees and try the high-gain again." Any response was lost, but Armstrong performed the yaw, which reduced the static. "Eagle, do you read Columbia?' Collins prompted.

"We read you," Aldrin confirmed.

With communications re-established, and barely 4 minutes remaining until the manoeuvre was due, Duke repeated the go-ahead, "Eagle, Houston. We read you now. You're Go for PDI."

"Roger," acknowledged Aldrin.

Armstrong and Aldrin turned their heads in their 'bubble' helmets and grinned at each other – they were going to get the chance to land! In order to eliminate the

[3] What no one realised was that the program driving the antenna was flawed, with the result that at certain times what was expected to be a clear line of sight to Earth was blocked by the structure of the vehicle.

[4] For the LM, yaw was a rotation around the thrust axis.

overhead of using the Push-To-Talk mode, Aldrin placed his microphone on VOX. This would enable Houston to hear everything Aldrin said, but Armstrong would be heard only when he closed the spring-loaded switch on the electrical connector of his suit or, later, on the hand controller.

"Flight, FIDO," called Greene. "MSFN shows we may be a little low."

"Rog," Kranz acknowledged.

"It's no problem," Greene added.

In the division of labour between Eagle's crewmen, Aldrin would operate the computer and monitor the systems, serving as Armstrong's eyes inside to enable Armstrong to direct his attention outside. The 16-millimetre Maurer camera was installed in Aldrin's window to record the descent on Ektachrome colour film. It had a 10-millimetre-focal-length lens that spanned 55 degrees horizontally and 41 degrees vertically on the rectangular frame. Two minutes prior to PDI, he started it running. "Sequence camera is coming on." At 6 frames per second, its 140-feet of thin-base film would last for 15 minutes, which was just sufficient for the nominal descent profile.[5]

Puddy recommended new angles for the steerable antenna. Duke relayed them, and a moment later Aldrin called, "I think I've got you on the high-gain now."

With 1 minute remaining, Kranz made a final round of checks.

"Have you got us locked up there, TELCOM?"

"It's just real weak, Flight," replied Puddy. Although the high-gain link was now continuous, the signal was still weak.

"Okay, how are you looking? Are your systems Go?"

"That's affirm, Flight," Puddy replied.

"How about you, Control?"

"We look good," replied Carlton.

"Guidance, are you happy?"

"Go, both systems," replied Bales, meaning that both the PGNS and AGS were satisfactory.

"FIDO, how about you?"

"We're Go," replied Greene. Then he repeated the indication from tracking by the Manned Space Flight Network, "We're a little low, Flight; no problem."

After 22 seconds of silence on the flight director's loop, Kranz warned, "Okay all flight controllers, 30 seconds to ignition."

When sending telemetry, Eagle's steerable high-gain antenna did not have the capacity to send television. The television networks had decided to play the audio downlink over technically accurate animations provided by NASA, following the nominal timeline.[6] Eagle was to start with the descent engine facing the direction

[5] There was a spare Maurer body, and Aldrin had tested both cameras during his inspection earlier in the mission; the spare was not needed (and was jettisoned with the trash after the moonwalk).

[6] This was long before the advent of computer-generated imagery, so the animations now appear quaint!

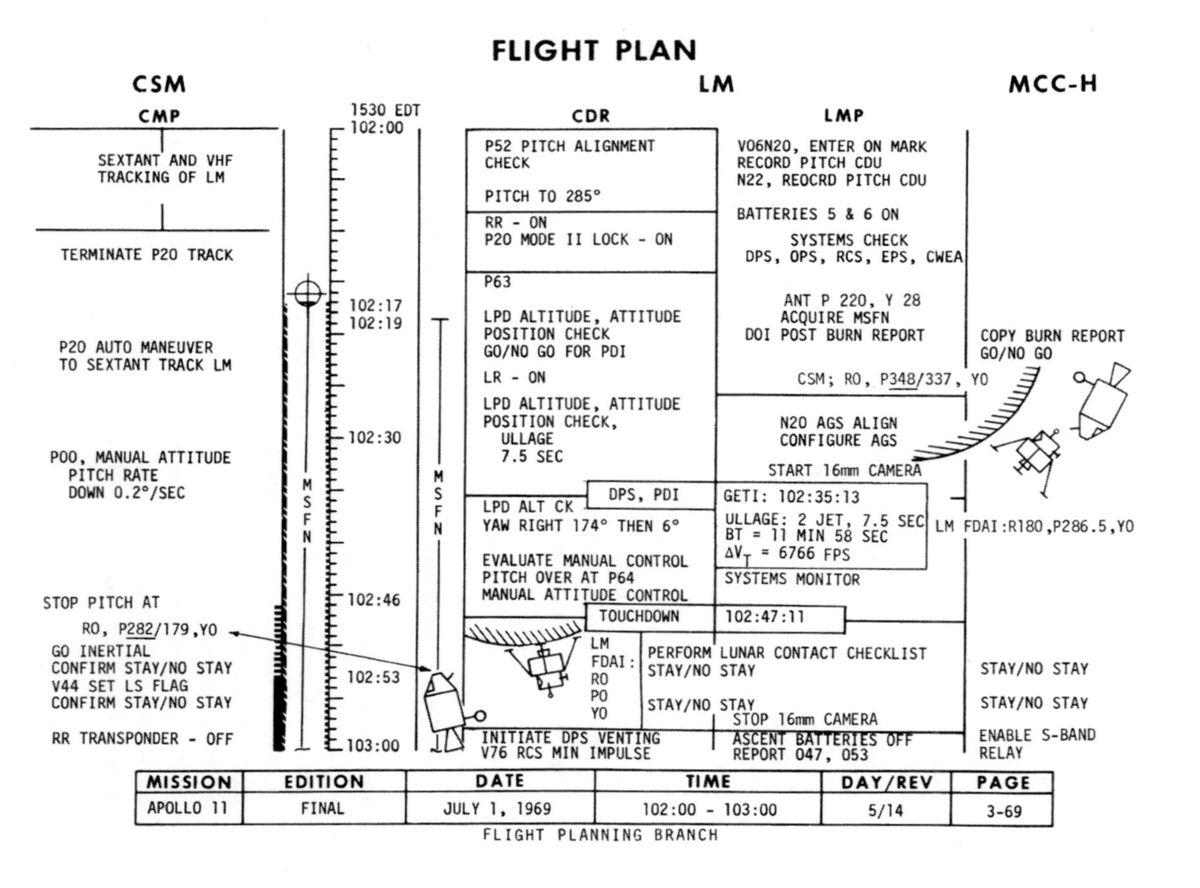

FLIGHT PLAN

CSM — CMP

SEXTANT AND VHF TRACKING OF LM

TERMINATE P20 TRACK

P20 AUTO MANEUVER TO SEXTANT TRACK LM

P00, MANUAL ATTITUDE PITCH RATE DOWN 0.2°/SEC

STOP PITCH AT
RO, P282/179, YO
GO INERTIAL
CONFIRM STAY/NO STAY
V44 SET LS FLAG
CONFIRM STAY/NO STAY

RR TRANSPONDER - OFF

1530 EDT 102:00 — 102:17 — 102:19 — 102:30 — 102:46 — 102:53 — 103:00

MSFN

LM — CDR

P52 PITCH ALIGNMENT CHECK

PITCH TO 285°

RR - ON
P20 MODE II LOCK - ON

P63

LPD ALTITUDE, ATTITUDE POSITION CHECK
GO/NO GO FOR PDI

LR - ON

LPD ALTITUDE, ATTITUDE POSITION CHECK,
ULLAGE
7.5 SEC

DPS, PDI

LPD ALT CK
YAW RIGHT 174° THEN 6°

EVALUATE MANUAL CONTROL
PITCH OVER AT P64
MANUAL ATTITUDE CONTROL

TOUCHDOWN

LM FDAI: RO PO YO

INITIATE DPS VENTING
V76 RCS MIN IMPULSE

LM — LMP

V06N20, ENTER ON MARK
RECORD PITCH CDU
N22, REOCRD PITCH CDU

BATTERIES 5 & 6 ON

SYSTEMS CHECK
DPS, OPS, RCS, EPS, CWEA

ANT P 220, Y 28
ACQUIRE MSFN
DOI POST BURN REPORT

CSM; RO, P348/337, YO

N20 AGS ALIGN
CONFIGURE AGS

START 16mm CAMERA

GETI: 102:35:13

ULLAGE: 2 JET, 7.5 SEC
BT = 11 MIN 58 SEC
ΔV_T = 6766 FPS

SYSTEMS MONITOR

102:47:11

PERFORM LUNAR CONTACT CHECKLIST
STAY/NO STAY

STAY/NO STAY

STOP 16mm CAMERA

ASCENT BATTERIES OFF
REPORT 047, 053

MCC-H

COPY BURN REPORT
GO/NO GO

LM FDAI: R180, P286.5, YO

STAY/NO STAY

STAY/NO STAY

ENABLE S-BAND RELAY

MISSION	EDITION	DATE	TIME	DAY/REV	PAGE
APOLLO 11	FINAL	JULY 1, 1969	102:00 - 103:00	5/14	3-69

FLIGHT PLANNING BRANCH

The page from the Apollo 11 flight plan dealing with Eagle's descent.

of travel to serve as a brake, and only when it had overcome most of its horizontal velocity would it reorient itself in order to make a steep descent that transitioned to hovering just above the surface for the selection of the best location, over which it would make the final vertical descent.

POWERED DESCENT

Armstrong armed the descent propulsion system (DPS) and Aldrin depressed the PROCEED key on the DSKY. As the thrusters provided ullage to settle the fluids in their tanks, Puddy had intended to accurately measure the propellant quantities, but the telemetry was inadequate and he had to resort to subtracting from the initial load the amount *estimated* to have been used during the DOI burn, which introduced an unfortunate uncertainty into his prediction of the total firing time available to the engine.

"Ignition," announced Armstrong when the computer decided that Eagle was at the PDI point. "10 percent."

"Just about on time," noted Aldrin.

It was just after 3 pm in Houston. Frustrated with the television commentary, Jan Armstrong had retired to her bedroom to listen to the powered descent on the squawk box, with Bill Anders joining her to provide technical exposition. Prior to launch, she had impressed on Slayton that if there was a problem she wanted her squawk box feed to continue. She did not want a repeat of Gemini 8, which Neil had commanded, when her audio had been cut as soon as it was realised the ship was in trouble and, even worse, on going to Mission Control to find out what was going on she had been refused entry.

Columbia was 120 nautical miles behind and above, but it would catch up and by the nominal landing time would be 200 nautical miles farther west, and near to or below the local horizon. Collins's role was to monitor the link between Eagle and Houston, and be prepared to act should intervention prove necessary.

The mood in the Mission Operations Control Room was intense concentration, and great expectation. The main wall screen showed a plot of the nominal powered descent profile, with a travelling symbol tracing Eagle's progress. Bales noted that at ignition the radial velocity component had been off by 20 feet per second. Being more than halfway towards the 'abort limit', this was concerning.[7] But he reasoned

[7] Post-mission analysis established that several interrelated factors contributed to the position–velocity error at PDI – including uncoupled attitude manoeuvres such as station-keeping, hot-fire thruster testing, and venting of the sublimator cooling system – but most of these perturbations were more or less self-cancelling. The principal error was the propagation forward of the impulse imparted at undocking due to the incomplete venting of the tunnel; this was not a mistake by Collins, it was an oversight in planning. Due to the 'vertical' attitude of the stack at undocking, the perturbation was to the *radial* component of Eagle's velocity.

that if it was a navigational issue the magnitude of the error would remain constant because it reflected a failure of the initial conditions, whereas if it was a guidance issue the error would probably increase; time would tell.

When the computer throttled up to 100 per cent 26 seconds into the burn, there was a silent high-frequency vibration and the astronauts' feet settled onto the floor, leaving them in no doubt that they had a good engine.

"PGNS is holding," Aldrin confirmed for Armstrong, being heard by Houston because he was on VOX.

The 10-degree yaw had helped communications, but because the spacecraft's attitude was fixed with respect to the surface as it travelled westwards, the line of sight to Earth was changing and the antenna was again being blocked. "Columbia, Houston," Duke called. "We've lost them. Tell them to go aft omni." On receiving Collins's relay, Aldrin opted to override the automatic pointing. He selected Slew mode and specified the pitch and yaw pointing angles appropriate to this phase of the descent profile. The signal improved.

"Eagle, we've got you now," Duke called.

"Rate of descent looks good," said Aldrin, speaking to Armstrong.

"Everything's looking good here," Duke said, by way of an advisory. Noting that Aldrin had the steerable antenna in Slew, Duke passed up a recommendation for how it should be pointed after Eagle had yawed 'windows up'.

"PGNS good? AGS good?" Armstrong prompted Aldrin.

"AGS and PGNS agree very closely," Aldrin confirmed.

The AGS was operating passively, ready for use in the event of PGNS failure. Although (as its name suggests) the AGS was for aborts, if the PGNS were to fail so close to the surface that an abort was deemed risky, the AGS would be used to continue the descent in order to land and then perform an emergency liftoff under more controlled conditions.

"How are you looking, Guidance?" Kranz prompted.

The residual in the radial velocity had remained constant, indicating it to be the result of a simple navigational error. "The residual is still 20 foot per second," Bales replied. "It looks good."

"No change, is what you're saying?" Kranz asked.

"No change," Bales confirmed. "That's down track, I know it." The PGNS was aiming for where it *thought* the target was; the fact that it had no way of knowing it was off course meant that it would land slightly downrange.

"Rog," acknowledged Kranz.

Armstrong confirmed to Houston, "RCS is good. No flags. DPS pressure is good. Two minutes."

"Altitude's a little high," warned Aldrin. They were about 47,000 feet.

Having re-established tracking by the Manned Space Flight Network following a brief hiatus, Greene said, "Flight, FIDO. We've reinitialised our filters, and we do have an altitude difference."

"Rog," acknowledged Kranz.

Since the post-DOI ranging test, the rendezvous radar had been in Auto Track mode.

A view northwest over Maskelyne from an altitude of 50,000 feet by Apollo 10 during it reconnaissance of ALS-2.

"Want to get rid of this radar?" suggested Armstrong.

"Yeah," agreed Aldrin.

"To Slew?"

"Slew," Aldrin confirmed.

This item on the checklist was a carry-over from the Apollo 10 mission, on which, because the plan had been to abort at the PDI point, the rendezvous radar had been set to continuously update the computer with the position of the CSM. At this point in Eagle's descent, however, this data was not only unnecessary, it would soon prove to be a distraction to the computer.

Aldrin noticed a fluctuation in the alternating current voltage. The concern was that the landing radar would need a stable AC power supply. However, there was no fluctuation in the telemetry and the problem was concluded to be an issue with the onboard meter.

"You're still looking good at 3 minutes," Duke advised.

"Control," Kranz called. "Let me know when he starts his yaw manoeuvre."

"Roger," acknowledged Carlton.

"How's the MSFN looking now, FIDO?" Kranz asked.

"We're Go," Greene replied.

"How about you, Guidance?"

"It's holding at about 18 feet per second," Bales replied, referring to the radial velocity residual. "We're going to make it, I think."

"Rog," Kranz acknowledged.

On making his final downrange position check, Armstrong observed that they flew over Maskelyne-W fully 2 seconds early. At their current horizontal speed of 1 nautical mile per second, this meant that they were significantly 'long'. Because the landmark checks at 3 minutes and 1 minute in advance of PDI had been on time, he was puzzled. At PDI, his attention had been inside, checking the performance of the engine, and he had not noticed precisely where they were at that moment. With the vehicle yawed 10 degrees to improve the line of sight of the high-gain antenna, it was difficult to estimate the crossrange error. The target ellipse was 11 nautical miles long and 3 nautical miles wide, with its major axis oriented along the direction of flight. Although they would land beyond the centre of the ellipse, he was certain they were 'in the ball park'. "We went by the 3-minute point early," he told Aldrin.

Aldrin was continuing to check their trajectory. One minute earlier they were slightly high, but the guidance system was steering towards the nominal trajectory. "The rate of descent is looking real good. Altitude is right about on."

Armstrong told Houston of their overshoot. "Our position checks show us to be a little 'long'."

"He thinks he's a little bit 'long'," Duke informed Kranz.

"We confirm that," Bales pointed out.

"Rog," Kranz acknowledged. Knowing the western end of the ellipse was rougher terrain than the target, Kranz mused that Armstrong might have difficulty in finding a spot on which to land, and this, in turn, alerted Kranz to the likelihood that the hovering phase of the descent might prove to be protracted.

Having begun the powered descent 'windows down' for landmark checking, Eagle now had to rotate around the thrust axis in order that the landing radar at the rear of the underside of the vehicle would face the surface. "Now watch that signal strength, because it's going to drop," Armstrong warned Aldrin as he initiated the yaw. With the steerable antenna in Slew mode, Aldrin would have to manually adjust it as the vehicle turned.

"Okay all flight controllers, I'm going to go around the horn," Kranz called as the 4-minute mark approached.

"We're yawing, Flight," Carlton informed Kranz, as requested.

"Boy, I tell you, this is hard to do," Armstrong observed.

"Keep it going," urged Aldrin.

Owing to the fact that Armstrong had neglected the checklist item to select a rapid rate of yaw, the manoeuvre began sluggishly and was erratic. Realising his error, he correctly set the Rate Switch and restarted the manoeuvre at the planned rate of 5 degrees per second. The torque from the disturbed propellants sloshing in the tanks not only made the yaw ragged, but also induced rates in the other axes, which caused much more thruster activity than he had expected. Despite Aldrin's attempt to keep the steerable antenna pointing at Earth during the turn, communications became intermittent. Kranz told his team to make their recommendations based on their most recent data, but when telemetry was restored before he could begin his poll he allowed them another few seconds.

Finally, Kranz took his poll, "Retro?"

"Go!" replied Deiterich.

"FIDO?"

"Go!" replied Greene.

"Guidance?"

"Go!" replied Bales.

"Control?"

"Go!" replied Carlton.

"TELCOM?"

"Go!" replied Puddy.

"GNC?"

"Go!" replied Willoughby.

"EECOM?"

"Go!" replied Aaron.

"Surgeon?"

"Go!" replied Zieglschmid.

"CapCom we're Go to continue PDI," Kranz announced.

Duke relayed the advisory, "Eagle, Houston, you are Go to continue powered descent."

"Roger," Aldrin acknowledged.

Eagle was now at 40,000 feet.

"Everybody, let's hang tight and look for the landing radar," said Kranz. The static cleared up. "Okay we've got data back."

The landing radar utilised four microwave beams to measure altitude in terms of echo-location and the rate of change of altitude by the Doppler effect. It was not expected to be very accurate above 35,000 feet. If it failed to function, a mission rule mandated an abort. However, in the event of difficulty bringing the radar on-line Kranz intended to permit the descent to continue to enable the problem to be investigated and, if it persisted, order the abort at 10,000 feet. He had selected this altitude because, in the absence of the radar, the spacecraft's navigation was based on Manned Space Flight Network tracking, which was calculated against a *mean* lunar surface measured with respect to the radius of the Moon at the landing site, which might as much as 10,000 feet in error; if the spacecraft were to pass below this altitude without radar it might well hit the surface. As Eagle yawed, the radar on its base began to get 'returns' from the surface.

"Radar, Flight," called Bales. "It looks good."

"Rog," Kranz acknowledged.

Because the yaw manoeuvre had run late, by the time it was complete Eagle was somewhat lower than intended at radar lock-on.

"Lock-on," Aldrin told Armstrong.

"Have we got a lock-on?"

"Yes," Aldrin confirmed. When the radar began to supply continuous data, a light on the control panel went out. "Altitude light's out."

When the altimetry became available, the PGNS was showing them to be at an altitude of 33,500 feet. The radar said they were somewhat lower. Aldrin reported this to Houston. "Delta-H is minus 2,900 feet."

The computer began by considering the orbital data from the Manned Space Flight Network to be accurate, and the radar altimetry to be suspect. But if the radar was functioning properly, its data would be more accurate. If the radar data differed significantly from the computer's navigation, the computer was to try to converge towards a compromise altitude. If the computer thought it was at 32,000 feet and the radar read 28,000 feet the computer could not simply accept this and revise its aim for the landing site, because the radar would be tracing the topography of the surface and would fluctuate. Instead, the computer would split the difference and use 30,000 feet, and iterate until it had properly 'corrected' its altitude, at which time it would recalculate its descent trajectory for the target. If they had found themselves in excess of 10,000 feet *higher* than the PGNS estimated, this would have required an abort, because if they had continued they would have run out fuel before reaching the surface.

With Eagle pitched at 77 degrees at this point in the descent, not quite on its back, its forward windows faced Earth which, because the spacecraft was east of the lunar meridian, was to the west of the zenith. Glancing out, Aldrin saw it as a half-disk of blue and white. "We have got the Earth right out our front window," he observed. As the descent continued, and Eagle progressively changed its pitch angle to face its direction of motion, transitioning to a hover, the home planet would drift out of the top of the windows.

Aldrin asked the computer to calculate and show the delta-H. As a precaution against loss of communication at this juncture, he had a chart with which to judge for himself whether the radar data was valid. Armstrong sought confirmation that Houston was also monitoring this, "Houston, are you looking at our delta-H?"

"That's affirmative," replied Duke.

"Looks good, Flight," Bales called on the flight director's loop.

"Is he accepting it, Guidance?" Kranz asked.

"Standby," replied Bales.

As Bales studied the radar data, the yellow Master Alarm in Eagle started to flash, a tone sounded in Armstrong's and Aldrin's headsets and the DSKY lit the yellow 'PROG' light.

Armstrong keyed his PTT and, with tension evident in his voice, announced, "Program alarm."

Aldrin queried the computer, which flashed '12-02'.

"It's a 12-02," Armstrong elaborated for Houston.

"12-02," confirmed Aldrin.

Armstrong and Aldrin turned their heads in their 'bubble' helmet to glance at each other; neither man had seen this alarm during simulations.

"What is it?" Armstrong asked Aldrin.

As the computer specialist, Aldrin knew in general terms what a program alarm meant, but had no way of deciding whether this was a hardware or a programming issue. "It's in core," he mused.

Although Armstrong knew that their telemetry would enable Houston to show the status of the hardware, he was also aware that if the situation were to turn sour he might have to abort without Houston's input.

Already psyched up by the task at hand, the alarm further boosted everyone's adrenaline. "When I heard Neil say '12-02' for the first time," reflected Duke, "I tell you, my heart hit the floor." The alarm caused consternation on Management Row. Gilruth, Phillips and Low sought insight from Kraft, but while he knew that some program alarms mandated an abort he was by no means an expert, and was unable to offer an explanation in this case.

Paules was the first to react, "12, 12-02 alarm." After a pause, "Yeah, it's the same thing we had." He was referring to the simulation in which Koos had caught them out – although in that case it had been a 12-01. Bales switched his attention from the radar and conferred with Jack Garman, an expert in the computer, on his support team. Garman, now fully familiar with all the program alarms, said, "It's executive overflow – if it doesn't occur again we'll be fine."

"Flight, Retro," called Deiterich while Bales and Garman were conferring.

"Go, Retro," prompted Kranz.

"Throttle down, 6 plus 25," announced Deiterich, drawing Kranz's attention to the time (measured in minutes and seconds since the start of the powered descent) at which Eagle was to throttle down.

"6 plus 25," acknowledged Kranz, annotating his console log.

In simulations Armstrong had been primed to abort, but now he was primed to push on. Nevertheless, he was concerned by the lack of a response from Houston, "Give us a reading on the 12-02 program alarm."

"We're Go on that, Flight," Bales finally announced, having established that, despite the alarm, the guidance system was performing its assigned tasks. But as he would reflect later, "In the Control Center any more than 3 seconds to reach a decision during powered descent is too long; and this took us about 10 to 15 seconds."

Duke replied to Armstrong, "Roger. We gotcha. We're Go on that alarm."

"If it doesn't recur, we'll be Go," Bales added.

"Rog," acknowledged Kranz, noting the confidence in Bales's voice. "Did you get the throttle down, CapCom?" Having missed it, Duke passed this information up to the spacecraft.

Eagle's altitude was now down to 27,000 feet. Bales, returning his attention to the landing radar, announced, "He's taking in the delta-H now."

"Rog," acknowledged Kranz.

"Flight, FIDO," called Greene. "We're converging on delta-H."

"Rog," acknowledged Kranz.

"Flight, Control," called Carlton. "We're on velocity."

"Rog," acknowledged Kranz.

Having received a Go on the 12-02, indicating that the computer was healthy, Aldrin again queried delta-H, and the alarm recurred. "Same alarm," he called, "It appears to come up when we have a 16/68 up." Keying Verb 16 with Noun 68 told the computer to display the altitude and velocity, the range to the landing site, and the time remaining in the braking phase (in essence, the time to the pitch-over manoeuvre). Aldrin was speculating that his checking of the delta-H convergence might be prompting the executive overflow. Aldrin was correct, but the true issue was that the rendezvous radar was needlessly interrupting the computer, leaving it little time to devote to computations in addition to its navigational tasks.

"Roger. Copy," acknowledged Duke.

This time Bales responded promptly, "It's okay." In the hope of relieving the load on the computer, he offered, "We'll monitor his delta-H, Flight."

"Rog," acknowledged Kranz.

Bales agreed with Aldrin's line of thought. "I think that's why he's getting it."

"Okay," said Kranz.

"Eagle, Houston," called Duke. "We'll monitor your delta-H."

"Delta-H is beautiful," Bales observed.

"Delta-H is looking good to us," Duke relayed.

"All flight controllers, hang tight," Kranz prompted, "We should be throttling down shortly."

At 6 minutes 25 seconds into the powered descent, the computer throttled down the DPS engine from 100 per cent to 55 per cent.

"Throttle down on time," announced Armstrong.

"Confirm throttle down," Carlton noted.

"Rog, confirmed," replied Kranz.

"Roger," Duke responded to Armstrong. "We copy throttle down."

"You can feel it in here when it throttles down better than the simulator," said Aldrin, tongue-in-cheek.

"Rog," acknowledged Duke.

The fact that the computer throttled down the engine on time indicated that it was unaware it was coming in 'long', as otherwise it would have delayed the transition in order to compensate and thereby re-establish its aim for the target.

"AGS and PGNS look real close," noted Aldrin.

"Flight, Control," called Carlton. "Everything looks good."

"Rog," acknowledged Kranz.

"Flight, FIDO," called Greene. "We're looking real good."

"Rog, FIDO, good," replied Kranz.

The spacecraft's altitude was now down to 21,000 feet, and it had slowed to 1,200 feet per second.

"At 7 minutes, you're looking great to us, Eagle," Duke called.

"TELCOM," Kranz prompted, "how're you looking?"

"It looks good, Flight," replied Puddy.

"Rog," acknowledged Kranz.

With the pitch angle now down to 60 degrees and the rate of change increasing, Aldrin announced, "I'm still on Slew, so you may tend to lose the high-gain as we gradually pitch over." Then he had second thoughts, "Let me try Auto again now, and see what happens."

"Roger," Duke acknowledged.

"We're going to try the steerable again, Don," Kranz warned TELCOM.

"Copy, Flight," replied Puddy.

"It looks like it's holding," reported Aldrin. With a clear line-of-sight and the steerable dish locked on, communications improved markedly.

"Roger," acknowledged Duke. "We've got good data."

"Are we on the steerable, Don?" Kranz asked.

"That's affirmative, Flight," replied Puddy. "And it's holding in there pretty good."

"Rog," acknowledged Kranz. His concern over telemetry drop-outs abated. It seemed that he would not, after all, face the decision as to whether communications had degraded to the point of requiring an abort.

The spacecraft's altitude was now down to 16,300 feet, and it had slowed to 760 feet per second.

"Okay, everybody hang tight," Kranz said. "Seven and a half minutes."

"Flight, Guidance," called Bales. "His landing radar's fixed to velocity; it's beautiful."

"Flight, Control. Descent 2 fuel," Carlton announced. Having closely studied the redundant propellant gauging systems, he recommended monitoring the 'low level' sensor in gauging system number 2.

"Descent 2 fuel crit," said Kranz.

"Descent 2 fuel, On," corrected Carlton. "I didn't want to say 'critical'."

"Rog," acknowledged Kranz.

Duke relayed the advisory, taking care not to be ambiguous, "Eagle, Houston. Set Descent 2 fuel to Monitor."

"Roger, 2," acknowledged Armstrong.

"Flight, FIDO," called Greene. "It's looking real good."

Pat Collins, listening to her squawk box, nervously clenched her fist.

Eagle's altitude was now down to 13,500 feet. Having elected not to use 16/68 to avoid further 12-02 program alarms, Aldrin asked for the time remaining in the braking phase, "Could you give us an estimated pitch-over time, please, Houston?"

"Stand by," said Duke. "You're looking great at 8 minutes."

"Thirty seconds to P64," called Bales, responding to Aldrin's request.

"Eagle, you've got 30 seconds to P64," relayed Duke. The P64 program would switch to the visual approach phase of the descent.

"Have we still got landing radar, Guidance?" Kranz asked.

"Affirm," replied Bales.

"Okay. Has it converged?" Kranz asked.

"It's beautiful," replied Bales.

"Has it converged?" Kranz repeated.

"Yes!" Bales replied.

"Flight, FIDO," called Greene. "We look real good."

"Rog," acknowledged Kranz.

"Eagle, Houston," called Duke. "Coming up 8 plus 30. You're looking great."

Having reached a point known as the 'high gate' at an altitude of 7,500 feet, Eagle's computer initiated P64, which rapidly reduced the pitch angle from 55 degrees down to 45 degrees. Thus far, most of the thrust had been devoted to slowing the horizontal velocity. As the pitch was further reduced, more of the thrust would be directed downwards. During the pitch-over, the radar on the base of Eagle swung from its 'Descent' position to 'Hover', where it would remain, and the horizon rapidly swung up into the bottom of the windows, giving Armstrong his first view of where the computer was heading, which at this altitude was a point some 3.5 nautical miles dead ahead, just on this side of the horizon.

"P64," called Aldrin.

"We copy," Duke acknowledged.

"Okay, they've got 64," Kranz announced over the flight director's loop. "All flight controllers, 20 seconds to Go/No-Go for landing."

"Eagle, you're looking great," Duke confirmed. "Coming up on 9 minutes."

The spacecraft was down to 5,200 feet and descending at 100 feet per second, which was as planned. Armstrong tested his hand controller in pitch and yaw, and then resumed 'hands off'. "Manual attitude control is good."

"Roger, copy," acknowledged Duke.

As Eagle descended through 4,000 feet, Kranz went around the horn, "All flight controllers, Go/No-Go for landing. Retro?"

"Go!" called Deiterich.

"FIDO?"

"Go!" called Greene.

"Guidance?"

"Go!" called Bales.

"Control?"

"Go!" called Carlton.

"TELCOM?"

"Go!" called Puddy.

"GNC?"

"Go!" called Willoughby.

"EECOM?"

"Go!" called Aaron.

"Surgeon?"

"Go!" called Zieglschmid.

"CapCom we're Go for landing."

"Eagle, Houston. You're Go for landing."

On hearing this, Jan Armstrong sat up on her heels at the foot of her bed. Pat Collins exclaimed, "Oh God, I can't stand it."

"Roger. Understand, Go for landing," acknowledged Aldrin. "3,000 feet." But then, "Program alarm." He keyed the DSKY for the code, "12-01."

"Roger," acknowledged Duke, "12-01 alarm."

"Same type," responded Bales immediately. "We're Go, Flight."

"We're Go. Same type," relayed Duke, the tension evident in his voice. "We're Go."

Armstrong had wanted to look for landmarks to determine how 'long' they were, but this alarm distracted him, and when he next looked out they were so low that he could not see any of the landmarks he had memorised, "So," he later reflected, "all those pictures Tom Stafford took on Apollo 10 to enable me to pick out where I was going and know precisely where I was, were to no avail."

"2,000 feet," called Aldrin.

Pat Collins nervously began to bite her lip.

As Aldrin had explained prior to launch, "During the landing, there is a fairly even division of labour. Neil will be looking more and more outside, his hand on the 'stick'. He is not able to look much at the instruments. This is where we must work as a finely tuned team, to ensure that he gets the information he requires to transfer whatever he sees into something meaningful. I'll relay this information. And at the same time I'll be looking at the various systems to make sure they're operating the way they should. However, here I am looking at five or six gauges, and, by telemetry, we've got teams of people looking at each gauge on Earth, so, really, I'm confirming what a lot of people are getting."

Left to itself, the computer would continue the descent until it either landed or crashed in the attempt, most likely as a result of unfavourable terrain. To find out where the computer was heading, Armstrong asked Aldrin for an angle for his Landing Point Designator, "Give me an LPD."

Aldrin interrogated the computer, "47 degrees."

The panes of Armstrong's two-layer window were annotated with a scale. The angle was measured downward, relative to directly 'ahead'. Positioning his head to align the scales, he sighted beyond the 47-degree mark to the position, a little more than 1 nautical mile away, where the computer was taking them. "That's not a bad-looking area," he observed to Aldrin.

Duke continued his advisories, "Eagle, looking great. You're Go."

As Eagle descended through 1,400 feet, the computer issued another program alarm. "12-02," called Aldrin.

"Roger," acknowledged Duke. "12-02."

"How are you doing, Control?" Kranz asked.
"We look good here, Flight," replied Carlton.
"How about you, TELCOM?"
"Go!" replied Puddy.
"Guidance, are you happy?"
"Go!" Bales replied.
"FIDO?"
"Go!" Greene replied.

To veteran reporters such as Reginald Turnill of the BBC, who had made the effort to learn something of the systems, this determination to push on regardless of the alarms began to look as if it would end with a crash.

"What's the LPD?" Armstrong asked.
"35 degrees," replied Aldrin. "750 feet, coming down at 23 [feet per second]."
Pat Collins now began to bite her finger.
"33 degrees," Aldrin called. "700 feet, 21 down."

With Aldrin acting as his eyes inside, Armstrong directed his attention outside. The computer was heading for a crater the size of a football field, surrounded by a field of ejecta excavated by the impact. He later reflected, "I was surprised by the size of the boulders, some of which were the size of small automobiles." The crater was 600 feet in diameter. "Pretty rocky area," Armstrong observed to Aldrin.

"600 feet, down at 19," Aldrin recited.

On the nominal descent, Armstrong was not to take control until Eagle was down to about 150 feet. However, in view of where it was heading, he could not let the computer continue to fly 'blind'. He considered trying to set down short of the crater or even among its ejecta in order to be able to inspect the boulders for the scientists, but ruled this out as being too risky and instead decided to follow his piloting instincts, and 'extend'. He selected the semi-automatic flight mode that would enable him to control attitude and horizontal velocity, while the computer – allowing for his commands – operated the throttle. At an altitude of 500 feet, at a point known as 'low gate' in the descent profile, he intervened. He cut the pitch angle from its current 20 degrees to about 5 degrees, thereby standing the vehicle essentially 'upright' to direct nearly all its thrust downwards in order to maintain the horizontal velocity of 60 feet per second and reduce the rate of descent from 19 feet per second to 9 feet per second. He then selected Attitude Hold, and let Eagle fly a shallow trajectory over the field of ejecta just north of the crater, while he looked for a clearer area further downrange.

"Attitude Hold!" called Carlton, on noting the mode change in the telemetry.
"Roger, Att-Hold," acknowledged Kranz.

At this point, as Duke recalled: "We were down to the last couple of minutes. Deke Slayton is sitting next to me. We're both glued to the screen on my console, and I'm just talking and talking and telling them all this stuff, and Deke punches me in the side and says 'Charlie, shut up and let them land'."

"I think I'd better be quiet, Flight," Duke said.
"Rog," acknowledged Kranz.

Because Armstrong had overridden the computer, Aldrin deleted the LPD angle from his cycle, and instead began to report their forward velocity: "400 feet, down at 9, 58 [feet per second] forward."

"The only call-outs now will be fuel," Kranz directed. Carlton, monitoring the propellant gauging system, would make the calls for Duke to relay. As the tension mounted, the flight controllers unconsciously grasped the handles of their display units; these were nicknamed 'comfort handles'.

"350 feet, down at 4," called Aldrin.

"P66," announced Carlton, reporting that the computer had switched from the approach phase to the landing phase.

"330, 6-1/2 down," called Aldrin. "We're pegged on horizontal velocity." At this point, there was a burst of static on the downlink.

Although Armstrong had not explained why he had intervened, it was evident from the fact that Eagle was passing downrange on an almost horizontal trajectory at high speed that he was taking evasive action. Kranz recognised that the locus of decision-making had transferred to Eagle. The vehicle was not yet into the 'dead man's box', but soon would be. The remainder of the descent would be up to Armstrong. Kranz also knew that as long as Armstrong thought he had a fair chance of making a landing he would press on. But, as Stafford had noted after the low pass by Apollo 10, the western end of the ellipse appeared to be much rougher than the aiming point.

On flying clear of the boulders around the big crater, Armstrong pitched Eagle back again in order to rapidly slow the horizontal velocity which, as a result of his evasive action, was now excessive for their altitude. On spotting a line of boulders up ahead, he neatly 'side stepped' off to the left – just as he had done when flying the LLTV, firstly by tilting Eagle in the direction he wished to go in order to use a component of the thrust to set up the requisite lateral velocity then, just before reaching where he wished to be, tilting in the opposite direction in order to cancel this translation, resuming the original orientation directly above his selected position. Although in such manoeuvres Eagle had the familiar sluggish response of the LLTV, he was delighted to find the LM easier to fly. To buy time, he began to use the toggle switch on the hand controller designed to adjust the rate of descent in increments of 1 foot per second; having been sceptical of this feature, Armstrong was delighted to find it very effective.

"Okay, how's the fuel?" asked Armstrong, as he continued to manoeuvre at an altitude of 300 feet.

"8 per cent," Aldrin replied.

Now well clear of the ejecta, Armstrong began to ease down.

"Okay, this looks like a good area here," Armstrong informed Aldrin.

Aldrin stole a glance outside and saw Eagle's shadow on the ground ahead. He was surprised since, being at an altitude of about 260 feet with the Sun low in the east, he had expected the shadow to be too far west to be readily visible; but there it was, distinctly showing the structure of the vehicle. "I got the shadow out there," he reported. Unfortunately, as a result of manoeuvring, Eagle was yawed slightly left, and the central pillar in front of the instrument panel blocked Armstrong's view of the shadow.

"250, down at 2-1/2, 19 forward," recited Aldrin.

"Okay, Bob. I'll be standing by for your call-outs shortly," Kranz prompted.

"Altitude/velocity light," noted Aldrin. This warning light indicated that the radar data had degraded. The logic was that the light illuminated when the output from the radar was unusable by the computer – it was lit prior to lock-on, went out with lock-on, and thereafter would come on to alert Aldrin to the fact that the radar had lost track of the surface. Because it had been deemed impractical to try to land by 'seat of the pants' flying, as there would not be the visual references to give a sense of altitude and rate of descent, the mission rules stated that if the radar were to fail they would have to abort. But they continued expectantly, and after 20 seconds the radar locked on again. Then Aldrin resumed his calls, "3-1/2 down, 220 feet, 13 forward."

As the downlink was lost to static, Jan Armstrong slipped her arm around son Ricky's shoulder. Joan Aldrin was standing in silence by the wall, grasping a door, her eyes moist, praying that Eagle would not crash. In Mission Control, Kranz had decided that he would not call an abort unless he was certain it was essential. As regards the mission rule that he had introduced requiring there to be telemetry for the powered descent to continue, he recalled, "Once we were close, I intended to let the crew go if everything appeared okay to them – I considered a low-altitude fire-in-the-hole abort riskier than landing without telemetry. I looked at a fire-in-the-hole abort the same way that I looked at a parachute when I was flying jets; that is, you use a parachute only when you've run out of options." Armstrong would later say that an abort involving (1) shutting down the DPS, (2) firing the pyrotechnics to sever all the structural and electrical connections between the stages, and (3) igniting the APS 'in the hole' in rapid succession, in close proximity to the lunar surface, "was not something in which I had a great deal of confidence". If the process were not to occur cleanly, it would jeopardise the ascent stage's departure. It had been done only on the unmanned test of LM-1 in 1968. In fact, this aversion to abort-staging had led to the mission rule that if a problem were to develop after the 5-minute point in the powered descent that did not mandate an in-flight abort, then every effort would be made to land in order to lift off several minutes later. However, if the DPS were to cut off once Eagle was within 200 feet of the surface, it would be doomed as it fell in the weak lunar gravity because by the time the abort-staging sequence was concluded, the APS would not be able to impart a positive rate of climb before the ascent stage struck the surface. Eagle was almost at this critical altitude.

"11 forward. Coming down nicely," said Aldrin.

"I'm going right over a crater," Armstrong pointed out. As he was not using his PTT, Houston did not hear this remark.[8]

"200 feet, 4-1/2 down, 5-1/2 down."

"I've got to get farther over here," Armstrong said, as he resumed manoeuvring.

[8] The large rock-strewn crater towards which the computer had been heading was named 'West', and the 75-foot-diameter crater over which they passed at this point would later be named variously 'Little West' or 'East' Crater.

"160 feet, 6-1/2 down," continued Aldrin.

There were 'level sensors' in each pair of propellant tanks, and Carlton had recommended that they use set 2. When either the fuel or oxidiser sensor in these tanks became exposed, it would illuminate the 'Descent Quantity' light on Eagle's control panel and generate the 'low level' signal in the telemetry. The signal meant there was now only 5.6 per cent of the initial propellant load remaining, which, in hovering flight with the throttle at about 32 per cent, meant the engine would cut off in 96 seconds. With 20 seconds reserved for the preliminary action of an abort during which the DPS would be throttled up to cancel the rate of descent and impart a positive rate of climb prior to abort-staging, the low-level signal meant that in 76 seconds Armstrong would be required either to abort or forgo the option of aborting and commit himself to touching down within the next 20 seconds. Borrowing pilots' slang, this decision point was known as the 'bingo' call.

"Low level," called Carlton over the otherwise silent flight director's loop. He started his stopwatch.

"Low level," echoed Kranz. This call "really grabbed my attention" he would later reflect, "mainly because in training runs we'd generally landed by this time".

"5-1/2 down, 9 forward," recited Aldrin. "You're looking good." After a burst of static, he was heard to say "120 feet."

Armstrong again slowed the rate of descent in order to manoeuvre to a flatter spot. Slope could be judged visually while hovering because, with the Sun low to the rear, a bright patch was probably sloping up because it was well illuminated, whereas a dark patch was probably sloping down and poorly illuminated. He had to find an evenly lit location that was free of rocks. The presence of rocks could be inferred from the shadows that they cast. As he recalled, "I changed my mind several times, looking for a parking place. Something would look good, and then as we got closer it really wasn't so good. Finally, we found an area ringed on one side by fairly good sized craters and on the other side by a boulder field; it wasn't particularly big, a couple of hundred square feet – about the size of a big house lot."

"100 feet, 3-1/2 down, 9 forward," recited Aldrin.

At the suggestion of Bill Tindall, the illumination of the Descent Quantity light did not trigger either the caution and warning light or sound an audible tone; it was a normal event after all, not something to risk distracting the crew so near the lunar surface. It was therefore some time before Aldrin noticed the amber lamp, "Five per cent. Quantity light."

Carlton was focused on his stopwatch. "Coming up on 60," he warned.

"Rog," acknowledged Kranz.

"Okay," continued Aldrin. "75 feet and it's looking good."

"60!" called Carlton.

"60 seconds," echoed Kranz.

Duke, who had been silent for some time, passed this on. Aldrin did not respond, opting instead to maintain his instrument readings for Armstrong.

Jan Armstrong sat forward, one hand over her mouth, her eyes a little brighter than usual. Joan Aldrin, tears in her eyes, was huddled against the frame of a door, one hand resting on a lamp shade, which was shaking.

Since Eagle might easily damage one of its legs (or possibly even tip over) if it were to land with a significant horizontal velocity, once Armstrong was directly over his chosen spot he focused on a point just in front as his visual reference and set about 'nulling' his lateral velocity components in preparation for a vertical descent. However, because he had no wish to drift backwards into an obstacle, he retained a very slow forward motion that tests had indicated the legs should be able to resist.

"Light's on," reported Aldrin. The illuminated altitude/velocity light indicated that the radar data had degraded again, but this time the drop-out lasted only a few seconds. "60 feet, down 2-1/2," he continued. After a pause, he added, "2 forward. That's good." And again, "40 feet, down 2-1/2."

Armstrong cut the throttle in order to descend. The exhaust plume was now in contact with the surface, but because the spacecraft had shed half of its mass since PDI the engine was delivering only about 1,000 pounds of thrust. Nevertheless, it stirred up the fine surface material. "Picking up some dust," Aldrin reported.

Unable to billow in the absence of an atmosphere, the dust travelled radially outward on 'flat' trajectories. The dust moving forward created the illusion that Eagle was drifting backward. Fortunately, the semi-transparent layer of 'ground fog' was so thin that some of the rocks poked up through it, and Armstrong was able to maintain his visual reference.

"30 feet, 2-1/2 down," called Aldrin. He saw the shadow of Eagle's right leg, the probe on its foot pad indicating that it was tantalisingly close to the surface. He also noticed that although shadows on the Moon were normally sharply defined, Eagle's shadow was softened by the dust passing just above the surface. "Faint shadow."

"And now for 30," called Carlton, monitoring his stopwatch.

"4 forward," Aldrin continued. "4 forward. Drifting to the right a little."

"30!" Carlton announced.

"30 seconds," echoed Kranz.

This was relayed by Duke with incredulity evident in his voice, "30 seconds."

In the Mission Operations Control Room, the flight controllers, managers and visitors began to breathe intermittently – some even ceased to breathe.

"20 feet, down a half," Aldrin called. "Drifting forward just a little bit. That's good." When one of the three 67-inch-long probes struck the surface it illuminated a blue lamp on the central control panel. Armstrong, his attention outside, did not see this, but Aldrin had the lamp in his peripheral vision. "Contact light!" Carlton had been about to call out 15 seconds as the start of a second-by-second countdown.

The final rate of descent was required not to exceed 3 feet per second, since (as factory testing had indicated) a faster sink rate could shock the legs sufficiently to damage them – possibly so much as to prevent a subsequent liftoff, during which the descent stage was to serve as the platform for the ascent stage. In practice, this meant that the vehicle was not to be allowed to fall in lunar gravity from a height exceeding 10 feet. The contact probe satisfied this requirement. Furthermore, the engine was to be shut down immediately the contact light lit, in order to preclude the possibility of back pressure from the plume in such close proximity to the surface damaging the engine, possibly causing it to explode. However, Armstrong was a second or so late, with the result that instead of falling the final 5 feet, Eagle settled onto the surface very gently at a sink rate of just 1.7 feet per second, with each of its pads pivoting to settle on the uneven surface. Although Armstrong had tried to cancel the lateral velocities and maintain a slight forward creep, it was later determined that Eagle had been drifting to the left at about 2 feet per second and the left leg was first to make contact, indicating that the vehicle had been tilted that way. As Aldrin reflected later, "I would think that it would be natural, looking out the left window and seeing dust moving left, that you'd get the impression of moving to the right and counteract by going to the left." As a result of the final manoeuvring, Eagle landed yawed around at an angle of 13 degrees left. On the uneven surface, its 4.5-degree backward tilt was well within the 10-degree tolerance.

"Shutdown!" announced Armstrong.

Turning their heads in their 'bubble' helmets Armstrong and Aldrin grinned at each other. Armstrong later reflected: "If there was an emotional high point, it was after touchdown when Buzz and I shook hands without saying a word." As Aldrin recalled the event, he was "surprised, in retrospect, that we even took time to slap each other on the shoulders".

Armstrong later insisted that the landing was everything he could have wished for, and the fact that it had been achieved with just seconds to spare had made it even more satisfying. In fact, he was not concerned by the narrow fuel margin, because this had always been so when flying the LLTV, which had severely limited flight time. A later analysis would show that when he began to manoeuvre, the fluids in the propellant tanks had sloshed around and because the level sensor in each tank was located on top of a 9-inch-tall rod the 'low level' signal had occurred 20 seconds prematurely. In fact, when Carlton's count reached the 15-second mark, the engine could have sustained 25 seconds of hovering prior to the 'bingo' point; the halving of the margin from 20 seconds at the 'low level' signal to 10 seconds at actual touchdown presumably being because Armstrong had departed from the nominal trajectory ahead of schedule in order to manoeuvre, thereby consuming propellant at an increased rate. Telemetry showed that Armstrong's heart rate had been 110 beats per minute at PDI, peaked at 156 during his final manoeuvres, and then rapidly dropped back to about 95.

Aldrin immediately started the post-shutdown checklist. "Engine Stop. ACA out of detent. Mode Control, both Auto. Descent Engine Command Override, Off; Engine Arm, Off; 413 is in."[9,10]

[9] The Attitude Control Assembly (ACA) was the hand controller used to fly the spacecraft. It was spring-loaded to stand in its central detent. The computer not only interpreted a displacement as a request for a manoeuvre but also remembered how the stick was being used. By nudging it out of detent after shutdown, Armstrong was essentially clearing it.

[10] The AGS used 'strap down' gyroscopes, which had a tendency to drift. Now that Eagle was on the surface, Aldrin loaded a specific value into address '413' of the AGS to tell that system to store its attitude information to ensure that if (1) an emergency liftoff became necessary, and (2) by sheer ill luck the PGNS were to malfunction beforehand, obliging them to use the AGS, then this system, by virtue of having stored its attitude immediately after landing, would be able to correct for any drift in its gyros.

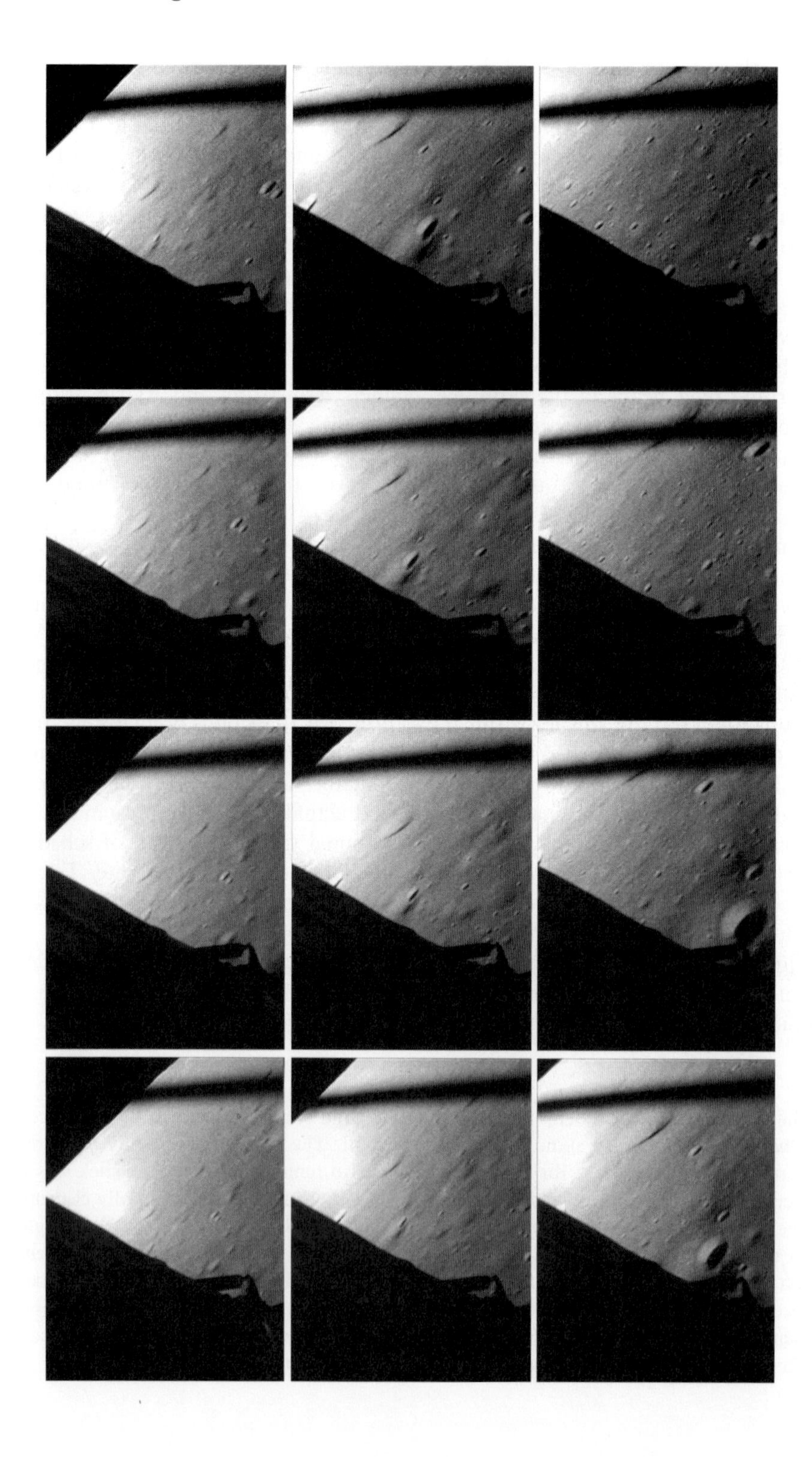

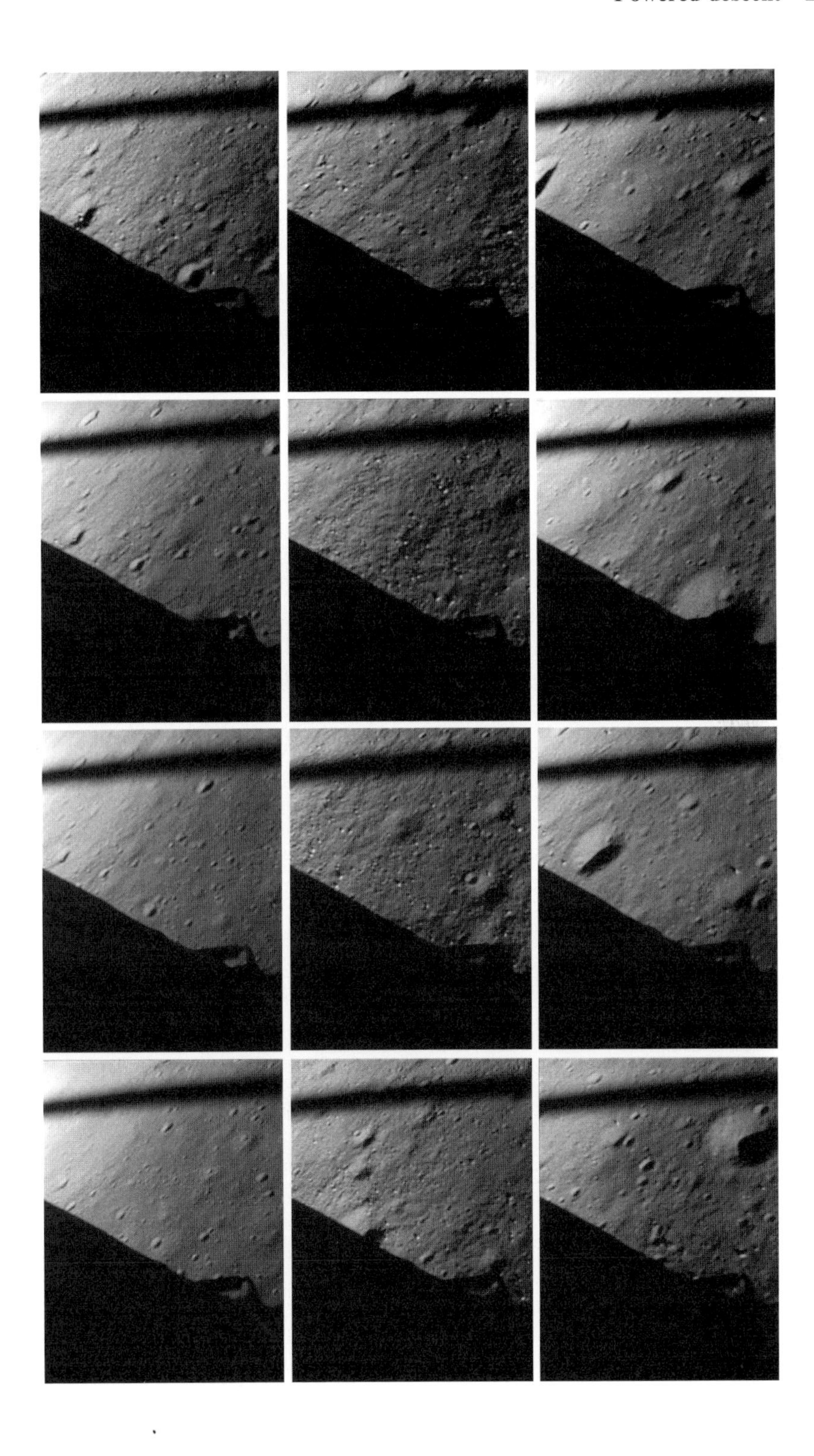

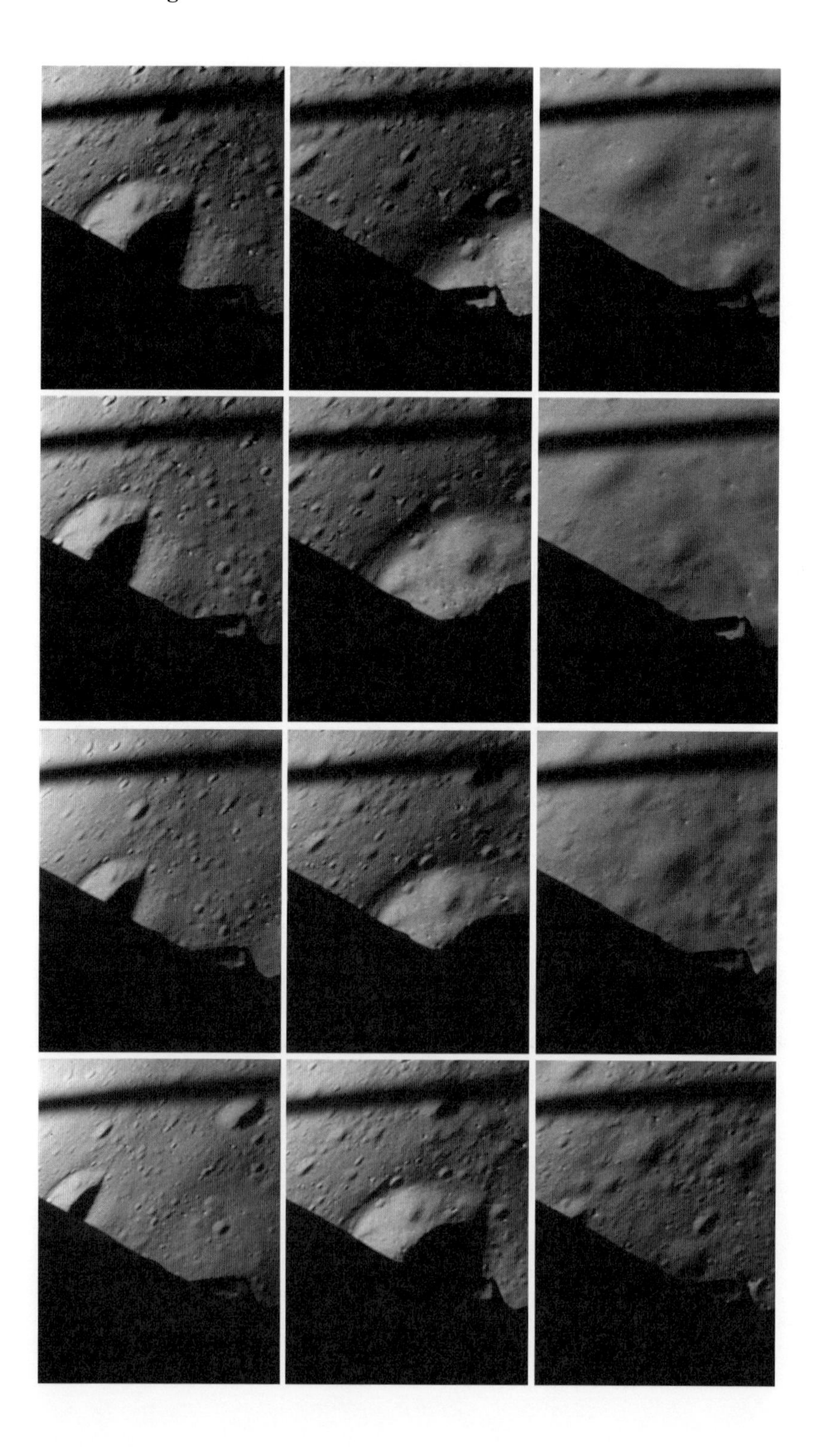

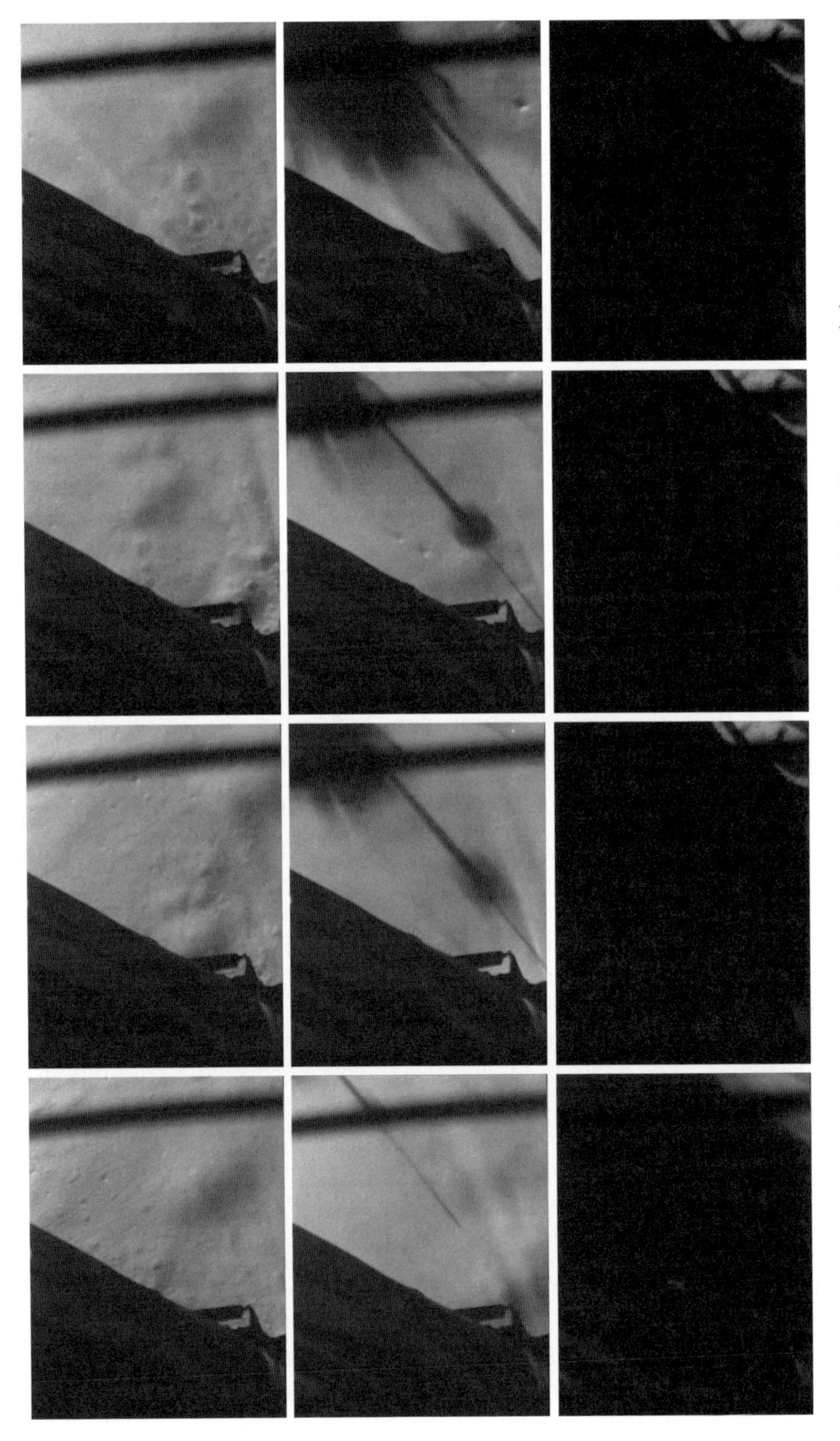

Frames from the 16-millimetre camera at 5-second intervals during Eagle's powered descent.

"Flight, we've had shutdown," confirmed Carlton.

"We copy you're down, Eagle," Duke called.

"Houston, Tranquility Base here," called Armstrong. "The Eagle has landed."

Duke had been alerted in order that he would not be caught out by a strange call sign, but he fluffed his reply. "Roger, Twank – Tranquility. We copy you on the ground." A moment later he continued, "You've got a bunch of guys about to turn blue. We're breathing again. Thanks a lot." With that, he slumped back in his chair and grinned at Slayton, who grinned back.

In the viewing gallery people stood to applaud, cheer, and wave small flags.

The powered descent had started at 102:33:07, and Armstrong called shutdown at 102:45:41 after a duration of 12 minutes 34 seconds – about half a minute over nominal. As he updated his console log, Kranz thought, 'My God, they've landed!'

At 'contact light' Pat Collins, head resting on her hands, broke into a smile for the first time in more than an hour. With the exception of Joan Aldrin, everyone in her house applauded at 'engine stop'; she had her head buried against the wall and was still shaking. Although Robert Moon went over to comfort her, she escaped to the solitude of her bedroom. Michael Archer, Joan's father, took daughter Jan, who was visibly shaken, to join her mother. After gathering her senses, Joan handed out a box of cigars. As she would reflect a few hours later, "My mind couldn't take it all in. I blacked out. I couldn't see anything. All I could see was a match cover on the floor. I wanted to bend down and pick it up, and I couldn't do it. I just kept looking at that match cover." With the 'landed' report, Jan Armstrong delightedly hugged son Ricky. A moment later, her sister Carolyn entered the room, leant against the wall and exclaimed, "Thank you, God."

In New York, Walter Cronkite, who was anchoring the CBS special, *Man on the Moon: The Epic Journey of Apollo 11*, had also been holding his breath. He removed his spectacles to wipe sweat from his forehead and, finding himself speechless, could only say, "Phew! Wow!" The Neilson ratings organisation later estimated that more than half of American households had had their television sets switched on during the landing. However, since all three networks were providing continuous coverage it was hard to avoid the event! Armstrong's parents were watching on their donated colour television. A baseball game in Yankee Stadium in New York was paused to permit the landing to be announced, and the audience delivered a rendition of *The Star-Spangled Banner*. Canon Michael Hamilton of Washington Cathedral noted, "The older people are getting a bigger bang out of this than the younger ones, who have grown up with astronauts and space; older people remember when it was just a dream." Of all the space program managers, the lunar landing must surely have meant the most to Wernher von Braun. It would not have been possible, however, without the challenge laid down by John F. Kennedy, on whose grave at Arlington National Cemetery a bouquet of flowers was deposited several hours later with the anonymous note 'Mr President, the Eagle has landed'. In Moscow, senior military officers and a dozen cosmonauts had gathered to monitor the American television coverage, and the landing prompted a round of applause. Alexei Leonov, who had hoped to make the first lunar landing for his country, later explained this praise of the American success as 'white envy'. On its final news bulletin of the day, Soviet television reported that the landing had succeeded, and that the Czar of the flight would soon step out onto the surface.

After watching Jan Armstrong give a press interview, Joan Aldrin went out to do likewise. A NASA Public Affairs Officer held an umbrella against the rain that had started to fall. Frustrated by banal questions such as "What are your plans for the moonwalk?" she burst out, "Listen! Aren't you all excited? They did it! They did it!" And with that she turned and strode back into the house.

ON THE SURFACE

Because four of Eagle's six batteries were in its descent stage, the ascent stage had power just for the several hours required to rendezvous. As this could begin only when Columbia was conveniently positioned, there was a brief window once per revolution. Although Columbia had just passed below the horizon, a rendezvous would be feasible if Eagle were to lift off within the next 12 minutes. Thus, even as those around them were celebrating the act of landing, the flight controllers on duty were studying their telemetry for any evidence of a problem that would oblige an immediate liftoff. In planning, it had been decided to provide two decision points, referred to as T1 and T2: the first, barely 2 minutes after landing, was to be made on the basis of 'first impressions'; the second was to be made just before the window closed, after a more thorough study. During training, Bill Tindall had expressed concern. By tradition, decisions were expressed as Go/No-Go, but in this case he saw scope for confusion: "Once we get to the Moon, does 'Go' mean stay on the surface, and does 'No-Go' mean abort from the surface? I think the decision should be changed to 'Stay/No-Stay' or something like that." His advice had been accepted.

"Let's go on," Armstrong said to Aldrin after they had completed the post-shutdown checklist, meaning that they should prepare for an immediate liftoff. Then he called Houston, "Okay, we're going to be busy for a minute."

"All flight controllers, about 45 seconds to T1 Stay/No-Stay," Kranz observed. Although the flight controllers had maintained discipline on their intercom loops, some of the others present were celebrating. "Keep the chatter down in this room," he ordered. "Okay, T1 Stay/No-Stay? Retro?"

"Stay!" replied Deiterich.

"FIDO?"

"Stay!" replied Greene.

"Guidance?"

"Stay!" replied Bales.

"Control?"

"Stay!" replied Carlton.

"TELCOM?"

"Stay!" replied Puddy.

"GNC?"

"Stay!" replied Willoughby

"EECOM?"

"Stay!" replied Aaron.

"Surgeon?"

"Stay!" replied Zieglschmid.

"CapCom, we're Stay for T1," Kranz directed.

"Eagle, you are Stay for T1," Duke relayed.

"Roger. Understand, Stay for T1," Armstrong acknowledged.

"Houston," Collins called. "How do you read Columbia on high-gain?" Now that he was no longer optically tracking Eagle, he had manoeuvred to enable his high-gain antenna to point at Earth.

"We read you five-by, Columbia. He's landed at Tranquility Base. Eagle is at Tranquility."

"Yes. I heard the whole thing," Collins noted. His wife was delighted that he had not been left out.

"Eagle, Houston. You are Stay for T2."

"Stay for T2," replied Armstrong. "We thank you." This decision committed Eagle to remaining on the Moon for 2 hours until Columbia came around again. Armstrong and Aldrin began to prepare for a normal liftoff, updating the guidance system to tell it that the vehicle was on the surface, in a specific orientation. This involved aligning the platform by a procedure that inferred local vertical from the gravity vector and azimuth from a star sighting. While Aldrin prepared to do this, Armstrong offered an explanation of the final phase of the descent. "Houston, that may have seemed like a very long final phase. The Auto targeting was taking us right into a football-field-sized crater with a large number of big boulders and rocks for about one or two crater diameters around it, and it required us going in P66 and flying manually over the rock field to find a reasonably good area."

"It was beautiful from here, Tranquility," Duke replied.

Aldrin gave his initial impression of the surface, "We will get to the details of what is around here later, but it looks like a collection of just about every variety of shape, angularity, granularity, about every variety of rock you could find. The colour varies pretty much depending on how you're looking relative to the zero phase point.[11] There doesn't appear to be too much of a general colour at all. However, it looks as though some of the rocks and boulders, of which there are quite a few in the near area, are going to have some interesting colours to them."

"Be advised there's lots of smiling faces in this room and all over the world," Duke congratulated.

"Well, there are two of them up here," Armstrong pointed out.

Although Columbia was below Eagle's horizon, Houston had set up a two-way relay. But a transmission from Eagle had to go both ways to reach Collins and this

[11] The down-Sun line was called the 'zero phase'. With the Sun low in the east, the shadows of rocks and craters were hidden when looking west, and coherent backscatter from cleavage planes in the fractured crystalline rocks produced a very strong solar reflection that tended to 'wash out' the scene.

imposed an extra time delay, with the result that anything that he chose to say in response to a transmission from Eagle was made almost 3 seconds late and what he said required 1.3 seconds to reach Earth, during which interval the CapCom might respond to Eagle. On hearing Armstrong's remark, Collins immediately chipped in, "And don't forget one in the command module", meaning that he was smiling too. But before this reached Earth Duke congratulated Eagle, "That was a beautiful job, you guys", with the result that Collins's remark created the impression that he was reminding Earth that he had done a beautiful job too!

"Columbia, say something," Duke prompted. "They ought to be able to hear you."

"Tranquility Base, it sure sounded great from up here," Collins called. "You guys did a fantastic job."

"Thank you," acknowledged Armstrong. "Just keep that orbiting base ready for us up there now."

"Will do," Collins promised.

Several minutes later, Armstrong called Houston, "The guys who said that we wouldn't be able to tell precisely where we are – they're the winners today. We were a little busy worrying about program alarms and things like that in the part of the descent where we would normally be picking out our landing spot. Aside from a good look at several of the craters we came over in the final descent, I have not been able to pick out anything on the horizon as a reference as yet."

"No sweat," replied Duke. "We'll figure out."

Bill Anders and Ken Danneberg, a friend of the Armstrong family, launched a $1 'pool' on where Eagle had set down.

The medics had expressed concern that when the astronauts ventured outside they might have to spend some time at the foot of the ladder adapting to the local gravity prior to moving off but, as had been pointed out in response, by then they would have had several hours to acclimatise by standing in the cabin. Armstrong made this point. "You might be interested to know that I don't think we notice any difficulty at all in adapting to one-sixth *g*. It seems immediately natural to us to move in this environment." He then reported what he could see through his window. "The area is a relatively level plain with a fairly large number of craters of the 5- to 50-foot variety, some ridges 20 to 30 feet high, I'd guess, and literally thousands of 1- and 2-foot craters. We see some angular blocks out several hundred feet in front of us that are probably 2 feet in size. There is a hill in view, just about on the ground track ahead of us. It's difficult to estimate, but it might be a half a mile, or a mile."

"It sounds like it looks a lot better than it did yesterday at that very low Sun angle," interjected Collins. "It looked rough as a cob then."

"It really was rough, Mike," Armstrong pointed out. "At the targeted landing area it was extremely rough with a crater and a large number of rocks that were probably larger than 5 or 10 feet in size."

"When in doubt, land 'long'," Collins observed.

"We did," Armstrong agreed. Picking up his observations, he continued, "I'd say the local surface is very comparable to that we observed from orbit at this Sun angle; about 10 degrees. It's pretty much without colour. It's a very white, chalky grey, as you look into the zero phase; and it's considerably darker grey, more like ashen grey

Immediate post-landing view from Aldrin's window down at Eagle's shadow

Immediate post-landing view out of Armstrong's window, looking southwest.

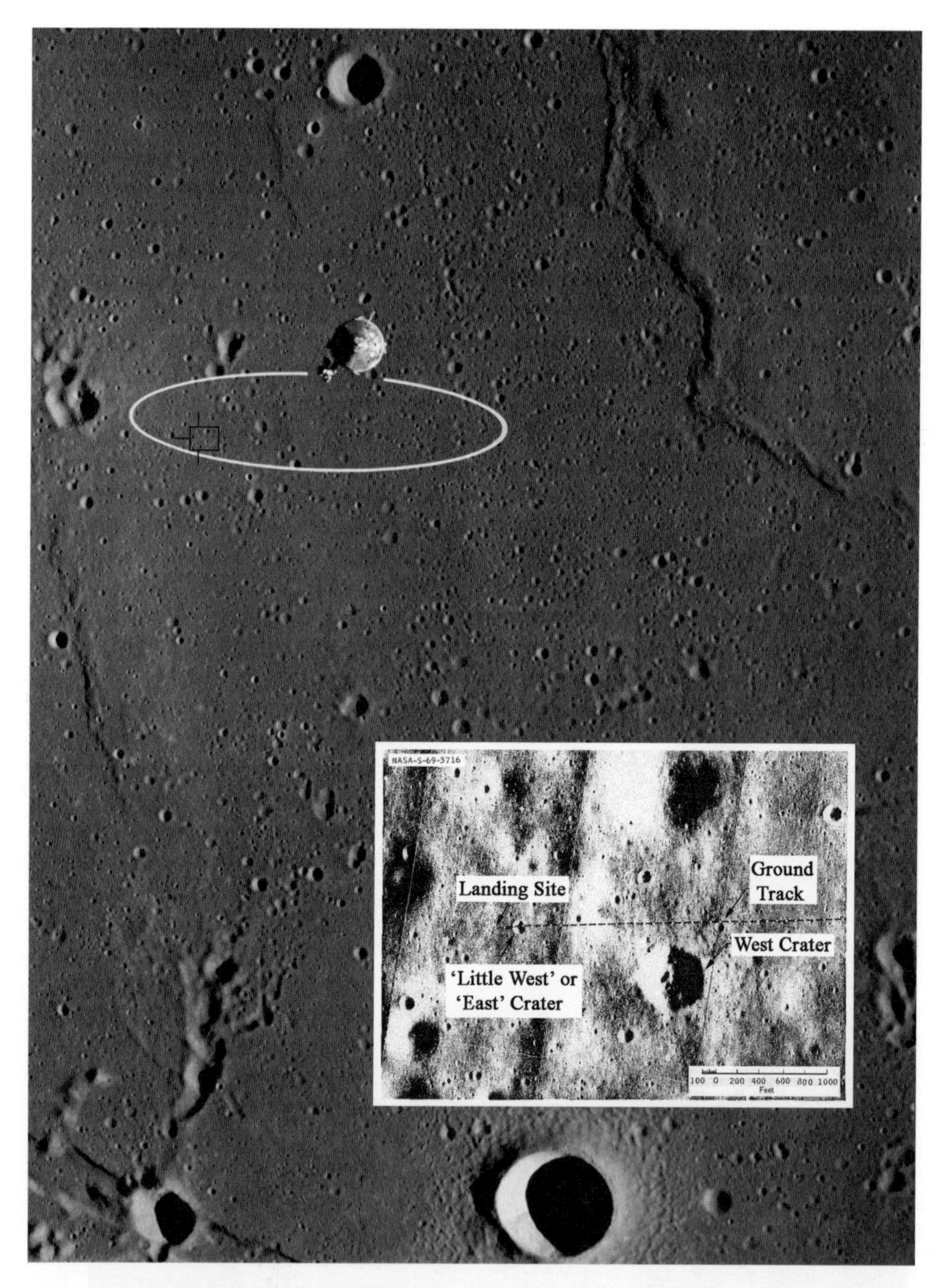

The target ellipse was 11 nautical miles long and 3 nautical miles wide, with its major axis along the intended direction of approach. Eagle came in south of track and landed at the point marked by the cross hairs. Inset: After passing over rocks north of West Crater, Eagle landed just beyond a smaller crater. (Based on S69-3716 and S69-3723)

as you look out 90 degrees to the Sun. Some of the rocks in close here that have been fractured or disturbed by the rocket engine plume are coated with this light grey on the outside; but where they've been broken they display a very dark grey interior – it looks like they could be country basalt."

While it was not necessary for mission success to determine in real-time where Eagle had landed, Collins was eager to find it by a P22 landmark tracking exercise in order to assist in the rendezvous. The sole reference was the large blocky crater that Armstrong had reported passing over. However, he had neglected to mention having later passed over a smaller crater. Nevertheless, the engineers were able to utilise tracking from the Manned Space Flight Network and telemetry from the onboard guidance systems to estimate the landing point by recreating the descent trajectory. As team member Lew Wade reflected, "All indications were that they were 'long', but the guidance systems didn't agree: PGNS put them a bit north of the planned site; AGS put them in the middle. The first Manned Space Flight Network report had them to the south." The various estimates were marked on a large-scale map of the ellipse, and all were within a 5-mile radius, with most clustered downrange of the big crater. Meanwhile, Gene Shoemaker's team of geologists, working in a science support room, evaluated the crew's descriptions of their surroundings (in particular that they were on "a relatively level plain with a fairly large number of craters") and reasoned that, as intended, Eagle had flown down the major axis of the ellipse; for some reason had come in 'long'; and Armstrong had 'extended' in order to avoid a blocky crater. Of two candidates for this crater, one was rejected, which left one. To cut a long story short, the issue would not be resolved until the 16-millimetre Maurer film of the powered descent was studied after the mission, whereupon it was realised that Shoemaker's team had identified the site to within 200 metres.

"We cannot see any stars out the window," Armstrong noted, "but out of my overhead window I'm looking at the Earth; it's big and bright and beautiful." At this point, Aldrin was preparing to take the star sighting for platform alignment. "Buzz is going to give a try at seeing some stars through the optics."

"Columbia, Houston," Duke called. "It's coming up on 2 minutes to LOS, and you're looking great going 'over the hill'."

While Collins passed around the far side of the Moon, Armstrong and Aldrin proceeded with their activities more or less independently of Houston. As soon as Collins reappeared around the eastern limb on revolution 15, Duke gave him one estimate of the landing site, which was "about 4 miles downrange". He provided selenographical coordinates, which Collins entered into his computer.

"Tranquility Base. Houston," Duke called. "You are Stay for T3."

"Roger." Armstrong acknowledged. This terminated the preparations for a lift off, and enabled Eagle to be powered down.

On concluding his momentous shift, Kranz shook Tindall's hand, found Koos to thank him for throwing the 12-01 program alarm into the final simulation, and then he accompanied Douglas Ward, his Public Affairs Officer, across the road to the News Center.

8

Tranquility Base

POST-LANDING ACTIVITIES

The flight plan called for the moonwalk to start about midnight Houston time. The idea had been to accommodate the possibility of the landing being delayed by one revolution, and still give Armstrong and Aldrin a period in which to wind down before starting the EVA preparations. However, as Armstrong reflected, "We had thought, even before launch, that: if everything went perfectly and we were able to touch down precisely on time; if we didn't have any systems problems to concern us; and if we found that we could adapt to the one-sixth gravity lunar environment readily; then it would make more sense to go ahead and complete the EVA while we were still wide awake – but, in all candour, we didn't think this was a very high probability." Nevertheless, some 2 hours after landing, Aldrin called Houston, "Our recommendation at this point is to plan the EVA, with your concurrence, to start about 8 o'clock Houston time; that's about 3 hours from now."

"Stand by," responded Duke.

"Well," said Aldrin, "we'll give you some time to think about that."

But half a minute later Duke was back, "We've thought about it; we'll support it. You're Go at that time." A moment later, he sought clarification, "Was your 8 o'clock Houston time a reference to opening the hatch, or starting the preparation for the EVA at that time?"

"That would be hatch opening," Armstrong replied.

"That's what we thought."

But Armstrong then decided to play safe, "It might be a little later than that; in other words, we'll start the preparations in about an hour or so."

Meanwhile, during his first pass overhead Collins had tried to see Eagle on the surface. As Columbia approached the general area of the landing site, he told his computer to slew the sextant onto the coordinates that Duke had given him. With the sextant maintaining the line of sight, he looked for either the momentary solar glint reflecting off the foil covering the vehicle, or the distinctive narrow shadow that it would project; but he saw neither. At his altitude of 69 nautical miles the 1.8-degree-wide circular field of view of the sextant covered an area of only a few square miles, and orbiting at almost 3,700 nautical miles per hour he was above 45 degrees elevation for less than 2 minutes 30 seconds, which was barely enough to study the designated location – it was certainly insufficient time to slew the sextant around to conduct a search.

"How did Tranquility look to you down there?" Duke enquired afterwards.

"Well," replied Collins, "the area looks smooth, but I was unable to see them." As an afterthought, he added, "It looks like a nice area, though."

Windler's Maroon Team took over in order to deal with the EVA preparations, with Owen Garriott as CapCom.

"Columbia," called Garriott, "Tranquility Base are prepared to begin the EVA early; they expect to begin depress operations in about 3 hours, at approximately 108 hours."

"Tell them to eat some lunch before they go," recommended Collins.

About 20 minutes later, Armstrong and Aldrin reached the point in their flight plan at which they could doff their helmets and gloves. Their next item was to be lunch which, since there was no provision for hot water in Eagle, would be a cold snack. First, however, Aldrin had an item of his own. A month previously, he had asked Dean Woodruff, the minister at the Webster Presbyterian Church, "to come up with some symbol that meant a little bit more than what most people might be thinking of [immediately after the landing]", and they had decided that Aldrin should celebrate the Sacrament of Holy Communion. When Aldrin told Deke Slayton of this, Slayton reminded Aldrin that Apollo 8's reading from *Genesis* had resulted in a law suit, and asked that he refrain from making overtly religious remarks.[1]

"This is the LM pilot," Aldrin called. "I would like to take this opportunity to ask every person listening in, whoever and wherever they may be, to pause for a moment and contemplate the events of the past few hours and to give thanks in his or her own way." Listening to her squawk box, his wife recognised the significance of his invitation for people to celebrate. He then switched off his microphone and drew from his personal preference kit a small silver chalice, a vial no larger than his little finger with a symbolic amount of wine, and a wafer representing bread loaf. Using the fold-down shelf in front of the DSKY as his altar he poured the wine, in the process observing that the fluid swirled around in the chalice for an inordinate time in the weak lunar gravity. He then read a card on which he had written the verse from the *Book of John* (15:5) traditionally used for this event: *I am the vine, you are the branches; he who remains in me, and I in him, will bear much fruit; for apart from me you can do nothing*. Communion was appropriate because it was a Sunday. Armstrong, who had been alerted in advance, watched in respectful silence. Unfortunately, NASA's hope that the ceremony would remain private was foiled by the Reverend Woodruff himself, who told a reporter that he had supplied Aldrin with a Communion kit.

[1] At the time of Apollo 11, the law suit brought by Madalyn Murray O'Hair regarding the reading from *Genesis* by the Apollo 8 crew was still pending.

When the time approached at which it had been estimated that the astronauts might initiate their EVA preparations, with there having been no communications since Aldrin's call, Garriott prompted, "We'd like some estimate of how far along you are with your eating, and when you may be ready to start your preparations."

"I think we'll be ready to start in about half an hour or so," Armstrong said. With that, the circuit reverted to silence.

In fact, preparing for the excursion proved to be rather more time consuming than expected. In training, the cabin had been 'clean', with just the apparatus for the exercise present, but in reality the cabin held everything they would require for the mission, and 90 minutes had elapsed before they were ready to begin the EVA preparations, and then activities that were assigned 2 hours actually took 3 hours. It was therefore fortunate that they had opted to forgo the early rest period.

Meanwhile, Charlesworth's Green Team took over, with Bruce McCandless as CapCom. He directed his attention to Columbia. Collins had inspected the second estimated position of the landing site, again in vain. While he would continue with these efforts, it dawned on him that although it was evident that Eagle was in the western half of the ellipse, the scatter in the estimated positions meant that Mission Control did not really know where it was.

EVA PREPARATIONS

The checklist for donning and checking the accoutrements required to venture out onto the surface was lengthy and excruciatingly detailed. First, each man unstowed his PLSS from its mount on the side wall and stood it on the floor by the forward hatch, then the bags containing the extravehicular gloves and helmet augmentation were retrieved and deposited on the floor next to the backpacks. Next, Armstrong retrieved the OPS packages from the stowage compartment on the side wall. In the event of the primary oxygen supply failing during the moonwalk, the OPS – which would be mounted on top of the PLSS backpack – contained sufficient oxygen to facilitate an emergency ingress and switch back to the cabin's supply. Finally, he retrieved the remote control units containing status displays for the PLSS and radio system that were to be worn on the chest, and the light-blue deeply treaded rubber overshoes for firm traction outside in the weak lunar gravity. After Aldrin assisted Armstrong to put on his overshoes, Armstrong did so for Aldrin. Aldrin transferred his PLSS to the circular cover of the main engine of the ascent stage, added the OPS, then turned around and reversed up against the backpack. Armstrong helped him to fasten the waist and shoulder harnesses, which clipped to rings on the suit, attach the chest unit to the shoulder harnesses, and run the oxygen umbilicals from the PLSS to the sockets on the front of the suit. Aldrin then helped Armstrong to do likewise. The next step was to conduct a communications check with the portable radios. Aldrin disconnected his cabin communications umbilical and plugged in the one from his PLSS, and Armstrong did the same. As the backpacks projected 10 inches to the rear, they greatly reduced the scope for movement in the cramped cabin. On Earth the backpack weighed 120 pounds but on the Moon it was only 20 pounds. However, it retained its inertia and if it were to even nudge against the lightweight structures it would deliver a significant impact.

"Houston," Collins called, "could you enable the S-Band relay at least one way from Eagle to Columbia, so that I can hear what's going on?"

"There's not much going on at the present time," McCandless noted, "but I'll see what I can do about the relay."

As part of the PLSS communications checks, the crew switched to VOX, and although this enabled people to listen in, the terse to and fro was not particularly illuminating for anyone without a checklist. McCandless participated only when addressed.

As 108 hours approached, Public Affairs Officer Jack Riley announced, "We do not at this time have a good estimate for the start of the EVA." The astronauts were 30 minutes behind on the checklist, and slipping further. Jan Armstrong, on the floor in front of the television, laughed. "It's taking them so long because Neil is trying to decide about the first words he's going to say when he steps out on the Moon." In the Collins home there was impatience rather than tension. Pat Collins sat with a plate of lasagne in the company of Clare Schweickart, Barbara Gordon and Mary Engle. Rusty Schweickart was attempting to follow the preparations by listening to the squawk box, but the astronauts' commentary was not particularly informative. Noting that her husband had requested a one-way relay to enable him to listen in to transmissions between Eagle and Houston while on the near side of the Moon, Pat suspected that he would be on the far side when the moonwalk started, and therefore miss it. Joan Aldrin was relaxing, listening to Duke Ellington records. On breaking off, she said to Audrey Moon, "I've thought sometimes, privately, and may even have said so to Buzz, that he was so caught up in the mechanics of all this that he really hadn't realised the significance of what he was doing; but now I really think he did!" She settled down to a plate of buffet nibbles and awaited developments.

Advancing through the preparations, Aldrin inserted the circuit breaker for the television system. "Houston, are you getting a signal on the television?" he asked. The MESA compartment was still in its stowed position, denying illumination to the camera inside, but the system was transmitting.

"The data we're receiving looks good," said McCandless, "and we're getting synchronisation pulses and a black picture."

With the Sun low in the east, the 23-foot-tall vehicle was casting a very long shadow, and the ladder on the forward leg was facing westward. "You'll find the area around the ladder is in a complete dark shadow," Armstrong pointed out, "so we're going to have a problem with the television, but I'm sure you'll be able to see the lighted horizon."

"We request that you open the television circuit breaker," McCandless called. "It's been on about 15 minutes now, with the MESA closed." Houston wanted it switched off to preclude the camera overheating in the insulated compartment. Aldrin was to have pulled the circuit breaker immediately following the test, but had missed a step on the checklist.

"Do we have a Go for cabin depress?" Armstrong asked.

When there was no response, Aldrin quipped about VOX letting Houston listen

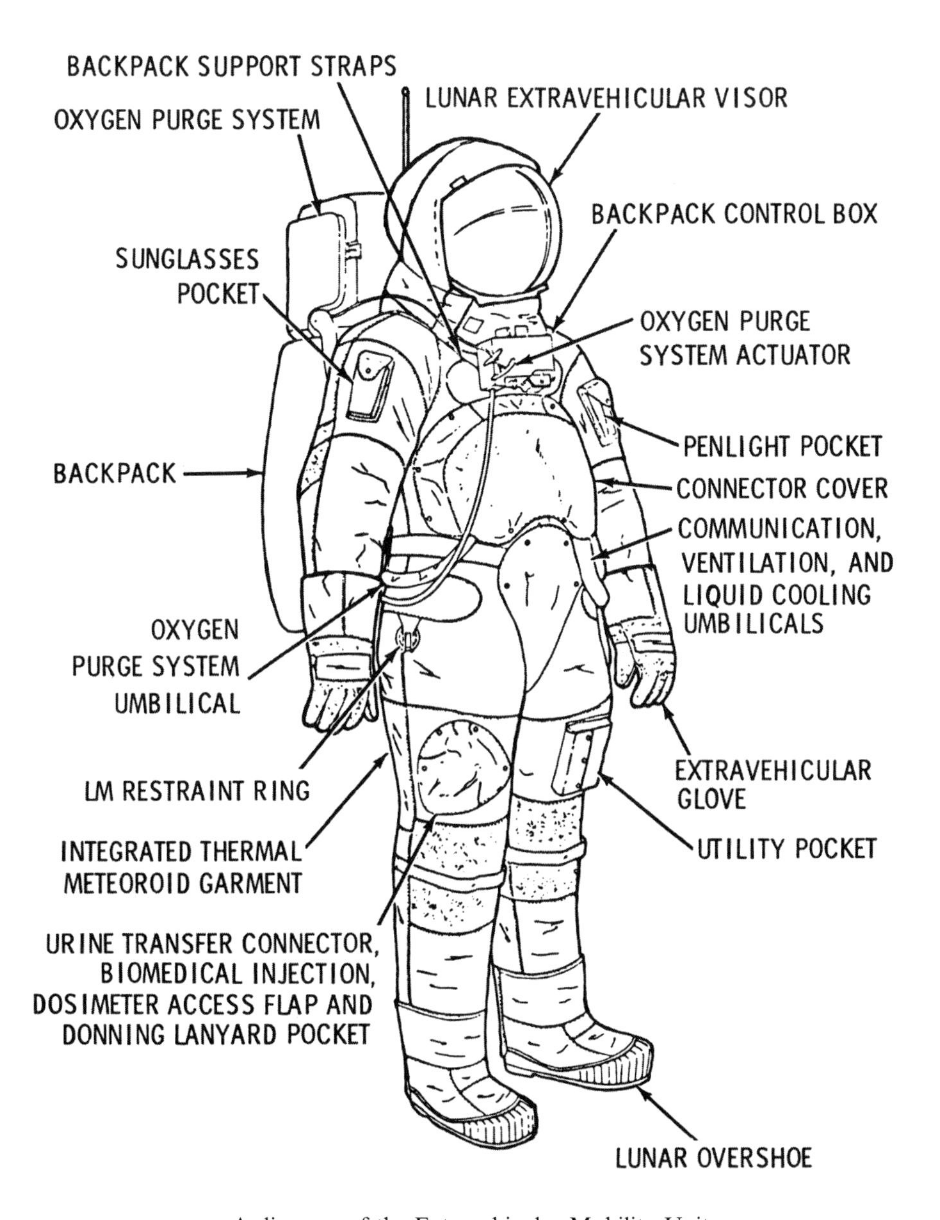

A diagram of the Extravehicular Mobility Unit.

in, "They hear everything but that!" He made the call himself just in case there had been a fault with Armstrong's radio. "Houston, this is Tranquility. We're standing by for a Go for cabin depress." Actually, the delay was because Charlesworth was polling his flight controllers. McCandless ought to have replied to Armstrong with a 'Standby'.

"You're Go for cabin depressurisation," McCandless replied finally.

In fact, they were currently some 22 minutes behind their 108:00 estimate, and had just received permission to proceed with the final phase of the preparations; it would be another half hour before they would be ready to start the depressurisation of the cabin. With the moonwalk imminent, the Apollo 12 LM prime and backup crews of Pete Conrad, Al Bean, Dave Scott and Jim Irwin joined McCandless. Gene Kranz returned and sat with Charlesworth to watch the telecast on one of the wall screens.

Having configured the cabin for exposure to vacuum, Armstrong and Aldrin were able to finish suiting. Once the cooling system had been checked, the water umbilical was run from the PLSS to the suit to circulate water through the liquid coolant garment. The umbilical to the cabin's oxygen system was removed, and the OPS umbilical was plugged into the vacant sockets. After he had applied an anti-fogging agent to the inside of Aldrin's 'bubble' helmet, Armstrong slipped it over Aldrin's head and locked it into position. He followed up with the polycarbonate shell incorporating the visor assembly that provided an inner and outer visor. The outer visor had a gold coating to reflect the harsh glare from the Sun which, in the absence of an atmosphere, was not only bright but also full-spectrum. Aldrin then repeated the procedure for Armstrong. Finally, they donned their extravehicular gloves, which had coverings of woven steel-fibre and rubberised finger tips for a modicum of dexterity. Taken together with the PLSS, OPS, helmet augmentation, extravehicular gloves and overshoes, the suit was described in NASA-ese as the extravehicular mobility unit. Aldrin wore a watch over the gauntlet-like sleeve of the right glove, but Armstrong decided to leave his watch inside as an emergency backup to the spacecraft's event timer. A checklist for the external activities had been stitched onto the gauntlet of each left glove, listing that man's specific tasks. Owing to uncertainty regarding the metabolic rate of an astronaut working on the lunar surface, and the fact that no one knew for certain how long the coolant water would last, the duration of the moonwalk had been set conservatively. However, the 'clock' on the life-support system started when the PLSSes were activated to pressurise the suits prior to depressurising the cabin.

"Now comes the gymnastics," Armstrong observed.

Aldrin carefully reached down and opened the valve built into the waist-high forward hatch, and then they monitored the pressure meter, which was initially at 4.8 psi. The depressurisation was slowed by the bacterial filter incorporated into the valve to protect the lunar environment from terrestrial biota, lest this be sampled, returned to Earth and misinterpreted as evidence of lunar life. As the pressure decreased, the rate of decrease slowed as it took time for the remaining gas molecules to find their way to the vent valve. "It sure takes a long time to get all the way down, doesn't it?" mused Aldrin. When the pressure fell below 0.2 psi he tried the hatch, but it would not budge.

"Neil, this is Houston. What's your status on hatch opening?" McCandless called 10 minutes into the procedure.

"Everything is Go here. We're just waiting for the cabin to blow enough pressure to open the hatch. It's about one-tenth on our gauge now," Armstrong replied. Fifteen seconds later, he announced, "We're going to try it." But even at 0.1 psi, the 32-inch-square hatch would not open. Aldrin suggested that they also open the overhead valve, but they decided to wait. The cabin had vented sooner in altitude chamber tests, but as a chamber never establishes a perfect vacuum this had not given an accurate time measurement. As the hatch was only a thin metal cover of little rigidity, Aldrin carefully peeled back one of its corners to break its seal. Ice crystals formed as the residual air rushed out. Aldrin then readily hinged the hatch in towards his feet. "The hatch is coming open," Armstrong announced.

"The valve's in Auto," Aldrin confirmed as he set the hatch valve to enable it to be operated from the outside.

The final task was to install the lunar equipment conveyor (LEC). This was a long nylon tether with a hook at each end to enable it to be linked into a loop and run around a pulley on a fixture located in the cabin. It had additional hooks on it to attach equipment for transfer to or from the surface. For Armstrong's egress, the tether was hooked to the tie-down strap of his neck-ring as a safety precaution. With the hatch open wide, Armstrong faced aft and, with one hand leaning on the cover of the ascent engine, carefully lowered himself until he was kneeling on the floor with his feet in the hatch. Then Aldrin provided cues to help him to reverse out. With the bulky PLSS on his back it was a tight fit, but the hatch was already as large as the vehicle's design could accommodate. Rusty Schweickart called to those in the Collins home that the hatch was finally open. Listening to Aldrin assisting Armstrong out, Schweickart remarked, "Don't bump into anything! Just find the ladder, Neil!" When CBS anchorman Walter Cronkite wondered why Armstrong was taking so long, Schweickart muttered that it was "because he doesn't have eyes in his rear end". Armstrong reversed out across the porch more or less flat on his stomach until his boots were at the far end, then he grasped the side rails and pushed himself up onto his knees. Aldrin pushed a bag of packaging through the hatch, and Armstrong dumped it over the side. By this point, they had been living off their PLSS resources for 25 minutes.

"Houston, I'm on the porch," Armstrong announced.

Fifteen seconds later, Aldrin started the Maurer 16-millimetre camera that he had earlier installed on the bracket in the top-right corner of his window. Since he had no viewfinder, he adjusted it as best he could to view Eagle's shadow, with a little of the illuminated lunar surface to each side. Armstrong entered the frame 30 seconds into the sequence.

"Stay where you are a minute, Neil," called Aldrin. In preparation for his own egress, Aldrin partially closed the hatch to enable him to cross to the left-hand side of the cabin, and he did not want Armstrong to snag the tether on the hatch.

"Can you pull the door open a little more?" Armstrong prompted, when told he could continue.

"All right," said Aldrin, and he opened the hatch fully, into the space where he had previously been standing.

As Armstrong prepared to descend the ladder, Columbia passed 'over the hill' and, deprived of his relay, Collins's eagerness to hear what Armstrong would say as he stepped onto the lunar surface was frustrated.

In case Armstrong had forgotten to deploy the MESA, Aldrin asked, "Did you get the MESA out?"

"I'm going to pull it now," Armstrong replied. He located the D-ring alongside the porch using his left hand, and tugged it. The pallet on the front-right quadrant of the descent stage hinged down until just short of horizontal. "Houston, the MESA came down all right."

"Roger," replied McCandless. "And we're standing by for your TV."

"The television circuit breaker is in," reported Aldrin. The transmission was by the high-gain antenna.

MOONWALK

On the flight plan, the moonwalk was scheduled to be made when the 210-foot-diameter dishes at Goldstone and the Parkes Observatory in New South Wales, Australia, would both have a clear line of sight in order to provide 100 per cent redundancy in these large antennas. However, advancing the schedule meant that the Parkes antenna, which could not dip all the way down to the horizon, could not receive until the Moon was well up; and in any case, wind gusts threatened to cut operations short. And since a problem at Goldstone was degrading the slow-scan television signal, the smaller 85-foot antenna at the Honeysuckle Creek Tracking Station 25 miles from Canberra, Australia, became the prime receiver for coverage of the egress. Built in 1967, Honeysuckle Creek had tracked previous missions but as Bernie Scrivener, the administrative officer, noted, "Somehow this seemed to be much more important; this was the day for which everyone on the station had worked and trained." Australia's Prime Minister, John Gorton, joined a large group of technicians to witness the event. At 11.15 am local time, the Moon rose and the signal was strong. The Westinghouse black-and-white camera provided 320 lines of resolution at a scan rate of 10 frames per second. As television technician Ed von Renouard reflected, "When I was sitting there in front of the scan converter waiting for a pattern on the input monitor, I was hardly aware of the rest of the world. I heard Buzz Aldrin say 'Television circuit breaker in'." The signal came in. "When the image first appeared, it was an indecipherable puzzle of stark blocks of black at the bottom and grey at the top, and was bisected by a bright diagonal streak. I realised that the sky should be at the top – and on the Moon the sky is black." Several weeks earlier, NASA had noticed that when the MESA was opened the camera affixed to it would be oriented upside down, and a switch had been installed at each ground station to flip the picture, and when von Renouard threw the switch "all of a sudden it made sense".

As one of the 10-foot-by-10-foot Eidophor screens in Mission Control flickered to life, it prompted cheers in the viewing gallery. "We're getting a picture on the TV," McCandless announced.

"You got a good picture, huh?" Aldrin asked.

"There's a great deal of contrast in it," McCandless replied, "and currently it's upside down on our monitor, but we can make out a fair amount of detail." In fact, the initial feed to Houston was by a land line from Goldstone, where the technician waited 30 seconds before flipping the image upright. Unfortunately, the problem with the conversion system made the image extremely contrasty.

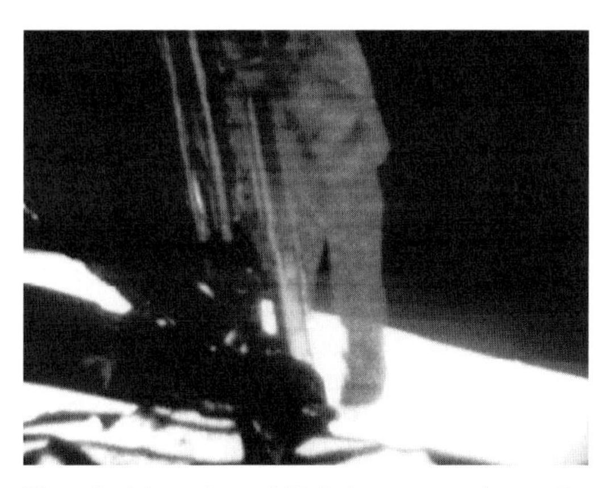

The television view of Neil Armstrong descending the ladder.

"Will you verify the aperture I ought to have on the camera?" Aldrin asked. He was referring to the Maurer.

"Stand by," McCandless replied.

Since the MESA rotation had stopped short of 90 degrees, the television view depicted the horizon tilted slightly down to the right. Watched by a mesmerised audience, Armstrong carefully descended the nine-rung ladder.

"Neil, we can see you coming down the ladder now," reported McCandless.

The specifications for the extravehicular suit had required it to protect against temperatures in the range –250°F to +310°F, the former representing deep shadow and the latter full sunlight plus heat radiated by the surface. Although the ladder was in shadow, Armstrong was able to hold onto it without ill effects. Indeed, as he would later recall, "At no time could I detect any temperature penetrating the insulated gloves as I touched things." Lacking an atmosphere, the lunar surface is exposed to micrometeoroid bombardment. The tough outer covering was judged sufficient to protect against microscopic motes impacting at cosmic speeds.

On touchdown, the narrow lower section of the strut was to act as a piston and slide into the wider main strut and, by virtue of crushing a honeycombed filler, absorb the shock of the vehicle falling the final few feet, but Eagle had landed so gently that there was no significant compression and the interval from the lowest rung on the ladder to the foot pad was at almost its maximum 3.5 feet – a fact that Armstrong discovered when he jumped backwards off the final rung and, sliding his hands down the rails to maintain his stability, landed with both feet within the 3-foot-diameter pad. He jumped back up onto the ladder to verify that ascent was feasible, then down again. Meanwhile, Aldrin opened the f-stop of the Maurer on his own initiative. Although this washed out the ground beyond the shadow, it much improved the view of Armstrong. Twenty seconds later, McCandless issued the recommended settings, "Buzz, f/2 and 1/160th second for shadow photography on the sequence camera."

At this point the Network controller in the Mission Operations Control Room noticed that the video from Honeysuckle Creek, which was being transmitted by microwave to Sydney and then relayed by an Intelsat geostationary satellite over the Pacific Ocean, was clearer, and so he switched to this. Although NASA fed the Honeysuckle picture to the commercial television networks, it used Goldstone's audio.

The first step

For Armstrong, being a test pilot, the significant achievement of the mission had been the act of landing on the Moon. But for the public, that was only the prelude to a man making an imprint of his boot on the lunar surface.

"I'm at the foot of the ladder," Armstrong announced. Standing on the foot pad, holding the ladder with his right hand and bent forward slightly in order to balance his PLSS, he inspected the surface to the left of the pad, as he had done in training. Although he was deep in Eagle's shadow, sunlight backscattered from the zero-phase angle illuminated the shadowed side of the vehicle, enabling him to see reasonably well. "The LM foot pads are only depressed in the surface about 1 or 2 inches, although the surface appears to be very, very fine grained; as you get close to it, it's almost like a powder." At home, his wife urged him to "be descriptive". Joan Aldrin clapped her hands in delight, "I can't believe this". In the Collins home, someone remarked in amazement, "This is science fiction!" Although the picture from the slow-scan camera was 'ghostly', it was remarkable (and, oddly, not envisaged by science fiction) that people on Earth were able to watch their representative take his first step onto the lunar surface 'live' on television. The Mission Operations Control Room was totally silent. In the News Center, journalists were watching on a theatre-sized Eidophor screen. Ironically, the town of Carnarvon, which hosted one of the tracking stations of the Manned Space Flight Network, had no television facilities. However, the Australian Broadcasting Corporation had arranged for the Overseas Telecommunications Commission to relay the moonwalk via satellite. In the local theatre the townspeople had installed a 14.5-inch set which people at the rear of the hall watched through binoculars!

Having described the strut and the adjacent surface, Armstrong announced, "I'm going to step off the LM now." On the translunar coast, Aldrin had asked him if he had decided what he would say on stepping onto the surface, and he had said he was still thinking it over. Having rejected quotations from Shakespeare and the Bible, and things that he deemed to be pretentious, it dawned on him as he stood at the foot of the ladder that there was, in fact, only one thing to say! Holding onto the ladder with his right hand, he placed his left boot firmly on the surface alongside the pad. "That's one small step for a man, one giant leap for Mankind."[2,3] The historic first words having been issued, the Mission Operations Control Room broke into hearty applause. Dave Scott, who flew on Gemini 8 with Armstrong, would later reflect that it was typical of the man to have deliberated for so long over what to say, and then expressed so much in so few words.

Armstrong released his grip on the handrail of the ladder and stepped fully off the foot pad. Walter Cronkite proudly told his CBS audience that a 38-year-old

[2] Due to Armstrong's manner of speech, he appears to have appended the 'a' to 'for', which came out as 'for-a', thereby giving the impression that he misspoke and uttered something meaningless!

[3] The time in Houston was 9.56 pm on Sunday, 20 July 1969.

American was now standing on the surface of the Moon. When Armstrong scraped his foot across the surface, he noticed that the dark powdery material coated his overshoe. "The surface is fine and powdery. I can kick it up loosely with my toe. It adheres in fine layers like powdered charcoal to the sole and sides of my boots." Although his boots only slightly impressed the surface, the material preserved the imprint of his boots very well. "I only go in a small fraction of an inch – maybe one-eighth of an inch – but I can see the prints of my boots and the treads in the fine, sandy particles."

Wearing the 180-pound extravehicular mobility unit, Armstrong's mass was 340 pounds, but in the weak lunar gravity he weighed only one-sixth of this and was light on his feet. He did not feel the weight of the suit, since its internal pressure made it self-supporting. Holding the ladder with both hands, he did several knee-bends and then took a few steps away from the foot pad, briefly leaving the view of the television camera. Some members of the medical community had expressed concern that the astronauts would have difficulty in rapidly adapting to lunar gravity, and had urged that time be reserved for acclimatisation, with an immediate recall if this were to prove difficult. However, this ignored the fact that they would have been exposed to lunar gravity inside the LM for several hours prior to egressing, during which time they would be acclimatising – if, indeed, a period of acclimatisation should prove necessary. Others had expressed concern that if an astronaut were to fall onto his back he might have difficulty regaining an upright stance. If Armstrong had slipped early on, this may have prompted his immediate recall. He was determined to allay such concern. "There seems to be no difficulty in moving around as we suspected," he continued. "It's even perhaps easier than the simulations at one-sixth gravity. It is actually no trouble to walk around."

Having moved back in order to see beneath the vehicle, Armstrong said, "The descent engine didn't leave a crater of any size. It has about 1 foot clearance off the ground. We're essentially on a very level place here. I can see some evidence of rays emanating from the descent engine, but a very insignificant amount." At this point, he unhooked the LEC from his suit, but kept hold of it. "Buzz, are we ready to bring down the camera?"

"I'm ready," replied Aldrin. "You'll have to pay out all the LEC. It looks like it's coming out nice and evenly."

After training had suggested that it would be difficult to carry the loaded rock boxes up the ladder at the end of the moonwalk, the LEC, dubbed the 'Brooklyn clothes line', had been devised. It had then been decided to use this to transfer out the Hasselblad camera. Armstrong was to use the LEC in hand-over-hand fashion to lower the equipment transfer bag. Now that he had stepped away from the LM, the illuminated terrain ruined his dark adaptation. "It's quite dark here in the shadow, and a little hard for me to verify that I have good footing," he pointed out. "I'll work my way over towards the sunlight here, without looking directly into the Sun." He did not want to enter full sunlight because he did not yet desire to lower his gold-coated visor. As he moved off to the southern edge of the shadow, Aldrin elevated the Maurer, and the western horizon appeared in the field of view. At this point, the movie camera exhausted the 8 minutes of film that had remained in the magazine with which Aldrin had set it running. At this point, too, having briefly switched the television feed back to Goldstone, then to Honeysuckle Creek and once again to Goldstone, Houston was informed that the 210-foot antenna at Parkes had finally acquired the signal and, since this had the best image quality to date, it was fed to commercial television networks for the remainder of the moonwalk.

The bodywork of the Hasselblad 500EL Data Camera was highly reflective for thermal control on the lunar surface, and it had an f/5.6 lens with a focal length of 60 millimetres. The shutter speed, aperture and focus were all manual; only the electric film-advance was automatic. The adjustments had been enlarged to enable them to be operated by gloved hands. The astronauts had memorised the exposure settings for different Sun angles. In order to facilitate precise measurements across a frame for 'data extraction', a glass plate bearing a grid of 25 crosses had been positioned immediately in front of the focal plane. It was originally intended that the camera would be operated hand-held, but during training Armstrong had suggested that a bracket be added to the chest pack in order to make the camera hands-free. Because the helmet would prevent use of the top-mounted view finder, the rotating mirror and viewing plate had been deleted, and the astronauts had learned to aim by trial and error. They had two cameras, but only one had the modifications for external use.

After removing the Hasselblad from the equipment transfer bag, Armstrong mounted it on his bracket. He draped the LEC beside the forward leg. Noticing that the Maurer had stopped, Aldrin attended to it. In addition to exchanging the film magazine, he moved the camera to a bracket on a bar running horizontally across his window, set the exposure for the illuminated terrain, and pointed it northwest in order to document the locus of most of the external activities. As Armstrong set out to snap pictures for a partial panorama of the site, McCandless prompted him to take the contingency sample. This had priority on the checklist for the early part of the excursion because, if a suit or PLSS problem were to oblige him to retreat before he could fill the rock boxes, or if it should prove impossible to transfer the boxes to the cabin, they might have to return to Earth with only this small sample. Armstrong said that he would collect the sample when he had taken his pictures. He shot nine frames, turning slightly each time to document the horizon running from due south, through west and around to due north. This done, he moved north about 12 feet, stepping out of Eagle's shadow into sunlight to enable him to see the ground as he took the sample, knowing that this position would be within the field of view of the Maurer. As he still had not lowered his gold-plated visor, he stood with his back to the Sun, drew a scoop from the pocket strapped onto his left thigh, and straightened the multi-segmented handle just as Aldrin restarted the Maurer. As Armstrong scraped the scoop across the surface, he discovered that although the surficial material was loose, it consolidated with depth and prevented the scoop from penetrating more than a few inches. Nevertheless, by dragging the tool across the ground several times he was able to fill the bag.

"It looks beautiful from here, Neil," Aldrin called, referring to the sampling.

Sections of the first panoramic sequence taken by Neil Armstrong, viewing south to west (top) and west to north.

Armstrong, presuming Aldrin's remark to be a comment on the moonscape, replied, "It has a stark beauty all its own. It's like much of the high desert of the United States. It's different, but it's very pretty out here." Then to Houston he said, "Be advised that a lot of rocks have what appear to be vesicles in their surfaces.[4] Also, I'm looking at one now that appears to have some sort of phenocrysts in it."[5] After detaching the transparent sample bag from the scoop he inserted the handle several inches into the ground. On inspecting the contents of the bag, he noted that although the surface generally appeared shades of tan, the sample was black. After sealing the bag, he kneaded it with his fingers, observing that although most of the material was very finely grained there were also fragments of rock. His next task was to put the bag into his thigh pocket, but the peripheral vision through the visor was so limited that he could not see the flap. "Is the pocket open, Buzz?"

"Yes, it is. It's not up against your suit though. Hit it back once more. More towards the inside. Okay. That's good."

"Is that in the pocket?"

"Yes. Push it down. It's not all the way in. Push it. There you go."

"The contingency sample is in the pocket," Armstrong informed Houston, to the relief of the scientists. In fact, they would have preferred the sample taken well away from Eagle because the exhaust plume had disturbed the fine material in the immediate vicinity and potentially contaminated that which remained. In addition, the oxidiser pressure had been relieved soon after landing by venting, and some of the nitrogen tetroxide might have coated the surface. But Armstrong had been told to remain close to the vehicle.

Second man out

After Armstrong had been on the lunar surface for about 15 minutes, Aldrin asked, "Are you ready for me to come out?"

"Just stand by a second. I'll move this over the handrail," replied Armstrong. He adjusted the position of the LEC on the strut to ensure that it would not hinder Aldrin's egress, and then he stood to the southwest in order to document Aldrin's egress. When Fred Haise alerted Joan to her husband's imminent appearance, she, as a former actress, observed, "It's like making an entrance on stage."

Before Aldrin left, he gave the Maurer camera a final inspection. On 27 February 1969 Maxime A. Faget, Director of Engineering and Development at the Manned Spacecraft Center, wrote to Owen E. Maynard, chief of the mission operations branch. Knowing that the television coverage of the moonwalk would not be of very high quality, Faget had had high hopes for the film record. But discovering what was

[4] Vesicles were a characteristic of igneous rock in which the melt contained bubbles of gas that left spherical holes in the solidified rock. Since this occurs more readily in lava that has been extruded onto the surface or is at shallow depth, it supported the inference that the landing site was a basalt lava flow. Armstrong would expand on this observation later in the excursion.

[5] Phenocrysts were crystals embedded in the finely grained matrix of an igneous rock.

intended, he was dismayed. "From the stand point of public information and historical documentation," he wrote, "I am terribly disappointed to find that although 560 feet of movie film has been set aside for lunar surface use, none will be exposed with the intent of providing a first-class visual appreciation of the astronauts' activity on the Moon during this singularly historical event. The impression of this occasion will be marred and distorted by the fact that the greatest frame rate [in automatic mode] is 12 frames per second. One can argue that 'suitable' (although jerky) motion rendition is produced by double-framing. Nevertheless, it is almost unbelievable that the culmination of a $20 billion program is to be recorded in such a stingy manner." The situation was actually worse than Faget had been led to believe, since with the camera running at its slowest rate of 1 frame per second a 140-foot reel of thin base film was sufficient for only 93 minutes, and because there would be no one available to replace the magazine it would not be possible to document the entire moonwalk.

"All set," called Armstrong. "You saw what difficulties I was having. I will try to watch your PLSS from here." As Aldrin's feet appeared in the hatch, Armstrong gave cues to assist him to reverse out along the porch.

Aldrin, "making sure not to lock it on my way out", partially closed the hatch in order to protect the cabin from the harsh thermal environment.

"A pretty good thought," chuckled Armstrong.

"That's our home for the next couple of hours, and we want to take good care of it," Aldrin added. "Okay. I'm on the top step." As he started down the ladder he provided commentary because one of his assignments was to evaluate the ability of a man to operate in the lunar environment, "It's a very simple matter to hop down from one step to the next." As her husband appeared on the television, Joan screamed with delight, rolled on her back and kicked her legs in the air, then sat up again and blew kisses.

"You've got three more steps and then a long one," Armstrong advised.

Aldrin continued his commentary as he prepared to jump down on to the foot pad, "I'm going to leave that one foot up there, and both hands down to about the fourth rung up."

"There you go," said Armstrong as Aldrin jumped.

Following Armstrong's lead, Aldrin tested jumping back up, but in this case his boot fell short of the lowest rung and dropped down again.

"About another inch," Armstrong noted.

Aldrin jumped up again, this time successfully. It was a matter of recalibrating his muscles for one-sixth gravity. "That's a good step," he noted wryly.

"About a 3-footer," said Armstrong.

"Beautiful view!" said Aldrin, as he looked left and right while standing on the foot pad.

"Isn't that something!" Armstrong agreed. "It's a magnificent sight out here."

Impressed by the contrast between the stark shadows and the barrenness of the illuminated surface, Aldrin said, "Magnificent desolation." Retaining hold of the ladder with both hands, he stepped backwards off the pad, then let go with his left hand and turned to face north. He was struck by the Moon's small size. To a man

Buzz Aldrin descends the ladder.

standing on the surface, the horizon was less than 2 miles away, making it very evident that he was on a sphere with the surface falling away in all directions. This had not been so evident when looking out of the window since, being higher, with the horizon further away, he had been able to see to a 'reasonable' distance. Standing freely, he noted that it was necessary to lean forward about 10 degrees to balance the mass of his backpack. However, this stance assisted in looking down, and he remarked that although the surface was finely grained and there were some rocks, there were some other objects that looked like clods of dirt. The loose material was different from terrestrial soil which, in addition to fragments of rock, contains the products of chemical weathering and organic humus – the lunar material is best described as a 'regolith', this being the term for material composed solely of rock fragments with a seriate distribution of sizes. Since a breccia is a consolidation of rock fragments bound in a matrix of finer material, geologists would subsequently introduce the term 'regolith breccias' for the compacted clods of regolith which, although they looked like rocks, readily fell apart when disturbed.

"This pad sure didn't penetrate far," Aldrin observed.

"No. It didn't," Armstrong agreed.

As had Armstrong, Aldrin stepped back and peered beneath Eagle, "There's absolutely no crater there at all from the engine." Although the plume had blown the dust out radially, it had not excavated the surface. However, there was a mark directly beneath the engine where the probe of the left landing gear had struck the surface.

Armstrong had also observed that whereas the probe on the left leg was bent beneath the vehicle, that on the right leg was bent outward. "I think that's a good representation of our sideward velocity at touchdown."

As per his checklist, Aldrin continued to familiarise himself, but since he had not set the voice-actuated keying control for his downlink at its maximum setting he was cutting out; at times, much of what he said was rendered unintelligible.

Meanwhile, Armstrong had tilted the MESA down past horizontal and pulled a lanyard to remove the thermal insulation blanket to expose the stowed apparatus. "Houston, I have the insulation off the MESA now, and the MESA seems to be in good shape." Turning his attention to the television camera, he announced, "I'm going to change lenses on you." As installed, the camera had a lens that provided an 80-degree field of view. He retrieved one with a 35-degree field of view from a slot of the MESA and put it on the camera, stowing the original lens. "Tell me if you're getting a new picture."

"That's affirmative," replied McCandless. "We're getting a new picture. You can tell it's a longer-focal-length lens. And for your information, all LM systems are Go."

"We appreciate that," Aldrin replied. "Thank you."

The commemorative plaque

Although each man had his individual checklist sewn onto the gauntlet of his left glove, several items were not listed. The unveiling of the commemorative plaque on the forward strut was such a late addition to the training that only Armstrong was

The design of the commemorative plaque on Eagle's forward leg, and a television view of its unveiling by Neil Armstrong (right of frame) and Buzz Aldrin.

familiar with it. With both men standing alongside the ladder, in view of the television camera, he described the plaque, "First, there's two hemispheres, one showing each of the two hemispheres of the Earth. Underneath it says 'Here men from the planet Earth first set foot upon the Moon, July 1969 AD. We came in peace for all mankind'. It has the crew members' signatures and the signature of the President of the United States." Measuring 9 by 7.625 inches with a thickness of 0.006 inch, it was made of #304 stainless steel with a brushed finish. The map and signatures were black epoxy in etched inscriptions. It was curved, conformal with the 4-inch radius of the strut but not actually in contact with it, being instead attached to the ladder by four spring clips, two on the third rung and two on the fourth rung. Armstrong hinged out and unlatched the sheet of stainless steel that had protected the plaque, in order to leave it exposed.

According to NASA Administrator Thomas O. Paine, the decision to make the plaque was a last-minute affair in which he and Wallis H. Shapley, an Associate Deputy Administrator, sketched the design, called in an artist to draw it properly, and sent the result to the White House for approval. However, Paine's account is contradicted by Jack A. Kinzler, an engineer at the Manned Spacecraft Center. This account stated that when Robert R. Gilruth phoned seeking ideas for how to celebrate the landing, Kinzler suggested a plaque to be left on the descent stage. Kinzler and colleague David L. McCraw produced a prototype that featured a US flag of red, white and blue paints baked into the etched figure in stainless steel, together with the signatures of the crew and the name of the landing site – on the assumption this would be named. Gilruth replaced the flag by two hemispheres with continental outlines devoid of national boundaries, to identify the planet of origin. Kinzler said that, "Once the plaque concept was approved, NASA headquarters took it over." When it was sent to the White House, Nixon changed the wording to past tense, and asked that his signature be added. Julian Scheer, head of the Public Affairs Office, has added a twist to the story, saying that NASA refused a suggestion by Nixon that 'under God' be inserted after the word 'peace'.[6] The plaque's design was made public shortly prior to launch. The astronauts had not been involved in the project, but felt it was tastefully done.

"Are you ready for the camera?" Armstrong asked.

Although Aldrin was scheduled to use the Hasselblad soon, he decided not to take it yet, "No. I'll get it later." He was closely following the checklist, and the next item was to relocate the television camera, "You take the television on out."

Having pulled a strap on the MESA to release the tripod on which he was to mount the television camera, Armstrong prompted, "Would you pull out some of my cable for me, Buzz?"

"How is the temperature on there?" Aldrin asked, as he drew the cable from a dispenser on the MESA. A temperature-sensitive patch on the television camera was designed to darken with increasing temperature; it was still white, indicating that the camera had not overheated while being tested prior to the deployment of the MESA.

"The temperature of the camera is showing 'cold'," Armstrong reported. He transferred the camera from the MESA to the tripod, and set off with it northwest, dragging the cable out as he went. On the way, something shiny on the bottom of a small crater attracted his attention. It was later concluded that this was a piece of glass, formed by the heat of a high-speed impact melting the regolith. In fact, there are two types of crater, 'primary' and 'secondary', with the primary produced by the impact of an object from space at cosmic velocity, and the secondary by the fall of ejecta issued by another impact. Since ejecta expelled faster than about 1.5 miles per second will escape the Moon, the speed of a secondary impact is necessarily at least an order of magnitude lower than that of a primary, and because kinetic energy is proportional to the square of the velocity, the energy of a primary strike for a specific mass greatly exceeds that of a secondary, and is sufficient to melt and fuse regolith.

Seeing Armstrong pause, Aldrin pointed out that there was more of the cable in the dispenser. "No, keep going. We've got a lot more, although it's getting a little harder to pull out."

"How far would you say I am, Buzz?"

"Oh, 40, 50 feet." Then Aldrin suggested that Armstrong give the audience a panoramic view. "Why don't you turn around and let them get a view from there, and see what the field of view looks like?"

"Okay," Armstrong agreed.

"You're backing into the cable," Aldrin warned, on seeing that Armstrong was at risk of entangling his feet in the cable. "Turn around to your right; that would be better."

"I don't want to go into the Sun if I can avoid it," Armstrong pointed out. Now that he was in full sunlight, he was taking care not to point the camera towards the Sun because flooding it with harsh sunlight would undoubtedly damage it.[7] He set

[6] These accounts are derived from interviews compiled by Glen E. Swanson in *Before this Decade is Out... Personal Reflections on the Apollo Program*, SP-4223, NASA, 1999.

[7] As indeed would happen at this point in the mission of Apollo 12.

the tripod down 55 feet northwest of Eagle. "I'll just leave it sitting like that, and walk around it." Once in position, he inspected the lines inscribed on the top of the camera body to indicate the angular field of view of the lens. "Houston. How's that field of view?"

"We'd like you to aim it a little bit more to the right," McCandless instructed. Armstrong adjusted the camera. "A little bit too much to the right! Can you bring it back left about 5 degrees?"

"Do you think I ought to be farther away, or closer?" Armstrong asked once it was lined up on Eagle, showing Aldrin, who, having advanced to the next item on his checklist, was configuring the MESA for sampling activities.

"You can't get much further away," Aldrin pointed out, having pulled out all the cable.

Television panorama

Armstrong now set about moving/aiming the camera to give the audience a series of views around the horizon. The image became a blur while the camera was in motion, and cleared when he set it down. "That's the first picture in the panorama," Armstrong announced. "It's taken just about north–northeast. Tell me if you've got a picture, Houston."

"We've got a beautiful picture, Neil," McCandless confirmed.

He moved the camera further around the horizon. "Okay. Here's another good one." The horizon was featureless, but there was a lot of detail in the foreground. "Now this one is right down-Sun, due west, and I want to know if you can see an angular rock in the foreground sticking up out of the soil."

"We see a large angular rock in the foreground," McCandless confirmed, "and it looks like there is a much smaller rock a couple of inches to the left of it."

"And about 10 feet beyond that is an even larger rock that's very rounded," said Armstrong. "The closest rock is sticking out of the soil about 1 foot; it's about 18 inches long and about 6 inches thick, but is standing on edge." Armstrong was spending a lot of time on this panorama because he believed on this first landing the geologists would welcome a view of the site.

"We've got this view, Neil," McCandless prompted.

Armstrong moved the camera again, "This is straight south."

"Roger," McCandless replied. "And we see the shadow of the LM."

"The little hill just beyond the shadow of the LM is a pair of elongate craters." There were two craters aligned in an east to west direction to the southwest of the vehicle. "Probably the pair together is about 40 feet long and 20 feet across, and they're probably 6 feet deep. We'll probably get some more work in there later."

Armstrong returned the camera to face Eagle, where Aldrin was still working at the MESA. He had attached to the edge of the MESA the teflon bag in which they were to place samples prior to stowing them in a rock box, and had raised a table on which a box was to be mounted for access; in their stowed positions the boxes were recessed into the MESA pallet. After some adjustments to the pointing, Armstrong left the television camera viewing Eagle and the area immediately to its front. There was an 8-foot-diameter deployable S-Band dish with its own tripod stowed in a compartment in the front-left quadrant of the descent stage, but since erecting this would take 20 minutes this was to be done only if Houston deemed the quality of the transmission using the smaller dish of Eagle's high-gain antenna to be unsatisfactory, which was not the case.

Having unstowed the Solar Wind Collector (SWC) from the MESA, Aldrin moved out to deploy it a short distance due north of Eagle. The experiment was a sheet of exceptionally clean aluminium on a staff that was to be positioned facing the Sun to soak up solar wind particles, particularly ions of helium, neon and argon (all of which were unreactive 'noble' elements in the Periodic Table). After Aldrin had extended the aluminium staff, he pulled out and locked the roller at the top, then drew down the 140-centimetre-tall and 30-centimetre-wide sheet from the roller and hooked it to a catch at the lower end of the staff.[8] He found it difficult to drive the staff into the ground because (as had been noted by Armstrong while collecting the contingency sample) the finely grained surface material became consolidated at a depth of 4 or 5 inches. The sheet was to be rolled up at the end of the moonwalk and returned to Earth. Because it was to be analysed by a laboratory in Switzerland, the experiment was also known as the 'Swiss flag'. On his way back from the television camera, Armstrong took several Hasselblad pictures of Aldrin with the experiment.

Aldrin observed that although the imprints left by their boot were generally only a fraction of an inch deep, their boots penetrated several inches where the loose material was piled up on the rims of small craters, and he wondered whether there was a correlation between the loose consistency and the change of slope. He had also noticed that when the toe of his boot penetrated the loose material at a shallow angle, it tended to displace a 'slab' of material as if it were solid, which, of course, it was not. A similar effect had been observed while pushing the surface material using the robotic arm of a Surveyor lander. Armstrong added an observation of his own, "I noticed in the soft spots where we leave foot prints nearly 1 inch deep, the soil is very cohesive, and will retain a slope of probably 70 degrees along the side of the foot prints." These were welcome 'soil mechanics' observations.

The flag ceremony

On 31 January 1969 Apollo Program Director Samuel C. Phillips asked Robert R. Gilruth of the Manned Spacecraft Center, Wernher von Braun of the Marshall Space Flight Center and Kurt H. Debus of the Kennedy Space Center to suggest symbolic activities that might be undertaken on the first lunar landing mission that would illustrate international agreements regarding the exploration of the Moon. The *Treaty on Principles Governing the Activities of States in the Exploration and Use of Outer Space* that was signed by the United States and the Soviet Union on 27 January 1967 (and, incidentally, witnessed by some of the astronauts, among them

[8] The SWC sheet was designed in metric units, so these have been used here to enhance fidelity.

Buzz Aldrin deploys the SWC.

Neil Armstrong) stated, in part, that the spacefaring powers agreed not to stake territorial claims on celestial bodies. When NASA proposed that the flag of the United Nations be raised, this was rejected by Congress, which directed that the US flag be flown. Phillips proposed that they also either raise the flag of the United Nations alongside the American flag, place decal flags of the member nations of the UN on the descent stage, or just deposit an appropriate information capsule on the surface.[9] However, Congress ordered that *only* the flag of the United States be raised. In order to preclude any manufacturer claiming to have made the flag used on the Moon, George M. Low ordered that a 3-foot-by-5-foot Stars and Stripes be purchased (at the average price of $3) from every official supplier, that their labels be removed, and that a secretary select a flag at random; the other flags would not go to waste, because if ever there was a mission to prompt the waving of a flag this would be it!

Having returned to Eagle, Armstrong and Aldrin retrieved the flag assembly from stowage in a thermal shroud by the left-hand ladder rail. They then set off northwest, in the general direction of the television camera, Aldrin carrying the lower part of the aluminium staff and Armstrong the upper part of the staff with the crossbar attached at its top by a locking hinge, incorporating the flag itself. Once they were in position, Armstrong rotated the crossbar into position and the two men grasped opposite ends of the telescoping rod in order to draw it out, but it became stuck just short of its full extension.

At this point Columbia appeared around the limb. "How's it going?" asked Collins. Joan Aldrin sympathised with him, "He doesn't know what's going on, poor Mike!"

"The EVA is progressing beautifully," McCandless replied. "I believe they are setting up the flag now."

"Great!" Collins said.

"I guess you're about the only person around that doesn't have TV coverage of the scene," McCandless consoled.

"That's all right," Collins insisted. "I don't mind a bit. How is the quality of the television?"

"Oh, it's beautiful, Mike. It really is," McCandless assured.

"Oh, gee, that's great!" said Collins. "Is the lighting half-way decent?"

"Yes, indeed," McCandless confirmed.

Having accepted that the crossbar would deploy no further, Armstrong set out to drive the lower section of the staff into the ground. As in the case of the staff of the SWC, the ground resisted penetration. Frustratingly, the surficial material gave little lateral support to hold the staff upright. On placing the flag assembly on top of the staff, Aldrin stepped back to salute and the flight control team stood, cheered and applauded.

[9] Engineers in Houston designed a 'stand' which, when deployed, would display the flags of the member states of the United Nations in the style of a tree.

The Mission Operations Control Room during the moonwalk.

Buzz Aldrin alongside the Stars and Stripes.

"They've got the flag up now," McCandless informed Collins, "and you can see the Stars and Stripes on the lunar surface!"

"Beautiful," replied Collins.

While Armstrong held the staff, Aldrin gripped the top and bottom of the flag and attempted to straighten it, in vain. They left it with a 'permanent wave' which, in retrospect, gave it a more natural appearance than if they had been able to draw it out totally flat. To finish off, Armstrong snapped two pictures of Aldrin standing by the flag.

Moving to his next checklist assignment, Aldrin set about evaluating modes of mobility. To enable the engineers to monitor his progress, he was to perform this exercise in front of the television camera. When asked, McCandless verified that he was in the field of view. He tested (1) a 'loping gait' in which he alternated his feet; (2) a 'skipping stride' that always led with the same foot; and (3) a 'kangaroo hop' in which both feet acted together. The conventional walking gait proved to be the most effective. On Earth he could easily halt his motion with a single step, but on the Moon it took several steps to slow down because the ratio of mass-to-weight had changed by a factor of 6. Similarly, changing direction while in motion had to be done in stages, stressing the outside leg in order to force the turn. As Aldrin paraded in front of the television camera, his wife laughed so much that her eyes wept. Pat Collins, watching with Barbara Gordon and Sue Bean, was amused by his antics. Jan Armstrong, who was ticking off items on her list, doubted they would achieve all of their assigned tasks in the time available.

Meanwhile, Armstrong had dismounted the Hasselblad and placed it on the MESA in order to start to prepare the equipment with which he was to collect what field geologists call a 'bulk' sample of the loose ground mass with embedded rock fragments.

A long-distance phone call

"Tranquility Base, this is Houston," McCandless called formally. "Could we get both of you on the camera for a minute, please."

"Say again, Houston," said Armstrong.

After repeating the request, McCandless added, "Neil and Buzz, the President of the United States is in his office now and would like to say a few words to you."

"That would be an honour," Armstrong said.

Richard Nixon had been watching them on television with Frank Borman in his private office in the White House. After the flag had been raised, Nixon went next door to the Oval Office to place a telephone call to the lunar surface. With a camera set up in the Oval Office, the television networks presented this historic call in split-screen fashion. Deke Slayton had alerted Armstrong that at some time during the moonwalk (the obvious moment being just after the flag was raised) they might receive a "special communication", which they both took to mean a call from Nixon. However, it came as a surprise to Aldrin.

"Go ahead, Mr President," said McCandless.

"Neil and Buzz," Nixon began, "I'm talking to you by telephone from the Oval Room at the White House, and this certainly has to be the most historic telephone call ever made. I just can't tell you how proud we all are of what you have done. For every American, this has to be the proudest day of our lives. And for people all over the world. I am sure they, too, join with Americans in recognising what an immense feat this is. Because of what you have done, the heavens have become a part of man's world. And as you talk to us from the Sea of Tranquility, it inspires us to redouble our efforts to bring peace and tranquillity to Earth. For one priceless moment in the whole history of man, all the people on this Earth are truly one; one in their pride in what you have done, and one in our prayers that you will return safely to Earth."

"Thank you, Mr President," Armstrong acknowledged. "It's a great honour and privilege for us to be here representing not only the United States but men of peace of all nations, and with interest and a curiosity and a vision for the future. It's an honour for us to be able to participate here today."

"And thank you very much," added Nixon, "and I look forward – all of us look forward – to seeing you on the *Hornet* on Thursday."

"I look forward to that very much, sir," replied Armstrong, signing off.

Both astronauts had remained in place throughout the call. Aldrin remained silent and left it to his commander to make the responses.

In the Collins house, Rusty Schweickart said there would be scientists around the world urging the astronauts to push on and collect some rocks. Indeed, Nixon was later criticised by some in the scientific community for having 'wasted' the limited time available to the astronauts. Aldrin, following his checklist, shuffled around repeatedly scuffing the surface with his boot to observe how the material dispersed. When sand on a terrestrial beach is scuffed, it disperses in an arc with some of the grains travelling further than others. On the Moon, in the absence of air-drag to sort the particles by size, all the grains landed at the same radius, which depended upon the impulse imparted and the weak lunar gravity. As this phenomenon marked a striking difference between training and reality, Aldrin found it fascinating. On returning to Eagle, Aldrin was struck by the sharpness of the vehicle's shadow. On standing in sunlight and projecting his arm into the shadow, it seemed to vanish. Furthermore, as he recalled later, "The light was sometimes annoying, because when it struck our helmets from a side angle it would enter the face plate and make a glare that reflected all over it. As we penetrated a shadow we would get a reflection of our own face, which would obscure anything else. Once when my face went into shadow it took maybe 20 seconds before my pupils dilated out again and I could see details."

Sampling

Having unstowed the long-handled scoop from the MESA, Armstrong set out to collect the bulk sample from the general vicinity of the SWC, on ground that had been documented as part of that experiment. The television cable was white, but when it became coated with dust it was difficult to see, and because it retained a memory of having been coiled in its dispenser it refused to sit flat on the ground. On seeing that Armstrong's feet were becoming entangled with the cable, Aldrin called "Watch it, Neil! Neil, you're on the cable." Armstrong tried to manoeuvre clear of the cable, but the visor of his helmet limited his downward view and the thickness of the suit prevented him feeling its presence. Aldrin went to help him. "You're clear

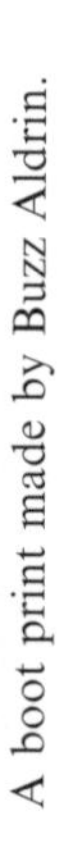

A boot print made by Buzz Aldrin.

now." As Armstrong would have to make many trips to the MESA to collect the bulk sample one scoop at a time, he used his scoop to lift the cable off the ground, Aldrin took it, dragged it aside, gathered the surplus and tossed it beneath the vehicle.

Leaving Armstrong to collect the sample, Aldrin began his first photographic task, which was to document the imprint that his boot made in the surface. After retrieving the Hasselblad from the MESA he went to a patch of ground that they had not yet disturbed and, using the camera hand-held, took a photograph of this. Then he made an impression with his right boot, stepped back and photographed the result. Moving further forward he put his boot on the surface again, and this time took the picture just as he lifted his foot; in so doing he noted that there was so much black material coating his overshoe that its light-blue colour was no longer visible. Moving on to his next task, he took a panorama from a location south of the tip of Eagle's shadow, covering 360 degrees in 11 frames, one of which captured Armstrong at the MESA. He then went to the southern side of Eagle and took a number of pictures to enable the Grumman engineers to assess the state of the vehicle, in the process capturing a view through the struts of the front gear of Armstrong once again at the MESA. As the next item on his checklist was to take 'after' shots of where the bulk sample was collected, he called, "How's the bulk sample coming, Neil?"

"It's just being sealed," replied Armstrong. Collecting the sample had proved to be more difficult than in training because, in the weak lunar gravity, the material readily spilled from the scoop as he carried it to the MESA, with the result that he lost part of each load. Whatever remained each time, he poured into the bag that Aldrin had prepared. The object of the exercise was to return sufficient material to satisfy the requirements of the many teams of scientists. Over about 15 minutes he drew 23 scoops. Since he did not wish to rely upon having time later to take fully documented samples, he had made an effort to collect a variety of small rocks for this sample. When he was finished, he placed the bag into the first rock box. The lid of the box was a precise fit, and included a razor edge in order to preserve the contents in vacuum once the box was taken into an atmosphere. As there was no lubricant on the hinge, sealing it took longer than expected, in part owing to the fact that in lunar gravity he did not have the same leverage as in training.

On reflection, Aldrin asked if Armstrong would rather take the 'after' pictures himself, because he knew precisely where he had sampled. "Do you want to get some particular photographs of the bulk sample area, Neil?"

"Okay," Armstrong replied. When Armstrong joined Aldrin, by now back in Eagle's shadow, Aldrin passed Armstrong the camera, who put it on his bracket even though he was to take just a few pictures. On impulse, he photographed the plaque on the forward strut, and since it was in deep shadow he shot it using a range of exposures. He then went to document the area from which he had collected the bulk sample. Aldrin followed him. Having finished his documentation, Armstrong took an impromptu picture of Aldrin, then returned the camera to the MESA.[10]

[10] This picture of Aldrin became the iconic Apollo 11 'Man on the Moon' image. It is on the front cover of this book.

"Buzz," McCandless called. "Have you removed the Close-up Camera from the MESA yet?"

"Negative," replied Aldrin.

The Apollo Lunar Surface Close-up Camera (ALSCC) was to provide extreme close-up stereoscopic pictures of lunar 'soil'. It was often referred to as the 'Gold camera' because it had been designed by Thomas Gold, an astronomer at Cornell University. As it was a late addition to the mission, the astronauts had very little time to train with it. The plan called for Aldrin to unstow it from the MESA, but Armstrong said he would do it. Aldrin therefore resumed his photographic task. Retrieving the Hasselblad, he took a panorama from a position northeast of Eagle, again covering 360 degrees in 11 frames, and then concluded his documentation of the vehicle.

"Houston, how does our time line appear to be going?" Aldrin enquired.

"It looks like you're about a half hour slow," McCandless replied.

Armstrong set off with the ALSCC. To take a picture (in fact, a stereo pair) he had to rest it on the ground with the Sun illuminating a window at its base, then pull a trigger to expose and advance the film.[11] It proved awkward to operate and, although designed to be self-standing, tended to fall over whenever he released it, which was frustrating because he then had to fetch one of the long-handled tools in order to raise its handle off the ground.

"Neil and Buzz, this is Houston," McCandless called, "Your consumables are in good shape at this time."

Deploying the instruments

On finishing his inspection of Eagle, Aldrin was ready to unstow the Early Apollo Surface Experiments Package (EASEP) from the scientific equipment (SEQ) bay, an activity which Armstrong was to document. "Neil, if you'll take the camera, I'll get to work on the SEQ bay."

"Okay," agreed Armstrong, taking the Hasselblad from Aldrin.

The compartment on the left-rear quadrant of the descent stage, opposite to the MESA, had two doors – a small door on the left that Aldrin simply hinged open, and a larger one that was hinged horizontally along its upper edge and was to be opened using a lanyard and pulley mechanism. Although the raised door failed to engage its lock, it remained in place. The base of the bay was at chest height. The Passive Seismic Experiment (PSE) was stowed in the left-hand compartment and the Lunar Ranging Retro Reflector (LRRR) on the right. For each, Aldrin had the option of drawing out a boom and using a pulley to lower the instrument onto the ground, but he chose instead to disconnect the hooks and extract them manually, finding this task to be rather easier than in training. Having extracted the PSE he moved off about 10 feet and put it on the ground, then returned to get the LRRR. Meanwhile, having taken the requisite pictures of Aldrin at work, Armstrong put down the

[11] In all, 17 stereo pairs were taken using the ALSCC.

A depiction of the pulley system available to Buzz Aldrin in retrieving the EASEP instruments from the SEQ bay.

ALSCC and moved to a point about 60 feet southeast of Eagle to shoot a 360-degree panorama in 11 frames. Aldrin closed and locked the doors of the SEQ bay to prevent the sunlight overheating the descent stage. He then asked Armstrong, "Have you got us a good area picked out?"

Although the terrain was pocked by craters, there was a reasonably level spot southwest of Eagle. "I think right on that rise out there is probably as good as any."

Holding the PSE in his left hand and the LRRR in his right, Aldrin hoisted the load – which in all weighed just 27 pounds in lunar gravity – and headed for the indicated area. After snapping several pictures of Aldrin carrying the instruments, Armstrong retrieved the ALSCC and followed.

"It's going to be a little difficult to find a good level spot here," Aldrin warned.

"The top of that next little ridge there," Armstrong prompted. "Wouldn't that be a pretty good place?"

Aldrin halted about 40 feet from Eagle, "Should I put the LRRR right about here?"

"All right."

Aldrin deposited the LRRR, and moved out 15 feet further out and put down the PSE.

Meanwhile, Armstrong had paused to study some of the larger rocks, "These boulders look like basalt," he ventured, "and they have probably 2 per cent white minerals in them – white crystals. But those things I reported as vesicles before, I now think they're small craters; they look like tiny impact craters where shot has hit the surface." He was correct. These light patches were where micrometeoroids had exposed clean crystals; they would later be named 'zap pits'. Armstrong then aligned the LRRR on an east–west axis, levelled it with respect to local vertical by means of an air bubble in fluid that had to be centred (observing that in the weak lunar gravity the bubble took a surprisingly long time to settle) and then tilted the mirror platform to face Earth. It is a common misconception that Earth, seen from the lunar surface, is always at the zenith. In fact, for an equatorial site 23 degrees east of the lunar meridian, Earth is correspondingly situated west of the zenith and revolves upon its axis. In its deployed state, the LRRR came to knee height. Its face incorporated an array of 100 fused silica 'corner-cube' mirrors that were to reflect a pulse of laser light straight back to its source. Although a laser directed by a large terrestrial telescope would start out as a narrow collimated beam, by the time the beam reached the Moon it would have dispersed to illuminate an area 2 miles in diameter, and as the instrument would be able to return only a tiny fraction of this the received 'signal' would be exceedingly weak. The first laser probe was made by the Lick Observatory near San Jose in California several hours later, but since the precise location of the landing site was not yet identified the first detection was not made until several days later.[12] As the reflected signal was difficult to discern when the site was in sunlight, the LRRR research was best undertaken during the lunar night.

The deployment of the PSE was rather more complicated. After orienting the instrument with respect to the Sun by ensuring that the shadow cast by a gnomon on the top of the package fell on a predetermined line, Aldrin set out to level it. The design had originally used a 'bubble' indicator (like the LRRR) but this had been replaced by a small ball in a cup. Aldrin shuffled the instrument on the uneven surface, pushing the loose material aside, but to his surprise the ball persisted in running around the periphery of the receptacle; on Earth it would have settled immediately. "That BB likes the outside. It won't go on the inside," mused Aldrin. Joining him, Armstrong speculated that the cup might be convex rather than concave. "Houston," Aldrin called, conscious that time was passing, "I don't think there's any hope for using this levelling device to come up with an accurate level."

"Press on," McCandless replied. "If you think it looks level by eye-ball, go ahead."

The instrument had a pair of 3-segment rectangular solar panels mounted on its sides to face east and west. One of the panels deployed automatically, and Aldrin deployed the other manually. As the mechanism unfolded the panels, their bottom corners came into contact with the ground and acquired a coating of dust. With its radio antenna deployed, pointing at Earth, the instrument rose to waist height. The initial transmission from the instrument was received by a 30-foot-diameter dish at Carnarvon in Australia. The seismometer was sufficiently sensitive to detect the astronauts walking about.[13]

At this point McCandless had some good news, "Neil, we've been looking at your consumables and you're in good shape. With your concurrence, we'd like to extend the duration of the EVA 15 minutes beyond nominal. We'll still give Buzz a hack at 10 minutes for heading in. Your current elapsed time is 2 plus 12."

"Okay," Armstrong replied. "That sounds fine."

"Buzz," McCandless prompted. "If you're still in the vicinity of the PSE, could you get a photograph of the ball?"

"I'll do that, Buzz," said Armstrong, who had the Hasselblad and had stepped beyond the EASEP to document the instruments with Eagle in the background for context. "Oh, shoot!" he exclaimed upon inspecting the PSE. "Would you believe

[12] Although McCandless was told that a laser reflection had been detected while Eagle was still on the surface, and he relayed this news to Collins, this was not so.

[13] The seismometer included a detector to measure dust accumulation and radiation damage to the solar cells, and an isotope heater to keep the electronics warm during the long lunar night. Despite operating temperatures that exceeded the planned maximum by 30°C, the instrument functioned normally through the maximum heating around lunar noon. With the power output from the solar arrays in decline about 5 hours before local sunset (on 3 August 1969) transmission was halted by command from Earth. It was turned on again on the next lunar day, but (on 27 August) near noon of this second lunar day the instrument ceased to accept commands and the experiment was terminated.

the ball is right in the middle now?" Lunar gravity had finally drawn the ball into the centre of the cup, which clearly was concave.

"Wonderful," Aldrin replied. "Take a picture before it moves!"

Winding up

On the plan, 30 minutes had been allocated to documented sampling, which was to be a two-man activity. The first task envisaged Aldrin hammering a core tube into the surface. Armstrong was to take pictures prior to sampling, with the tube in the ground, and following its extraction. They were then to collect a number of rocks, each of which was to be photographed *in situ*, carefully lifted, and inserted into an individual sample bag. Although the lunar material was to be put inside a vacuum-sealed rock box, some material was to be put into a can which, when sealed, would retain any readily volatised constituents that would otherwise be difficult to preserve when the rock box was opened in the laboratory. Finally, if time permitted, they were to collect a second core sample. But when McCandless announced that only 10 minutes was available for this sampling it was decided to forgo the documentation.

While Aldrin prepared a core tube at the MESA, Armstrong disappeared out of sight of the television. Although he had been surprised to discover that, on looking east, he could not see the boulders that surrounded the large crater where Eagle's computer would have tried to land, Armstrong was able to see the smaller crater over which he had passed just prior to landing. As this was only 200 feet away he decided to inspect it. Saying nothing of his intention, he set off, carrying the ALSCC.[14] On reaching the southwestern rim of the crater he shot a sequence of 8 frames across the pit ranging from up-Sun, around the northern horizon and on down-Sun to Eagle. The crater had a raised rim and an interior strewn with rocks. He yearned to enter it to collect a rock as a treat for the scientists, but the pit was 70–80 feet in diameter and 15–20 feet deep and, in any case, he had to rush back. In all, his excursion had lasted just over 3 minutes. He had no difficulty sustaining a 'loping' gait, which the timing indicated to have been at 2 miles per hour.

"Buzz," McCandless called while Aldrin was still at the MESA attaching the extension handle to the core tube, "You've got about 10 minutes left now prior to commencing your EVA termination activities."

"I understand," replied Aldrin. A minute later he took the core tube and went to sample some already documented ground near the SWC. This 'soil mechanics' study was to determine soil density, strength and compressibility as functions of depth. It would also reveal layering, either in terms of the chemical composition of the loose material or its physical characteristics, such as grain size. The plan called for the hollow tube to be driven to a depth of 18 inches. The staffs of the flag and the SWC had indicated that the surface material was consolidated at a depth of several inches, but Aldrin hoped that by hammering on the core tube he would be able to drive it in.

[14] Armstrong would later express surprise that he had lugged Gold's camera around with him for so long.

A view across the crater that Neil Armstrong visited, variously named ‘Little West’ and ‘East’ Crater.

A view looking back at Eagle from the rim of the crater.

The extension handle came up to waist height. At first Aldrin raised the hammer only to chest level, but then he increased this to head level in order to generate the additional force. A complication was that the tube gained little support from the material it penetrated, and he had to maintain a grip on the tool with one hand throughout. As he became more determined, he observed that the hammer was denting the top of the handle. "I hope you're watching how hard I have to hit this into the ground to the tune of about 5 inches, Houston," he said pointedly. In fact, his hammering drove the tube in only 2 inches beyond the depth to which he had inserted it by hand. Giving up, he withdrew the tube from the ground. The finely grained material coated the section that had penetrated the ground. "It almost looks wet," he noted. To his relief, the material did not dribble out of the open end. On his return, Armstrong snapped pictures of Aldrin at work, then accompanied him to the MESA to help him to cap the tube. A post-mission investigation concluded that the design of the aperture of the tube had inhibited penetration. In the expectation that the surface material would be loose to considerable depth, the core tube had been designed with an internal bevel to compact the material entering the tube as it was hammered into the ground, but because the lunar material at a few inches depth was close to its maximum density, it jammed in the aperture. This discovery made even more ludicrous the idea that the lunar surface was a dust trap that would swallow a spacecraft!

"Neil and Buzz," McCandless called. "We'd like y'all to get two core tubes and the Solar Wind." At Aldrin's suggestion, Armstrong completed capping the first core tube, and Aldrin took the second sample 15 feet beyond where he had taken the first. "Buzz," McCandless called as Aldrin hammered the second tube, "in approximately 3 minutes you'll have to commence your EVA termination activities." On realising that he was gaining no greater penetration than before, Aldrin withdrew the tube and returned to the MESA to cap it.

"Neil, after you have got the core tubes and the Solar Wind, anything else that you can throw into the box would be acceptable," McCandless called.

"If you want to pick up some stuff," Aldrin said to Armstrong, "I'll get the Solar Wind." Aldrin detached the collector sheet from its staff, rolled it up, and stuffed it into a bag. He discarded the staff. He deposited the SWC on the MESA next to the core tubes, ready for Armstrong to stow in the second rock box.

Meanwhile, Armstrong had used a pair of long-handled tongs to collect rocks for the 'suite' – a field geologist's term for a collection of rocks representative of a site, including both the typical and the exotic. This was essentially as planned, but without documentation and with the rocks going into a single large bag rather than into individual bags.

"Buzz," McCandless called. "It's time for you to start your EVA close-out."

"That's in progress," Aldrin replied.

As the moonwalkers began to wrap up in silence, Columbia once again flew 'over the hill' and out of communication.

"We'd like to remind you of the Close-up Camera magazine before you start up the ladder, Buzz," McCandless called.

"Have you got that over with you, Neil?" Aldrin asked.

Armstrong had dispensed with the ALSCC in order to collect samples. "No, the Close-up Camera's underneath the MESA." Having made an early report of what appeared to be vesicular rock and then retracted this claim, Armstrong had located some genuine examples, "I'm picking up several pieces of *really* vesicular rock out here, now."

"You didn't get any environmental samples, did you?" Aldrin asked, referring to the material they were to have sealed into cans.

"Not yet," replied Armstrong.

"Well, I don't think we'll have time."

"Neil and Buzz," McCandless called. "Let's press on with getting the Close-up Camera magazine and closing out the sample return containers."

Aldrin went to the MESA and, supporting himself with one hand, bent down to retrieve the ALSCC. After removing the film, he asked Armstrong to assist in inserting the magazine into his thigh pocket. "Anything more before I head on up, Bruce?"

"Negative. Head on up the ladder, Buzz."

"Remember the film off of that," Aldrin reminded Armstrong, referring to the Hasselblad.

"I will," Armstrong promised.

"I'll head on in, and get the LEC ready for the first rock box," Aldrin said. As he ascended the ladder he noticed that the dust coating his boots made the rungs seem slippery. Armstrong was to have tried to dust him off, but there was no time.

Armstrong carried the bulk sample box from the MESA out in front of Eagle and hooked it to the LEC, then added the Hasselblad magazine to the same hook. "How are you doing, Buzz?"

"I'm okay," replied Aldrin, who was now inside the cabin. "Are you ready to send up the LEC?"

The method for hoisting the box to the hatch required Armstrong to pull on the loop as if drawing washing along a clothes line. Watching, Joan Aldrin laughed, "God bless the rock box. I feel as if I've lived with that rock box for the last six months." As the scene played out, she was amazed, "This is like a Walt Disney cartoon, or even a television show – it's all too much to believe or understand." As the lanyard thrashed in the weak lunar gravity, the film pack detached and fell to the ground by the forward leg. With the leading edge of the box nudging the upper rim of the hatch, Aldrin asked Armstrong to slacken off the tension in the tether in order to lower the box sufficiently to enable it to enter. While Aldrin was stowing the box, Armstrong retrieved the Hasselblad magazine. Because this had fallen beside the foot pad, he decided not to fetch his tongs from the MESA, and instead gripped the ladder with one hand and, bending at the waist, leaned to lift the magazine which, as with everything else that came into contact with the lunar surface, was coated with fine black dust.

"This one's in. No problem," reported Aldrin, having stowed the first box in its receptacle in the cabin.

At this point McCandless asked Armstrong for an "EMU check". Although this was nominally a request that he read out the status of his PLSS systems, the flight surgeon was concerned that in manhandling the rock box and working the LEC, his heart rate had shot up to 160 beats per minute, and the EMU check was a hint that he should take a rest.

"How's it coming, Neil?" Aldrin asked a minute later.

Having placed the bag of rocks, core tubes and SWC into the second box and sealed it, Armstrong tethered the box and added the recovered magazine. "Boy," he observed, "that filth from on the LEC is kind of falling all over me while I'm doing this."

"All that soot, huh?"

To give Armstrong a rest, Aldrin suggested they revise the procedure for hauling up the box, "If you can just kind of hold it, I think I can do the pulling."

"Stand by a minute. Let me move back," said Armstrong. He backed away to tension the LEC. Once the box was up, Aldrin detached the LEC from the pulley and tossed the cable out through the hatch. The uploading of the boxes had taken rather longer than in training, and repeatedly working against the restraint system built into the shoulder joints of the suit was the greatest exertion of the moonwalk.

"How about that package out of your sleeve, did you get that?" Armstrong enquired.

This was a reference to a small canvas bag of mementoes Aldrin had carried in his shoulder pocket with the intention of leaving it on the surface prior to his ingress; he had forgotten. Armstrong proposed that Aldrin pass him the bag once he was on the porch, but Aldrin tossed the bag out through the hatch and it landed at Armstrong's feet. It contained a gold medallion bearing a representation of the 'olive branch' motif – one of four that Aldrin had in his personal preference kit, the others being destined for the astronauts' wives. There was also an Apollo 1 mission patch in memory of Gus Grissom, Ed White and Roger Chaffee, who died when their capsule caught fire on the pad on 27 January 1967. On returning from his visit to the Soviet Union, Frank Borman handed over two medals that his hosts had requested be left on the Moon. These honoured Yuri Gagarin, the first man to orbit Earth, who died in an aircraft accident on 27 March 1968, and Vladimir Komorov, who died on 24 April 1967 when the parachute of Soyuz 1 failed to open. A more formal memento was a text bearing statements issued by Presidents Eisenhower, Kennedy, Johnson and Nixon, a message from the Pope, and messages of goodwill from the leaders of 73 countries of the United Nations. Some messages were handwritten, others typed, in a variety of languages. It also included a listing of the leadership of the Congress in 1969, a list of members of the committees of the House and Senate responsible for NASA legislation, and the names of NASA management. It was photographed and reduced by a factor of 200, transferred to glass for use as a mask for etching by ultraviolet light onto a 1.5-inch-diameter silicon disk – the same technology as was used to etch integrated circuitry. The disk bore the inscription 'Goodwill messages from around the world brought to the Moon by the astronauts of Apollo 11'. Around the rim was 'From Planet Earth', and 'July 1969'. Although silicon was chosen for its ability to withstand the temperature extremes of the lunar surface, it was enclosed in an aluminium container to protect the delicate crystal from shock. If it had been intended to mark the placement of these items, the moment had been lost.

Meanwhile, Houston, oblivious to what was going on, was eager to confirm that everything that was to have been loaded was indeed on board. "Neil, did you get the Hasselblad magazine?"

Armstrong had just stepped onto the foot pad. "Yes, I did. And we got about, I'd say, 20 pounds of carefully selected, if not documented, samples."

"Well done."

Grasping the ladder with both hands for stability, Armstrong adopted a deep knee-bend, then jumped, and his feet landed on the third rung from the bottom of the ladder! It was a shame, he would reflect, that they had not been able to remain out for longer. He had hoped to inspect the boulders off to the north, which, while distance was difficult to judge, appeared to be several feet across.

BACK INSIDE

As soon as Aldrin had guided Armstrong in through the hatch, he reported, "The hatch is closed, latched, and verified secure."

The astronauts' boots, lower legs and gloves were coated with the black lunar dust, and because the LEC had been coated with this material as a result of being trailed across the ground, dust had entered the cabin while hauling up the rock boxes. Some scientists had suggested that iron-rich material on the lunar surface might have been so modified by its long exposure to the charged particles of the solar wind that it would burst into flame on coming into contact with oxygen, and had expressed concern that the last that would be heard from Eagle would be a recital of the checklist leading to cabin repressurisation! Indeed, for several minutes as the cabin pressure built up to 4.8 psi of oxygen, the crew did not respond to calls – but they had not been consumed by flame, they were switching their umbilicals from the PLSS to Eagle's communications system. On raising their helmets, they noted an odour that Armstrong compared to "wet ashes in a fireplace" and Aldrin to "spent gunpowder".

As Armstrong and Aldrin ran through their post-ingress checklist, Columbia, reappeared on revolution 19. McCandless brought Collins up to date, "The crew of Tranquility Base is back inside their base, repressurised, and are in the process of doffing the PLSSes. Everything went beautifully."

"Hallelujah," replied Collins.

In order to lighten the ascent stage, all items that were no longer needed were to be jettisoned. The extravehicular Hasselblad had been left on the MESA. The version for internal use was to be jettisoned, but first the astronauts used up their film by documenting the views from their windows showing the evidence of their activities. They were amazed at the number of boot prints. To finish up, they shot some interior views, obtaining an excellent picture of Armstrong looking deeply content.

Although the hatch had been open for 2 hours 31 minutes 41 seconds, they had spent longer on the PLSS systems due to having switched to the portable systems prior to opening the hatch, and remained on them for some time after its closure. As the rate at which they would consume coolant water was not accurately predictable

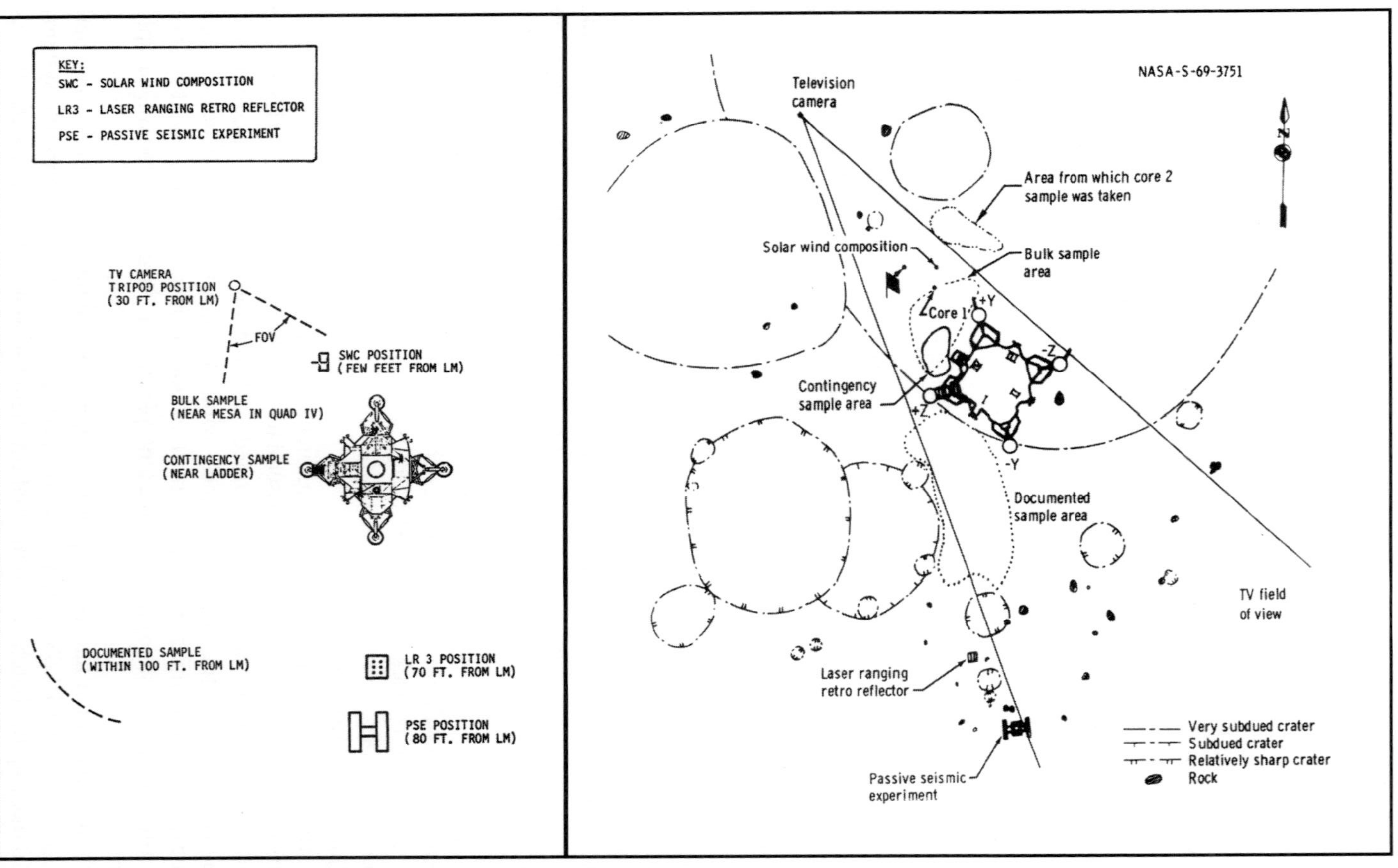

The lunar surface activity plan (left) and actual (right), both on the same scale.

and they had required a margin in case of difficulty in reverting to the cabin's life-support system, the nominal duration of the surface excursion had been set much shorter than the potential total duration of the backpack. To evaluate the cooling system of the PLSS, the water remaining in their tanks was poured out and weighed. It had been thought that Armstrong (by being outside for longer) would use some 5.4 pounds of coolant and Aldrin 5.1 pounds. In fact, Armstrong used a mere 2.9 pounds; Aldrin, however, used 4.4, this being due to his preference for intermediate cooling. As he reflected later, "I had cooler levels set on the (feed water) diverter valve because it just seemed to be pleasant that way. In retrospect, it appears that this leads towards a higher consumption of water. I wasn't fully aware that when I was on a higher flow I would be pumping more water overboard. It wasn't clear to me pre-flight that it would have such an effect on my water consumption. I certainly could have operated at lower levels without overheating." Nevertheless, both men consumed significantly less water than expected, and could have remained outside for a longer period.

As they tidied up, Aldrin discovered that the button had snapped off a circuit breaker on the panel on his side of the cabin, and that of an adjacent breaker had been pushed in; evidently, while wearing his PLSS he had nudged the panel. The breakers were of the standard push–pull configuration used in aircraft. When the button was pulled to open the circuit it exposed a white band, and when pushed to close the circuit it hid the band as a visual cue. Although it would be possible to close the circuit by inserting the tip of a pen to set the latch, there would be no way to open it again. As the damaged breaker would be used to feed power to the ascent engine, the priority was to determine its current state. Houston said the telemetry indicated that it was open. It was decided to wait until the circuit was required, then use a pen to push it in; if it were to fail to latch, the designers had a manual workaround to feed power to the engine.

As the astronauts finished the post-moonwalk meal of cocktail sausages and fruit punch, Deke Slayton called to offer his congratulations, "I want to let you guys know that since you're an hour and a half over your time line and we're all taking a day off tomorrow, we're going to leave you. See you later."

"I don't blame you a bit," replied Armstrong.

"That was a great day, guys. I really enjoyed it," said Slayton.

"Thank you," Armstrong said. "You couldn't have enjoyed it as much as we did."

"It was great," Aldrin added.

"We sure wish you'd hurry up and get that trash out of there, though," Slayton urged.

"We're just about to do it," Armstrong assured him.

To depressurise the cabin they put their helmets and gloves back on and verified their umbilicals to the cabin's life-support system. In view of the time it had taken to vent the cabin the first time, it was suggested that this time they should also open the valve in the overhead hatch. This reduced the depressurisation time to a mere 90 seconds. There were no communications while the hatch was open. Armstrong shoved out first one PLSS and then the other with sufficient force to clear the far end of the porch. The overshoes and helmet augmentation followed, together with the cabin arm rests, the body of the Hasselblad, a saturated lithium hydroxide carbon dioxide filter from the cabin system, urine bags and food packaging. The OPS were retained in case an external transfer to Columbia proved necessary.

Meanwhile, Columbia reappeared once again and McCandless called, "I guess we'll bid you a good night and let you get some sleep, Mike."

"Sounds fine," agreed Collins.

Having flown a very 'busy' Gemini 10, Collins relished the day that he would have alone in Columbia. "I really looked forward to a chance to relax and look out the window – to get some assessment of what it's all about." He enjoyed the 48 minutes per orbit during which he was on the far side of the Moon. He would later write, in the style of a diary, "I'm alone now, truly alone, and absolutely isolated from any known life. I'm it. If a count were taken, the score would be three billion plus two over on the other side of the Moon, and one plus God-knows-what on this side. I feel this powerfully – not out of fear, or loneliness – but as awareness, anticipation, satisfaction, confidence, almost exultation. I like the feeling."

"Repress complete," Armstrong announced, ending Eagle's period of silence.

"We observed your equipment jettison on the TV," said McCandless, "and the passive seismic experiment recorded the shocks when each PLSS hit the surface."

"You can't get away with anything anymore, can you?" laughed Armstrong in respect of their littering.

"No, indeed," McCandless agreed.

Several minutes later, McCandless was back, "We'd like to say, from all of us down here in Houston – and really from all of us in all the countries in the entire world – we think that you've done a magnificent job up there today."

"Thank you, very much," Armstrong replied.

"It's been a long day," Aldrin noted.

"Indeed," McCandless agreed. "Get some rest there and have at it tomorrow."

"Have you had enough TV for today?" Aldrin asked.

"Yes, it's been a mighty fine presentation."

"Okay. Signing off. See you again tomorrow." Aldrin pulled the circuit breaker to end the television transmission. The time in Houston was almost 3 am, and most people had already retired.

As McCandless went off-shift, he handed over to Owen Garriott of the Maroon Team, who posed a number of questions on various aspects of the surface phase of the mission. After accepting a deferment of a detailed description of the geology of the area, Garriott wrapped up with a question designed to assist in identifying their location, "You commented, Neil, that on your approach to the landing spot you passed over a football-field-sized crater containing blocks of rock 10 to 15 feet in size. Can you estimate its distance from your present position?"

"I thought we'd be close enough so that when we got outside we'd be able to see its rim back there, but I couldn't. But I don't think that we're more than a half mile beyond it – that is, a half mile west of it," Armstrong explained.

"So you would estimate your position as less than half a mile, approximately, west of this large crater," Garriott asked.

"That's correct," Armstrong agreed.

With that, Garriott wished the crew of Eagle good night.

Armstrong and Aldrin had a 7-hour rest period scheduled prior to initiating preparations to liftoff. As a safety precaution against airborne dust they donned their helmets and gloves. Aldrin settled on the floor across the cabin, with his legs bent since the cabin was not wide enough to stretch flat. Armstrong reclined on the circular cover of the ascent engine, leaning against the aft wall and with his feet suspended above Aldrin in a sling improvised by hanging one of the waist tethers from the fixture installed for the LEC. With the windows shaded and the vehicle powered down, the temperature dropped. "The thing that really kept us awake, was the temperature," recalled Aldrin. "It was very chilly in there. After about 3 hours it became unbearable. We had the liquid cooling system in operation in our suits, of course, and we tried to get comfortable by turning the water circulation down to minimum; it didn't help much. We turned the temperature of our oxygen system to maximum; that didn't help much either. We could have raised the window shades and let the light in to warm us but, of course, to do that would have destroyed any remaining possibility of sleeping." The telemetry system allowed monitoring just one set of biomedical sensors. This indicated that although Armstrong was unable to fall into deep sleep, the period of inactivity permitted him to 'wind down' after the most momentous of days.

Meanwhile in the Sea of Crises

As Armstrong and Aldrin were attempting to sleep, the Soviet Union's unmanned spacecraft Luna 15 tried to land and crashed. Sir Bernard Lovell, having tracked it using the 250-foot-diameter radio telescope at Jodrell Bank, estimated that it fell onto the Sea of Crises about 500 miles east of Tranquility Base. The Soviet news agency *TASS* reported, "The program of research in space near the Moon, and of checking the new systems of the automated station Luna 15, has been completed. At 6.47 pm Moscow time on 21 July, a retrorocket was switched on and the station left orbit and achieved the Moon's surface in the preselected area." If Luna 15 had managed to land and scoop up some material, it would, by virtue of not pausing in lunar orbit, have been able to return to Earth a day or so ahead of Apollo 11. Gerry Carr pointed out that the presence of the human pilot had undoubtedly saved Eagle from similar disaster. "Its computer was heading for a blocky crater, and Neil just intervened and moved it over a bit." Although Soviet television had not shown the moonwalk 'live', a short clip was included in the news the next day. The People's Republic of China, however, made no mention of the mission.

9

Home in triumph

LIFTOFF AND RENDEZVOUS

Flight day 6, Monday, 21 July, started at 121 hours elapsed time with Ron Evans making his wake-up call to Columbia. In fact, Collins was already awake and was having his breakfast. Because Columbia was only a few minutes from going 'over the hill', Evans launched straight into updates to the flight plan which promised to keep Collins "a little busy": while in the Moon's shadow he was to perform a P52 to align the inertial platform, and then, on starting the near-side pass of revolution 23, he was to track crater 130-prime in order to measure the plane of his orbit accurately prior to rendezvous.

Meanwhile, Evans called Eagle, "Tranquility Base, Houston."

"Good morning," replied Armstrong promptly.

When Evans asked how they had spent the night in the cramped cabin, Aldrin replied, "Neil rigged himself a really good hammock with a waist tether, and he's been lying on the ascent engine cover. I curled up on the floor." For breakfast they had bacon squares, peaches, sugar cookie cubes, and a drink made from pineapple and grapefruit.

In view of the program alarms during the descent indicating that the computer had been overloaded, and the fact that the 'duty cycle' would be 15 per cent greater during the ascent, Mission Control had decided to leave the rendezvous radar off until they reached orbit. As Columbia flew overhead on revolution 24, Eagle tested its rendezvous radar by tracking the CSM's transponder. The inertial platform of the PGNS, which had lost its reference when powered down at the 'T3' milestone following landing, was aligned by taking star sightings with the telescope.

"Eagle and Columbia, this is the backup crew," announced Jim Lovell. "Our congratulations for yesterday's performance, and our prayers are with you for the rendezvous."

"Thank you kindly, Jim," replied Armstrong.

"Thank you, Jim," Aldrin concurred.

"We're glad to have y'all looking over our shoulders," Collins added.

As they waited for the appointed time to lift off, Armstrong called, "Houston, Tranquility Base is going to give you a few comments with regard to the geology question of last night. We landed in a relatively smooth field of secondary craters, most of which have raised rims irrespective of their size, but that's not universally true because a few of the smaller craters don't have a discernible rim. The ground mass throughout the area is a very fine sand to a silt. I'd say the thing that would be most like it on Earth is powdered graphite. Immersed in this ground mass are a wide variety of rock shapes, sizes, textures, rounded and angular, and many with varying consistencies. As I said, I have seen what appeared to be plain basalt and vesicular basalt, others with no crystals, and some with small white phenocrysts of maybe 1 to less than 5 per cent. And we are in a boulder field where the boulders range up to 2 feet, with a few larger than that. Some of the boulders are lying on top of the surface, some are partially exposed and some are just barely exposed. In our traverse around on the surface, and particularly working with the scoop, we ran into boulders below the surface – probably buried under several inches of the ground mass. I suspect this boulder field may have some of its origin with this large sharp-edged rocky-rim crater that we passed over in the final descent. Yesterday I said that was about the size of a football field. I have to admit it was a little hard to measure, coming in, but I thought it might just fit into the Astrodome as we came by it. The rocks in the vicinity of this rocky-rim crater are much larger than those in this area. Some are 10 feet or so and perhaps bigger, and they are very thickly populated out to about one crater diameter beyond the crater rim. Beyond that, there is some diminishing. Even out in this area the blocks seem to run in rows with irregular patterns, and then there are paths between them with considerably less surface evidence of hard rocks."

"Thank you, very much," acknowledged Evans.

As they ran through the checklist for liftoff, Aldrin used his pen to push in the damaged circuit breaker.

"For your information," Evans reported, "the circuitry looks real fine on that ascent engine arm circuit breaker." With this confirmed, Armstrong set the switch on the Engine Selector to Ascent.[1]

Glynn Lunney polled his Black Team on Eagle's status, and Evans relayed the result, "Eagle's looking real fine to us."

When Columbia appeared on revolution 25, Evans called Collins with another estimate for the position of the landing site, this time the one determined by Gene Shoemaker's team of geologists, but as there was less than half an hour remaining to liftoff Collins was too busy to look.

"Our guidance recommendation is PGNS," Evans advised Eagle, "and you're cleared for takeoff."

"Roger. Understand," replied Aldrin. "We're number one on the runway."

Armstrong and Aldrin once again harnessed themselves to the floor to enable them to 'stand' upright while in flight.

The ascent propulsion system (APS) of Eagle was built by Bell Aerosystems, using

[1] The options were, in turn, Descent, Off and Ascent.

an injector plate that was supplied by the Rocketdyne division of North American Rockwell. In all, the engine stood 4 feet 6 inches tall. Its combustion chamber was actually located inside the cabin, within a cylindrical cover that rose up from the floor, and its short nozzle sat on top of the central part of the descent stage. The only moving parts were the ball valves to allow propellants to flow to the injector. For redundancy, the primary valve was supplemented by a backup with a bypass line. It was required only that the valves open, since the hypergolic propellants would ignite on coming into contact in the chamber. If the computer command did not reach the valves, the circuit could be bypassed. Armstrong had suggested adding an option for manually operating the valves, but there had been no time. And, of course, the engine would require to burn long enough to enable the spacecraft to achieve some kind of orbit. If liftoff were delayed, Eagle would have to adopt a lower orbit in order to catch up with Columbia more rapidly, but if liftoff were to be so delayed that this would be impracticable, it was to climb to a high altitude and go passive, while Collins lowered Columbia in order to complete an 'extra' orbit to get into position to chase Eagle. There were many variations on this theme, depending on the circumstances, and Collins had a book detailing the procedures for each – all of which had been rehearsed in simulations. However, because Collins was allowed to descend to no lower than 50,000 feet, all of his options presumed that Eagle managed to attain at least this altitude. In a simulation in mid-June, the APS had shut down during ascent and Eagle had made up the velocity shortfall using a lengthy firing of all four of its downward-facing 100-pound-thrust RCS thrusters which, although nominally independent of the primary propulsion system, in an emergency could be fed from the main tanks to produce a sustained burn. On the possibility of the APS not igniting, Armstrong pointed out before the mission, "When pilots *really* get worried, is when they run out of options and run out of time simultaneously." If it failed to fire, there were various procedures to try. Although Columbia would be able to remain in orbit for two more days before it had to set off for home, the issue would by then have been resolved, as Eagle's power and oxygen would last no longer than 24 hours. Collins would later admit that his "secret terror" was that he would have to leave his colleagues on the Moon. As he told a reporter, "They know, and I know, and Mission Control knows, that there are certain categories of malfunction where I just simply light my engine and come home without them."

The White House had prepared a speech for President Nixon in case this were to occur: "Fate has ordained that the men who went to the Moon to explore in peace will stay on the Moon in peace. These brave men, Neil Armstrong and Edwin Aldrin, know that there is no hope for their recovery. But they also know that there is hope for mankind in their sacrifice. These two men are laying down their lives in mankind's most noble goal: the search for truth and understanding. They will be mourned by their families and friends; they will be mourned by their nation; they will be mourned by the people of the world; they will be mourned by a Mother Earth that dared send two of her sons into the unknown. In their exploration, they stirred the people of the world to feel as one; in their sacrifice, they bind more tightly the brotherhood of man. In ancient days, men looked at stars and saw their heroes in the constellations. In modern times, we do much the same, but our heroes are epic men of flesh and blood. Others will follow, and surely find their way home. Man's search will not be denied. But these men were the first, and they will remain the foremost in our hearts. For every human being who looks up at the Moon in the nights to come will know that there is some corner of another world that is forever mankind." Prior to making this speech, the President was to have telephoned the wives of Eagle's crew. When it was evident that communications were nearing their conclusion, a clergyman was to have commended their souls in the manner of a burial at sea. Although macabre, it was only right and proper that such plans should be drawn up.

As the final minute ticked away, Armstrong issued a final reminder, "At five seconds to go I'm going to get Abort Stage and Engine Arm, and you're going to hit Proceed."

"Right," confirmed Aldrin.

"And that's all," Armstrong added wryly.

There was a 'thud' as pyrotechnics cut the structural and electrical connections between the two stages, and then the engine lit.

"We're off!" Armstrong announced.[2]

The television camera drew its power from the descent stage, but after the moonwalk Aldrin had pulled the circuit breaker to end its transmission through the high-gain antenna on the ascent stage. If they had erected the large self-standing antenna on the surface, it would have been possible to televise the liftoff. As they departed, Aldrin started the 16-millimetre Maurer camera that was mounted in his window in order to document their ascent to orbit.

It was just before 1 pm in Houston. Joan Aldrin was on the floor in front of the television. When Gerry Carr said Eagle had lifted off, she rolled onto her back and kicked her legs in the air (as she was wont to do), then stood up and leaned against the wall. "It's strange," she said, "from the beginning I have worried more about liftoff than touchdown." This was perhaps because she had confidence in the astronauts' ability to fly the vehicle, but at liftoff they were at the mercy of the hardware. "They're on their way home!"

After a prayer, Herman Clark, the Grumman quality control inspector who had checked out the APS of LM-5, had held his breath literally as well as figuratively as he waited for news; on hearing that the engine had lit he told himself, "Job well done, QC-wise", then resumed work checking another LM on the production line.

The APS plume shredded the foil that had covered the descent stage, sending fragments radially outwards. "Look at that stuff go all over the place," exclaimed Armstrong. The plume did not stir up any dust, but Aldrin saw it blow the flag to the ground. The engine delivered just 3,500 pounds of thrust, but the ascent stage was light and it climbed rapidly, imposing a load of about 0.5 *g*. As it ascended, Aldrin remarked on its shadow, which raced over the surface. After 10 seconds of vertical rise, Eagle pitched over 45 degrees in order to begin to build a horizontal component to its velocity.

"One minute and you're looking good," Evans advised.

[2] Eagle had been on the surface for 21 hours 36 minutes.

The LM normally controlled its pitch and roll by using opposed upward- and downward-aimed thrusters to rotate the vehicle. However, as the upward-aimed thrusters would reduce the effect of the APS, it had been decided that during the ascent only the downward-aimed thrusters would be used, and firing the fore and aft thrusters in pairs to implement the progressive pitch-over produced a pronounced oscillation. "It's a very quiet ride," Aldrin noted. "There's just a little bit of slow wallowing back and forth; not very much thruster activity."

"You're looking good at 2 minutes," Evans continued. "PGNS, AGS, and the MSFN all agree."

"We're within 1 foot per second, AGS to PGNS," Aldrin pointed out. If the PGNS navigation were to falter, the AGS would take over.

"Go at 3 minutes. Everything's looking good," called Evans.

"We're going right down US One," Armstrong pointed out. US Highway One was the main north–south route running between New York and Florida, and the astronauts had borrowed the name for a distinctive linear valley that astronomers knew as Hypatia Rille.

"We've got Sabine off to our right," Aldrin announced. And a moment later, "There's Ritter." These were two large craters just beyond the western shore of the Sea of Tranquility. "And there's Schmidt. Man, that's impressive looking, isn't it?" During their earlier orbital sightseeing, these craters had been in darkness.

The APS burned for the planned 7 minutes and inserted the spacecraft into orbit 166 nautical miles west of Tranquility Base. Its progress towards and across the terminator into darkness was monitored by the Madrid station of the Manned Space Flight Network. This showed a perilune of 9.4 nautical miles, an apolune of 46.7 nautical miles and a velocity of about 5,537 feet per second. The PGNS gave 9.5 by 47.3, and the AGS gave 9.5 by 46.6 – indicating that, all things considered, the redundant systems were in excellent agreement. Such a low orbit would readily be perturbed by the mascons, but this did not matter because the initial perilune would be raised during the rendezvous. The ascent had halved the 10,837-pound mass of the ascent stage, reducing it to 5,885 pounds.

Meanwhile, at home

Jan Armstrong listened to the ascent on her squawk box alone, while having a late breakfast. When she heard her husband call shutdown right on time she knew that, with Eagle in orbit, if it were to encounter any difficulties Columbia would be able to rescue it.

Back in space

In weightlessness, Armstrong and Aldrin expected the specks of dust which had settled on the floor of the cabin to float around and pose a health hazard, but most of it remained in place, seemingly because it had developed an electrostatic charge and become 'clingy'. Free of the risk of inhalation or eye contamination, they were safely able to remove their helmets. The first in-orbit task was to do a P52 to check the inertial platform prior to initiating the rendezvous.

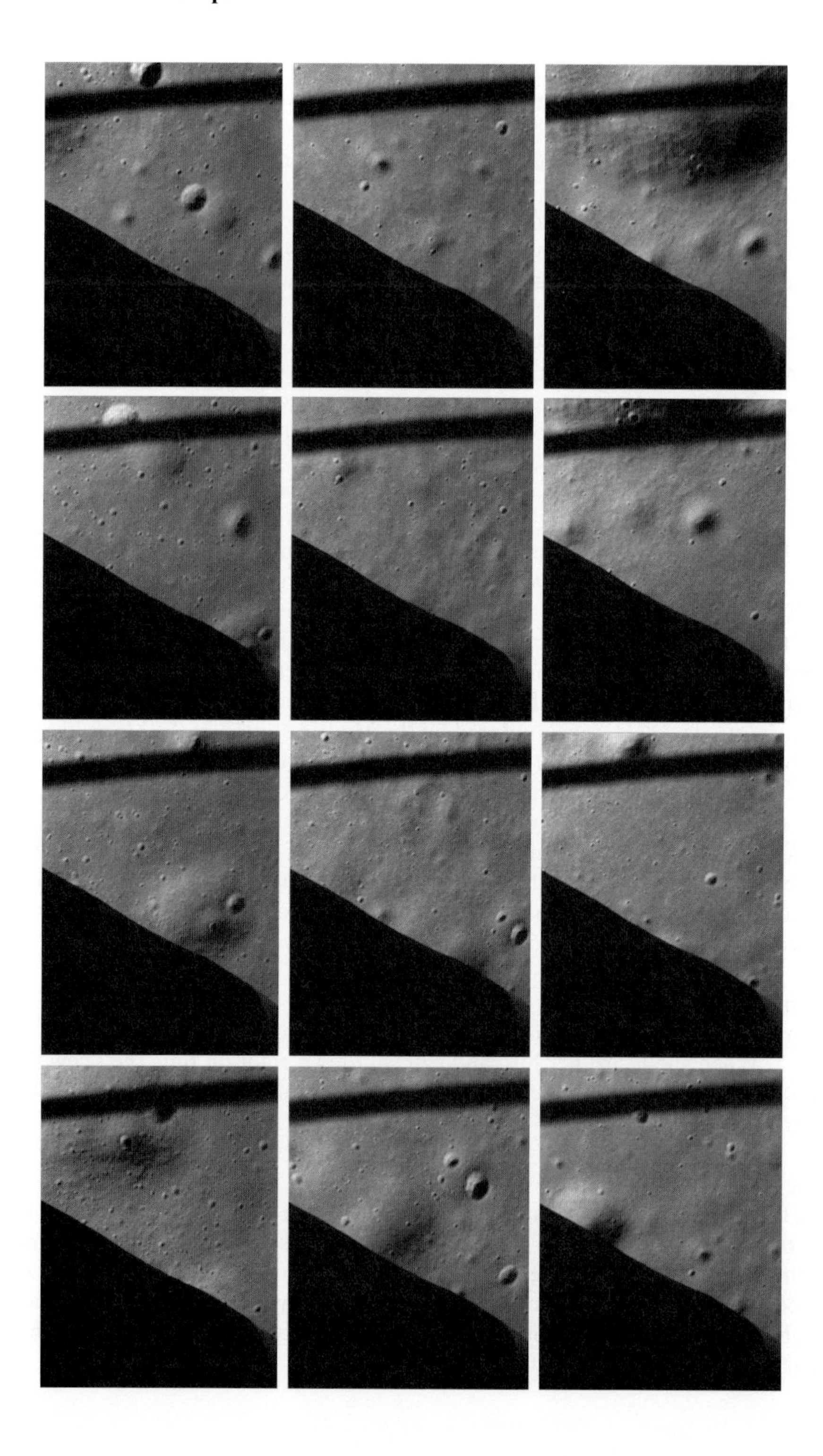

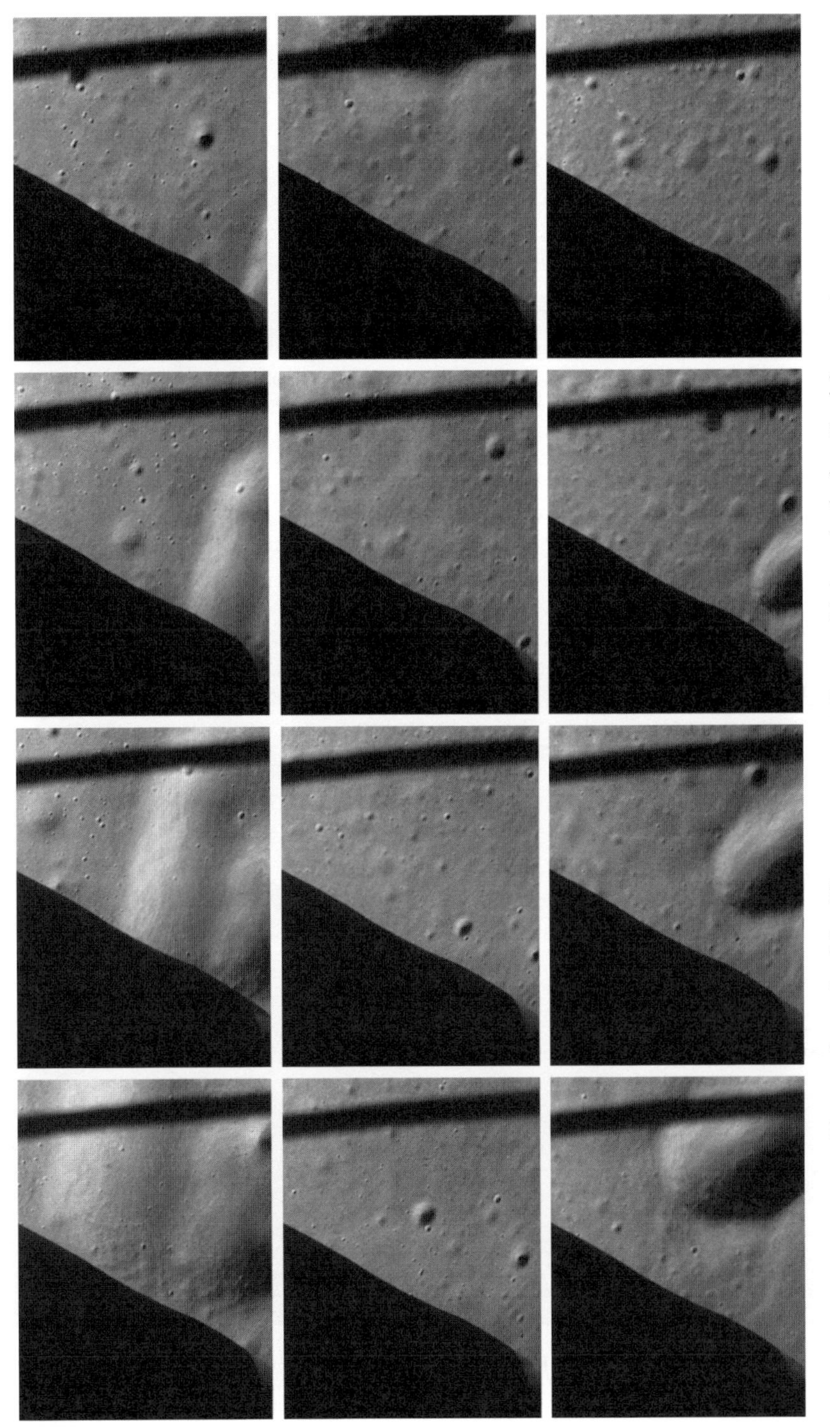

Frames from the 16-millimetre camera at 5-second intervals during Eagle's ascent.

The craters Sabine and Ritter over which Eagle flew during its ascent.

Columbia was oriented to point its radar transponder towards Eagle. Its VHF-ranging apparatus acquired Eagle at a range of 250 nautical miles, but the lock was intermittent. However, when Collins spotted the flashing beacon in the darkness he centred the cross hair of his sextant on it and instructed the computer to note the sighting. Eagle was to be the active partner in the rendezvous but if at any time it was unable to make a manoeuvre, Collins was ready to perform the 'mirror image' manoeuvre approximately one minute later, and take over the active role. The first rendezvous manoeuvre was calculated by Houston. The apolune of the initial orbit was on the far side. On nearing apolune, Eagle was to use its RCS thrusters to add 51.5 feet per second to lift the perilune. This coelliptic sequence initiation (CSI) manoeuvre gave an orbit of 46.1 by 49.5 nautical miles. On emerging from behind the Moon on revolution 26, the separation between the vehicles was 100 nautical miles and, since Eagle was lower and travelling faster, this was reducing at 99 feet per second. As a result of the perturbations by the mascons, Columbia's orbit was 56.8 by 63.2 nautical miles. About one hour after CSI, the constant differential height (CDH) manoeuvre was a radial burn of 45 seconds for a delta-V of 9.2 feet per second to make the orbit coelliptical with and ideally 15 nautical miles below the orbit of Columbia. As Eagle caught up, Columbia 'rose' progressively above the horizon. About 40 minutes after CDH, with the separation at 38 nautical miles, the elevation reached 27 degrees and Eagle made the terminal phase initiation (TPI) manoeuvre, a 25-foot-per-second burn designed to set up an interception. Eagle had been tracking Columbia by radar, but now Armstrong was to track it visually and make lateral adjustments in order to hold his target 'fixed' against the stars, thereby ensuring that although the vehicles were tracing different arcs around the Moon, he made a *straight-line* (i.e. inertial) approach to Columbia. Shortly after the burn, Eagle passed 'over the hill'. If things went to plan, by AOS on revolution 27 it would be station-keeping alongside Columbia.

The terminal phase finalisation (TPF) manoeuvre involved a series of small burns to refine the line of approach and, when the range was down to a few thousand feet, to brake in order to come to a halt within 100 feet of Columbia. It was a procedure perfected by the Gemini missions. The timing had been set to enable the terminal phase to occur in darkness, with Armstrong visually observing the flashing beacon on Columbia against the stars. The braking phase would occur just after emerging from the Moon's shadow, to enable him to see and manoeuvre towards his target. This is the reason that this part of the mission, like so many other critical events, had to occur out of communication on the far side of the Moon.

Thus far, Columbia had been oriented to face its lower equipment bay towards the approaching spacecraft to enable Collins to use the sextant, but now he turned to point his apex towards the newcomer, and watched it from the left couch using the optical reticle in window 2, with the 16-millimetre Maurer camera in window 4 to document the remainder of the approach. He was delighted to see that Eagle remained centred in his reticle. Early in mission planning, Aldrin had noted that if the ascent stage was oriented 'upside down' as it finished the approach, this would obviate solar glare. As Eagle drew to a halt about 50 feet from Columbia, Collins took a series of Hasselblad photographs to document the state of the ascent stage.

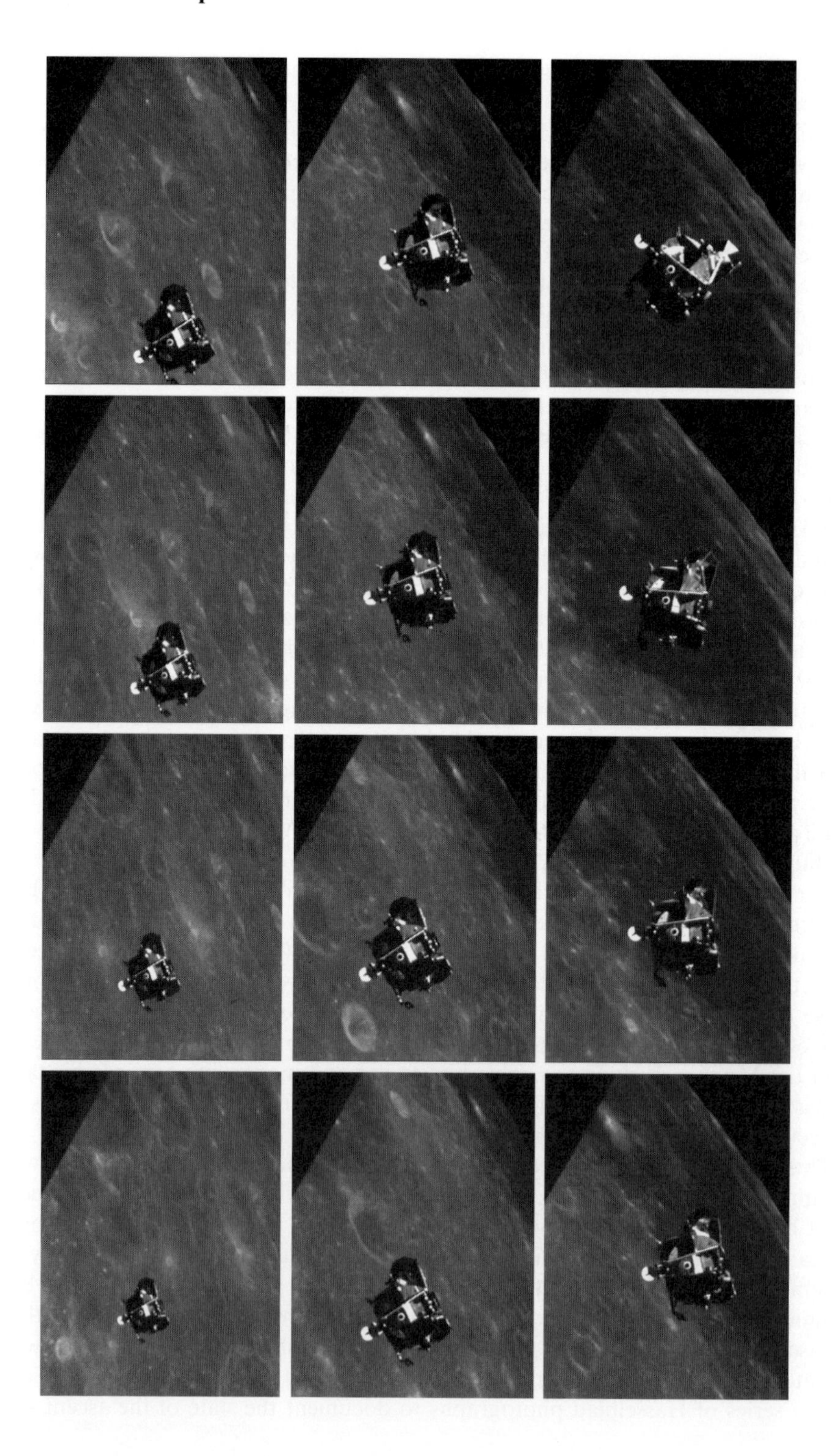

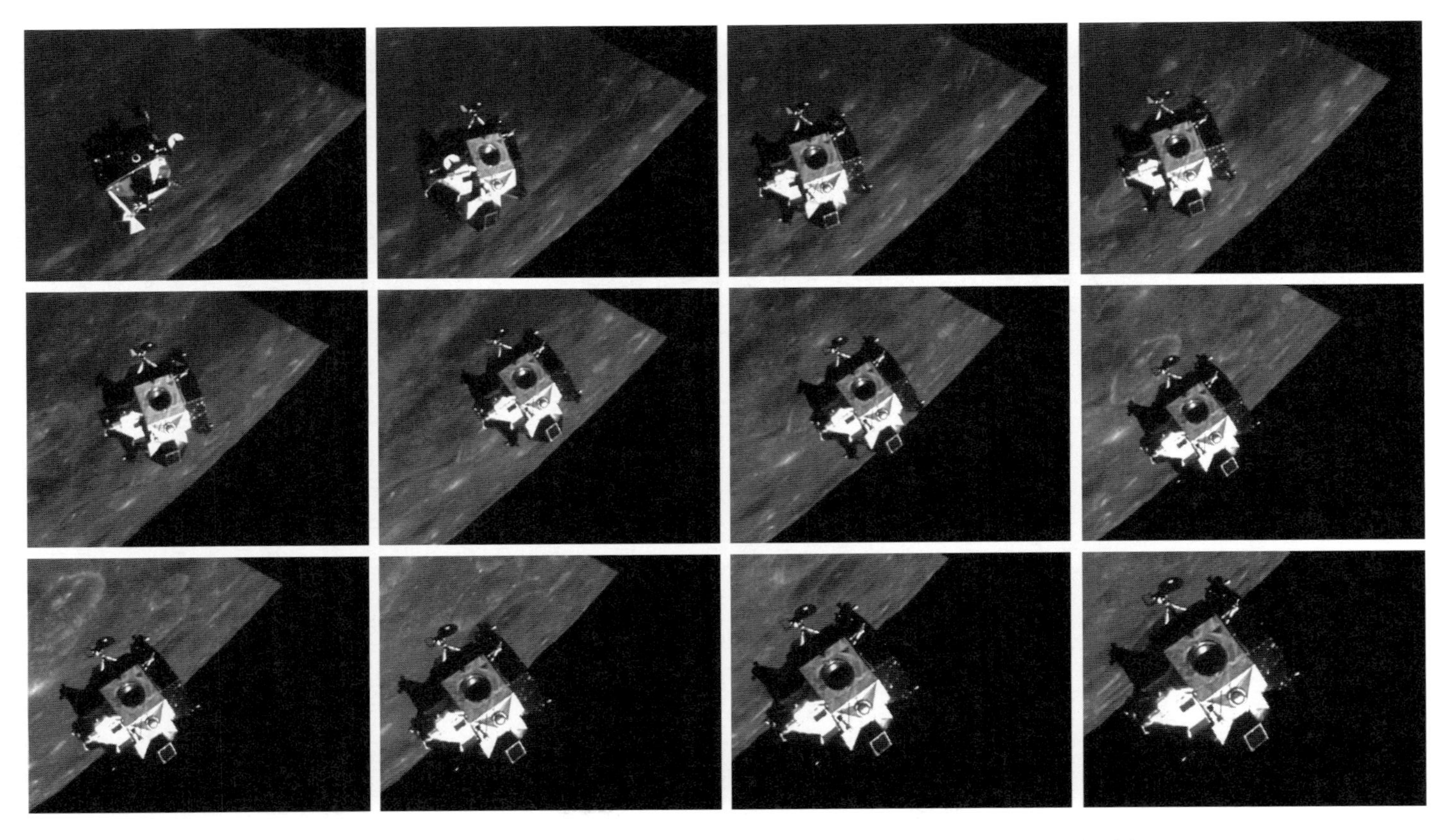

Frames from the 16-millimetre camera during Eagle's rendezvous with Columbia.

During the terminal phase, Eagle had drawn ahead of Columbia, and therefore as the vehicles jointly flew around the limb he was able to snap his companion with Earth on the lunar horizon.

"Eagle and Columbia, Houston. Standing by," Evans called to announce that communications had been restored.

"We're station-keeping," Armstrong replied, prompting sustained applause in the Mission Operations Control Room.

An exciting docking

Although Eagle had performed the rendezvous, it would be more convenient for Collins to undertake the act of docking since he had a better view. As Armstrong's attention was directed through the roof window in orienting his vehicle to face its docking system towards Columbia, he did not notice that the attitude indicator on the instrument panel showed the imminent onset of 'gimbal lock'. As this condition occurs when two or more of the three gimbals of the inertial platform become coaligned, rendering it ineffective, he switched from the PGNS to the AGS for an inertial reference. With Eagle's drogue facing the probe on the apex of Columbia stationed 25 feet away, he engaged Attitude Hold. Collins repeated the procedure he had used during the docking manoeuvre 5 days earlier. When the capture latches had engaged the drogue he commanded the probe to retract, and was concerned to see the ascent stage, now a lightweight 5,785 pounds, rapidly swing 15 degrees to the right (as he viewed it). He promptly fired his thrusters to counteract this dynamic activity, and restored the alignment just before the 8-second pneumatic retraction drew the two collars together and triggered the 12 latches for a 'hard' docking.

"That was a funny one," Collins called to Eagle. "You know, I didn't feel the shock and I thought things were pretty steady so I went to Retract, and that's when all hell broke loose." He had not felt the moment of contact, and was unaware that he was slightly off axis, with the probe scraping the side of the drogue. At contact, Armstrong thrusted towards Columbia. The resulting rotation prompted the flight control system, which was set in Minimum Deadband, to fire thrusters in an effort to hold attitude. To rectify this, Armstrong switched to Maximum Deadband and acted manually to re-establish the desired attitude. During the probe retraction sequence, therefore, *both* vehicles were actively attempting to regain a stable configuration! As Collins noted several hours afterwards, "No sooner had I fired that goddamned [gas] bottle, than wow, away we went." He really ought to have waited to establish that the vehicles were in a stable configuration but, as he pointed out, "I was in the habit of, you know, as soon as contact is made, I look at it, it looks okay, and I fire the bottle right away." When NASA released its latest increment of the transcript an hour after the docking, the remark 'all hell broke loose' had been elided; it was reinstated only when one of the reporters played his own audio tape to the Public Affairs Officer.

If the latches had failed to engage due to the collars being misaligned, Collins would have re-extended the probe, realigned, and tried again in the hope that the mechanism had not been damaged by the 'thrashing'. If this failed, or if for some reason the tunnel could not be opened, he would have retained Eagle in the 'soft' docked configuration on the end of the extended probe, and Armstrong and Aldrin would have made an external transfer using their OPS (which is why these were retained after the moonwalk) with each man dragging one of the rock boxes. There was a rail on Eagle to enable them to translate from the forward hatch to a position from which they could reach across to a rail beside Columbia's side hatch. Collins was already suited in case an external transfer proved necessary, and the centre couch was stowed to create an isle into the lower equipment bay to accommodate two suited newcomers and their cargo. There was also an external lamp on the LM to provide illumination in case the transfer was made in darkness. In the event, of course, none of these contingency measures were necessary.

Reunited

After docking, Armstrong and Aldrin released the harnesses and unstowed a small vacuum cleaner with which they were to clean the lunar dust from their suits, but the remarkable ability of the particles to adhere made this ineffective, and they instead resorted to brushing themselves by hand and then hoovering up the floating motes. Aldrin remarked that he did not think the tool would "be much of a competitor to the leading vacuum cleaner brands". Despite all efforts, their suits remained dark with dust. Meanwhile, Collins pumped up his cabin pressure to ensure that when the tunnel was opened the air flow would be into rather than out of Eagle's dirty cabin. Once the tunnel had been pressurised and both hatches were open, the weary moonwalkers re-entered Columbia. As each man emerged from the tunnel with a beaming grin on his face, Collins felt like grabbing him and kissing his forehead, but refrained. As he recalled later, "We shook hands, hard, and that was it. I said I was glad to see them, and they said they were happy to be back."

"Apollo 11, Houston," called Evans just before the spacecraft went 'over the hill' on revolution 27. "You're looking great. It's been a mighty fine day."

"Boy, you're not kidding," Collins replied.

On the far side of the Moon, Aldrin returned to Eagle to retrieve the rock boxes and the other items to be returned to Earth. As each of the 2-foot-long shiny metal sample return containers floated out of the tunnel, Collins zipped it into a white bag for stowage in a receptacle in the lower equipment bay. As he was later to recall, "I handled them as if they were absolutely jam-packed with rare jewels, which, in a sense, they were."

Armstrong gave Collins a small white cloth bag, "If you want to have a look at what the Moon looks like, you can open that up and look, but don't open the other bag."

Collins unzipped the cloth to find a small plastic bag containing dark powder. "What's in the bag?"

"The contingency sample."

"Any rocks?"

"Yes there's some rocks in it too. You can feel them, but you can't see them, they are covered with that dark –" Armstrong searched for the best word for the clingy powder " graphite."

Meanwhile, at home

At about this time, Jan Armstrong, who had Dee O'Hara with her, received a brief visit from NASA Administrator Thomas O. Paine, who expressed his admiration for the astronauts' achievements. "I still can't believe they have really been there," she admitted to him.

Meanwhile, Lurton Scott arrived at the Collins home with the 'hot news' that after their return to Earth the astronauts and their families were to receive parades in New York, Chicago and Los Angeles, whereupon they would have dinner with President Nixon. Pat Collins was stunned. On suggesting that the fuss would be "all over by Christmas", she was told that as a result of Apollo 11 her life would never be the same.

Table: Apollo 11 lunar orbit limb-crossing times

Revolution	AOS GET (h:m:s)	LOS GET (h:m:s)
0	–	075:41:23
1	076:15:29	077:41
2	078:23:31	079:48:30
3	080:33:21	081:45
4	082:32	083:44
5	084:30	085:42
6	086:28	087:40
7	088:27	089:38
8	090:25	091:36
9	092:23	093:35
10	094:21	095:34
11	096:20	097:32
12	098:18	099:30
13	100:16	101:28
14	102:15	103:27
15	104:15	105:25
16	106:11	107:23
17	108:09	109:21
18	110:08	111:19
19	112:06	113:17
20	114:04	115:16
21	116:02	117:14
22	118:00	119:12
23	119:58	121:10
24	121:56	123:09
25	123:55	125:07
26	125:53	127:05
27	127:52	129:03
28	129:51	131:02
29	131:49	132:59
30	133:46	134:57
31	135:34	–

Source: The flight plan and mission transcript.

EARTHWARD BOUND

While Apollo 11 was behind the Moon, Lunney handed over to Kranz to manage the transearth injection (TEI) manoeuvre. To reduce the spacecraft's mass for this burn, the urine bags and fecal garments, food packaging and other trash, including the probe, were dumped into Eagle. By the time they reappeared around the limb on revolution 28 they had closed Columbia's hatch and were venting the tunnel. At 129:40, shortly prior to finally departing Eagle, Aldrin, as per instructions, had switched the coolant system from the primary to the secondary loop. This left the PGNS without cooling. The PGNS and AGS had been precisely coaligned, and the high-gain antenna pointed at Earth. Eagle was left in Attitude Hold to enable the rate at which the two guidance systems diverged to be monitored, as a means of determining the degradation of the PGNS. It was an engineering experiment to find out how long the system would continue to function without coolant. Predictions ranged from 1 hour to several hours.[3] With Eagle in this state, they were eager to jettison it as soon as possible. In the flight plan this was scheduled for 131:52, but on realising that they were well ahead Duke suggested they do so at their convenience, and several minutes later, at 130:09, Collins fired the pyrotechnics to truncate the tunnel and discard the heavy docking collar with the ascent stage.

"There she goes", observed Armstrong appreciatively as Eagle floated away at a speed of several feet per second.

"It was a good one," added Collins.

Collins then reoriented Columbia and fired its thrusters for 7 seconds to achieve a 2-foot-per-second retrograde burn to move well clear of the discarded vehicle. He discussed with Duke the possibility of advancing the TEI burn by one revolution, but the flight dynamics team wanted more Manned Space Flight Network tracking to measure the post-separation orbit as a preliminary to calculating the manoeuvre. With the matter resolved, Duke said, "It looks like it's going to be a pretty relaxed time here for the next couple of hours."

"I imagine that place has cleared out a little bit since the rendezvous. You can find a place to sit down almost, huh?" said Collins.

"The MOCR's about empty right now. We're taking it a little easy. How does it feel up there to have some company?"

"Damn good, I'll tell you," Collins assured.

"I'll bet," said Duke. "I bet you'd almost have been talking to yourself up there after 10 revolutions or so."

[3] The platform began to be unusable after 4 hours, and the computer failed just over 3 hours later. Both items had operated for considerably longer than had been predicted. The other systems were still functioning. The last contact with Eagle was at 137:55, when the battery output dipped below that required for the AGS to maintain the vehicle's attitude within the antenna's requirements for communication with Earth. Although Eagle was released in an almost circular orbit, perturbations by the mascons would soon have caused it to strike the surface, but it is not known when or where this occurred.

"No, it's a happy home here," insisted Collins. A moment later he added, "It'd be nice to have company. As a matter of fact, it would be nice to have a couple of hundred million Americans up here to let them see what they're getting for their money."

"Well, they were with you in spirit anyway – at least that many. We heard on the news today that after you made your landing the *New York Times* came out with the largest headlines they've ever used in the history of the newspaper."

"Save us a copy," said Collins.

Shortly thereafter, Apollo 11 passed 'over the hill'. While on the far side, the crew had lunch.

"If you guys want some news, I can read it up," Duke offered, after they had reappeared on revolution 29.

"We'd be pleased to have it," replied Armstrong.

"To start off, congratulatory messages on the mission have been pouring into the White House from world leaders in a steady stream all day, including one on behalf of the Soviet cosmonauts." Collins hoped the celebrations would not prove to be premature! "Mrs Robert H. Goddard said today that her husband would've been so happy, 'He wouldn't have shouted or anything. He'd just have glowed. That was his dream, sending a rocket to the Moon.' People all around the world had many reasons to be happy about the Apollo 11 mission. The Italian police reported that Sunday night was the most crime-free night of the year. And in London, a boy who had the faith to bet $5 with a bookie that a man would reach the Moon before 1970 collected $24,000; now that's pretty good odds! You're probably interested in the comments your wives have made. Neil, Jan said about yesterday's activities, 'The evening was unbelievably perfect. It is an honour and a privilege to share with my husband, the crew, the Manned Spacecraft Center, the American public, and all mankind, the magnificent experience of the beginning of lunar exploration.' She was then asked if she considered the Moon landing the greatest moment in her life. She said. 'No, that was the day we were married.' And Mike, Pat said simply, 'It was fantastically marvellous.' Buzz, Joan apparently couldn't quite believe the EVA. She said, 'It was hard to think it was real until the men actually moved. After the Moon touchdown, I wept because I was so happy.' But she added, 'The best part of the mission will be the splashdown.' In other news – and there *was* a little bit – another explorer, Thor Heyerdahl, had to give up his attempt to sail the papyrus boat, *Ra*, across the Atlantic. The storm-damaged boat was abandoned about 650 miles from Bermuda. The speed of the craft had been reduced to about 25 miles a day, and Heyerdahl said the object of the voyage had not been to provide an endurance test for the crew."

"Thank you, Charlie," Armstrong acknowledged.

"You're welcome."

With the crew of Apollo 11 reunited, Walter Cronkite drew to a conclusion the CBS network's unprecedented 32-hour continuous coverage.

Several minutes before Apollo 11 went 'over the hill' for the final time, Kranz polled his flight control team and Duke relayed, "Apollo 11, Houston. You're Go for TEI."

"Thank you," acknowledged Collins.

"Go sic 'em," Duke urged.

On passing beyond the limb, the astronauts settled down to check and recheck the configuration for the burn.

"Okay. I'm ready to proceed now," Armstrong concluded.

"Do it!" confirmed Aldrin.

"Say, you guys," Collins mused, "is there anything you want to do?"

"We've got to make a star check yet," Armstrong pointed out.

In attempting to take the star sighting, Collins made a mistake that prompted the computer to report Operator Error, giving rise to considerable swearing on his part and some laughter from his colleagues. At sunrise, 2 minutes prior to the burn, he visually verified their attitude with respect to the lunar horizon as a final check that the computer knew what it was doing, "Yes. I see a horizon. It looks like we are going forward." This prompted further laughter.

"Shades of Gemini," observed Armstrong.

"It is important that we be going forward," Collins laughed. The spacecraft had to be pointing forward in order to boost out of lunar orbit. "There is just one really bad mistake you can make here."

"Let's see," recited Aldrin cheekily, "the motor points this way and the gases escape that way, thereby imparting a thrust that-a-way."

After the separation manoeuvre Columbia's orbit was 54 by 63 nautical miles, and by TEI it was 20 nautical miles ahead of Eagle and 1 mile below it. The mass of the CSM prior to the TEI burn was 36,691 pounds. The SPS engine was to ignite at 135:23:42 and deliver 20,500 pounds of thrust for 2 minutes 28 seconds, in the process consuming 10,000 pounds of propellant. Prior to the manoeuvre Columbia was travelling at 5,355 feet per second. It was to be accelerated by 3,285 feet per second to a speed of about 8,640 feet per second, or about 5,900 nautical miles per hour. Confidence was high. Such a manoeuvre had been successfully performed by both Apollo 8 and Apollo 10, and their own engine had worked perfectly twice while settling into lunar orbit. It was a very robust configuration but if it were to make a 'hard' start (as had occurred to other types of engine) the astronauts would be stranded in lunar orbit at best, and possibly killed outright in an explosion.

Collins took the left couch for the manoeuvre, with Armstrong in the middle and Aldrin on the right. The thrusters of two of the quads on the service module fired for 16 seconds to settle the propellants in the main tanks, then the SPS ignited.

"Burn! A good one," called Collins. "Man, that feels like *g*, doesn't it?"

"How is it, Mike?" Aldrin asked.

"It's really busy in roll, but it's holding in its deadband. It's possible that we have a roll-thruster problem, but even if we have, it's taking it out; there's no point in worrying about it," Collins reported. "One minute. Chamber pressure is holding right on 100 psi." "The gimbal looks good. The total attitude looks good. It's still a little busy." They passed the 2-minute mark. "When it hits the end of that roll deadband, it really comes crisply back." "Chamber pressure is falling off a little bit, to 96 or so. Now it's going back up. It's oscillating just a tad."

"Ten seconds left," called Armstrong.

"We don't really care about the chamber pressure any more," said Collins. "Brace yourselves. Standing by."

"It should be shutdown, *now*," Armstrong announced.

Collins allowed the SPS to run for 2 seconds longer than the planned duration and hit shutdown – just as the computer cut the engine. On checking the residuals, which proved negligible, Collins was delighted. "Beautiful burn; SPS, I love you; you are a jewel! Whoosh!"

On completing the post-TEI checklist, Armstrong asked, "What time's AOS?"

"I haven't the foggiest," Aldrin replied.

"It's 135:34," Collins indicated.

"That's right now!" Armstrong noted. "Has anybody got any choice greetings they want to make to Houston?"

"Hey," said Aldrin, "I hope someone's getting the picture of the Earth coming up." Their final Earthrise was a welcome sight. All loose objects had been stowed for the manoeuvre, but they hastily retrieved the Hasselblad.

Acquisition of signal was to be handled by Honeysuckle Creek. If the burn had not occurred then the spacecraft would not appear until 135:44. The detection of the carrier signal at the scheduled time told the story, but Houston had to wait for the high-gain antenna to slew around and lock on before they could communicate.

"Hello Apollo 11," Duke called. "How did it go?"

"It's time to open up the LRL doors, Charlie," Collins replied, referring to the Lunar Receiving Laboratory in which they were to be quarantined.

"Roger," acknowledged Duke. "It's well stocked."

Coasting

On its approach to the Moon, Apollo 11 had been within the Moon's shadow and the astronauts been able to see only its dark leading hemisphere, but their line of departure was over the illuminated trailing hemisphere. In contrast to when they left Earth, and were preoccupied by retrieving Eagle from the spent S-IVB stage, they were now free to act as tourists and, on the basis that there was little point returning with unexposed film, they snapped away. They were amazed by the rate at which the Moon shrank before their eyes. But, as after the translunar injection manoeuvre, they slowed down as the range increased, and by the time they were 1,000 nautical miles out they had slowed to 6,698 feet per second.

"How does that tracking look?" Collins enquired.

"We've projected your onboard vector forward and it looks real good," replied Duke. "We've only got about 24 minutes of tracking now. It's really too early to tell on the radar." The Manned Space Flight Network did not yet have sufficient data to plot an accurate course.

"Understand," said Collins.

"Have you finished taking pictures?" Duke asked.

"We're just finishing up, Charlie," Collins reported.

The post-TEI tasks were to do a P52 to verify the alignment of the platform, upload a new state vector and switch to the REFSMMAT for the transearth coast (which, as for the translunar coast, was relative to the ecliptic) and then set up the PTC roll. As Armstrong and Aldrin had had no real sleep for 40 hours and were exhausted, Kranz wanted them to retire as soon as possible. Deke Slayton called in order to emphasise this point, "This is the original CapCom. Congratulations on an outstanding job. You guys have really put on a great show up there, but I think it's time you powered down and got a little rest; you've had a mighty long day. I'm looking forward to seeing you when you get back. Don't fraternise with any of those bugs *en route* except for the *Hornet*." The USS *Hornet* was the prime recovery vessel.

"Thank you, boss," Armstrong replied. "We're looking forward to a little rest and a relaxing trip back."

"You've earned it," Slayton assured.

"It's shift change time, here," Duke announced shortly thereafter. "The White Team bids you good night. We'll see you tomorrow."

"Good night, Charlie," Armstrong replied. "Thank you."

"Adios," Duke signed off. "Thanks again for a great show, you guys."

Windler's Purple Team took over for the 'graveyard' shift, and then handed over to Charlesworth's Green Team. The flight plan allotted 10 hours to the sleep period, but because the wake-up time was not critical it was decided to let the crew sleep through and, in fact, with each man gaining 8.5 hours of solid sleep, this was the best 'night' of the mission. Aldrin was the first to stir, at noon on Tuesday, 22 July, with his colleagues joining him several minutes later. At that time Apollo 11 was 32,253 nautical miles from the Moon and receding at 4,303 feet per second.

"Good afternoon, Houston," announced Collins, making the first call of flight day 7.

McCandless responded straight away with a list of chores, and the news that they would have to make a midcourse correction. If the trajectory resulting from the TEI manoeuvre would not enter the 36-nautical-mile-wide entry corridor, the flight plan had options for three midcourse corrections. Collins would later write that returning to Earth was as demanding as "trying to split a human hair with a razor blade thrown from a distance of 20 feet". Midcourse correction 5 was set for TEI + 15 hours. The only issue of concern in Mission Control was tropical storm Claudia, which was threatening to encroach on the prime recovery zone.

When at 148:07:22 Apollo 11 crossed the neutral point, 33,800 nautical miles from the Moon, Collins jokingly asked that off-duty flight dynamics officer Phil Shaffer be advised that the spacecraft "gave a little jump" as it left the Moon's sphere of influence.

Collins halted PTC, oriented the spacecraft for midcourse correction 5 and then took a star sighting to verify the platform. At 150:30, when 169,000 nautical miles from Earth and travelling at 4,076 feet per second, thrusters on all four quads were fired in unison for 11 seconds for a 4.8-foot-per-second retrograde burn designed to centre the trajectory in the entry corridor. Afterwards, Collins reinstated the PTC roll and the crew settled down for lunch, during which McCandless put questions drawn up by the geologists.

Table: Apollo 11 propulsive manoeuvres

Manoeuvre type	Propulsion system	GET (h:m:s)	Duration (s)	delta-V (fps)
TLI	S-IVB	002:44:16.2	347.3	10,441.0
Post-TD&E evasion	CSM/SPS	004:40:01.0	3.4	19.7
MCC-2	CSM/SPS	026:44:57.92	2.91	20.9
LOI-1	CSM/SPS	075:49:49.6	362.1	2,917.5
LOI-2	CSM/SPS	080:11:36.0	17.0	158.8
Post-undocking separation	CSM/RCS	100:39:50	8.2	2.6
DOI	LM/DPS	101:36:14.1	29.8	76.4
PDI	LM/DPS	102:33:04.4	712.6	6,775.8
Ascent	LM/APS	124:22:00.0	439.9	6.070.1
CSI	LM/RCS	125:19:34.7	47.0	51.5
CDH	LM/RCS	126:17:46.0	18.1	19.9
TPI	LM/RCS	127:03:30.8	22.8	25.3
TPF	LM/RCS	127:45:54	28.4	31.4
Post-jettison separation	CSM/RCS	130:30:00	7.1	2.2
TEI	CSM/SPS	135:23:42.0	150.0	3,278.8
MCC-5	CSM/RCS	150:29:54.5	10.8	4.8

Source: Post-Launch Mission Operation Report, 24 July 1969 (M932-69-11)

"For $64,000, we're still trying to work out the location of your landing site, Tranquility Base." McCandless provided the coordinates of a position and asked them to check it on their charts, but as these had been stowed he described it. "The position I gave you is slightly west of that large crater – I would guess about 0.2 kilometre to the west – and we were wondering if Neil or Buzz had observed any additional landmarks during descent, lunar stay, or ascent that would confirm or disprove this." Armstrong said they were sure that the descent and ascent movies would resolve it. A few minutes later, in discussing how he had collected rocks, Armstrong volunteered that after they had deployed the EASEP he had taken "a stroll to a crater behind us that was maybe 70 or 80 feet in diameter and 15 or 20 feet deep". On hearing this, Shoemaker's team were certain that they had correctly identified the landing site. As Armstrong had been out of the field of view of the television camera at that time, his excursion to the crater came as a surprise to all concerned, because the plan had been for them to remain within about 100 feet of the LM. As Shoemaker noted, "Had Neil told us about the small crater behind the LM, we could have pinpointed them right then, within 40 to 60 feet." The geologists were ecstatic when Armstrong said that he had taken pictures of the interior of the crater, which "had rocks in the bottom of pretty good size, considerably bigger than any that were out on the surface". His brief observations provided the very welcome insight that because there was no bedrock outcropping from the walls of the crater, the impact that excavated it had, as he put it, "not gotten below the regolith", and this set the lower limit for the thickness of the regolith on that part of the Sea of Tranquility. Although the presence of the crater undoubtedly complicated the landing, the fact that it provided a 'window' into the subsurface made its inspection the most significant single observation of the entire surface activity.

Charlesworth handed over to Kranz, and Duke took over the CapCom console. By design, the flight plan was sparse. The highlight, a telecast, was scheduled to start at 8 pm in Houston. The 85-foot-diameter antenna at Goldstone was used to receive the signal because the 210-foot was occupied communicating with an unmanned spacecraft that was approaching the planet Mars.

"Are you picking up our television signal?" Armstrong enquired, as they began the transmission with a view of the Moon.

"That's affirmative," said Duke. "The focus is a little bit out. We see the Earth in the centre of the screen, and see some landmasses in the centre, at least I guess that's what it is."

"I believe that's where we just came from," Aldrin corrected.

"It is, huh?" said Duke, embarrassed. "Well, I'm really looking at a bad screen here. Stand by one. Hey, you're right."

"It's not bad enough, not finding the right landing spot, but when you haven't even got the right planet!" Collins teased.

"I'll never live that one down," Duke mused.

Collins used the zoom to make the disk shrink in order to emphasise that they were leaving it behind, "We're making it get smaller and smaller here to make sure that it really is the one we're leaving."

"All right. That's enough you guys," Duke insisted.[4]

"That's enough of the Moon, Charlie," announced Armstrong. "We're getting set up now for some inside pictures." The camera turned to Armstrong, who showed off the rock boxes in the lower equipment bay. "We know there's a lot of scientists from a number of countries standing by to see the lunar samples, and we thought you would be interested to see them. These two boxes are the sample return containers. They're vacuum-packed containers that were closed in a vacuum on the lunar surface, sealed, and then brought inside the LM and put inside these fibre-glass bags, zippered, and resealed around the outside and placed in these receptacles in the side of the command module. When we get onto the ship, I'm sure these boxes will immediately be transferred for delivery to the Lunar Receiving Laboratory. In them are the samples of the various types of rock, the groundmass of the soil, the particle collector for the Solar Wind Experiment and the core tubes that took depth samples of the lunar surface."

"Well, we appreciate it," Duke replied.

Next, Aldrin demonstrated the preparation in weightlessness of a sandwich with ham paste spread, and then he spun the empty can to illustrate the action of the gyroscope. Picking up this theme, Collins demonstrated for "all kids everywhere" the behaviour of blobs of water. After showing Duke a view of Earth, Armstrong signed off with the observation that, "No matter where you travel, it's always nice to get home."

[4] On subsequent missions, crews would tease Duke about this misidentification.

A view of Earth during the transearth coast, showing specular reflection.

After the telecast Collins teased Duke concerning the lax schedule, "The White Team's really got a busy one tonight, huh, Charlie?"

"Boy," said Duke, taking the bait. "We're really booming along here with all this activity. I can barely believe it."

"What are you doing? Sitting around with your feet up on the console drinking coffee?"

Duke laughed, "You must have X-ray eyes. You sure can see a long way."

A few minutes later, Collins was back, "I'm just wondering how everything is going on the home front?"

"Everything's going fine. All the gals are having a little party tonight, as far as I know."

"Glad to hear it," Collins said.

After another several hours, Duke announced, "Apollo 11, it's good night from a sleepy White Team."

Aldrin replied, "Thank you, very much. We're not as sleepy tonight as we were last night."

After the Black Team, the Green Team took over and again opted to leave the crew in peace until telemetry showed them to be active, and then Garriott, standing in for McCandless, put in the call to begin flight day 8, Wednesday, 23 July: "Are you up and at 'em yet?"

"Well, we're up, at least, Owen," Armstrong replied.

Of course, Garriott launched straight into the flight plan updates, the main one of which was the cancellation of midcourse correction 6, which meant that they could remain in PTC, but as this had become "a little bit ragged" Collins was asked to refine it. Houston updated their state vector and recommended a P52 to check the platform alignment.

When communications lapsed, Garriott called, "Apollo 11, Houston."

"Go ahead," replied Armstrong.

"I just wanted to make sure you fellows hadn't gone back to sleep again!" But he continued, "I have a bit of news here if you'd like to find out what's been happening in the last 12 to 14 hours."

"Go ahead," Armstrong invited.

"Hot off the press: Juan Carlos was formally designated yesterday, Tuesday, to become General Franco's successor as the Chief of State of Spain and, eventually, King. He will be sworn in today as the successor designate after taking an oath of loyalty to the law and the National Movement, which is Spain's only legal political organisation. He'll be called the Prince of Spain." When this elicited no response Garriott went on, "Here, the House Ways and Means Committee agreed yesterday to tax changes affecting oil companies, also banks and utilities, which could add as much as 2 billion dollars per year to the federal revenue. The committee also voted tentatively to change the accounting procedures for telephone, electric, gas, and oil pipeline companies, and to reduce the tax benefits of the mutual savings and loan institutions. So, it looks as if tax reform may be on the way." Garriott allowed for a response, but there was none. "South Korea's first super highway, linking Seoul with the Port of Inchon, has been named the Apollo Highway to commemorate your trip.

I think we mentioned last night that President Nixon has already started on his round-the-world trip. Today, he's in San Francisco on his first stop. Next he will go to the USS *Hornet*, from which he'll watch the return of your spacecraft. He plans to visit seven nations during this trip." He paused again, then continued, "The West Coast residents in Seattle, Washington; Portland, Oregon; Vancouver, British Columbia; and San Francisco, California, all plan to make themselves visible to you by switching on their lights between 9 pm and midnight tonight. There is clear weather predicted there, so you may be able to see."

"Good show," interjected Armstrong.

Garriott continued. "In Memphis, Tennessee, a young lady who is presently tipping the scales at 8 pounds 2 ounces was named Module by her parents, Mr and Mrs Eddie Lee McGhee. 'It was my husband's idea', insisted Mrs McGhee. She said she had balked at the name 'Lunar Module McGhee' because it didn't sound too good, but apparently they've compromised on just 'Module'." This prompted some laughter on the downlink.

When Collins said his breakfast of "sliced peaches, sausage patties, two cups of coffee and I forget what else" was "magnificent as usual", Garriott said he was "way overdue for a meal" himself, and soon thereafter he handed over to McCandless.

Final telecast

At 6 pm in Houston, Apollo 11, having halted its PTC roll in order to maintain the high-gain antenna pointing towards Earth, made its final telecast. Since each man wished to make a specific point, they used cue cards.

"Good evening," began Armstrong. "This is the Commander of Apollo 11. One hundred years ago, Jules Verne wrote a book about a voyage to the Moon. His spaceship, 'Columbiad', took off from Florida and landed in the Pacific Ocean after completing a trip to the Moon.[5] It seems appropriate to us to share with you some of the reflections of the crew as the modern-day Columbia completes its rendezvous with the planet Earth and the same Pacific Ocean tomorrow. First, Mike Collins."

"This trip of ours to the Moon may have looked, to you, simple or easy," said Collins to introduce his theme. "I'd like to assure you that *that* has not been the case. The Saturn V rocket which put us into orbit is an incredibly complicated piece of machinery, every piece of which worked flawlessly. This computer above my head has a 38,000-word vocabulary, each word of which has been carefully chosen to be of the utmost value to us, the crew. The switch I have in my hand now has over 300 counterparts in the command module alone. In addition to that, there are myriads of circuit breakers, levers, rods and other associated controls. The SPS engine, the large rocket engine on the aft end of our service module, must have performed flawlessly

[5] In fact, Armstrong was in error because Columbiad was the name of the giant cannon that fired Verne's spaceship to the Moon; the ship did not have a name, always being referred to simply as "the projectile".

or we would have been stranded in lunar orbit. The parachutes up above my head must work perfectly tomorrow or we will plummet into the ocean. We have always had confidence that all this equipment will work properly, and we continue to have confidence that it will do so for the remainder of the flight. All this is possible only through the blood, sweat and tears of a number of people. First, the American workmen who put these pieces of machinery together in the factory. Second, the painstaking work done by the various test teams during assembly and retest after assembly. And finally, the people at the Manned Spacecraft Center, both in management, in mission planning, in flight control and, last but not least, in crew training. This operation is somewhat like the periscope of a submarine. All you see is the three of us, but beneath the surface are thousands and thousands of others, and to all of those, I would like to say: Thank you, very much."

"Good evening," called Aldrin. "I'd like to discuss with you a few of the more symbolic aspects of the flight of our mission, Apollo 11. As we've been discussing the events that have taken place over the past two or three days here on board our spacecraft, we've come to the conclusion that this has been far more than three men on a voyage to the Moon; more, still, than the efforts of a government and industry team; more, even, than the efforts of one nation. We feel that this stands as a symbol of the insatiable curiosity of all mankind to explore the unknown. Neil's statement the other day upon first setting foot on the surface of the Moon, 'This is a small step for a man, but a giant leap for mankind', I believe, sums up these feelings very nicely. We accepted the challenge of going to the Moon; the acceptance of this challenge was inevitable. The relative ease with which we carried out our mission, I believe, is a tribute to the timeliness of that acceptance. Today, I feel we're fully capable of accepting expanded roles in the exploration of space. In retrospect, we have all been particularly pleased with the call signs that we very laboriously chose for our spacecraft: Columbia and Eagle. We've been particularly pleased with the emblem of our flight, depicting the US eagle bearing the universal symbol of peace from the Earth, from the planet Earth, to the Moon; that symbol being the olive branch. It was our overall crew choice to deposit a replica of this symbol on the Moon. Personally, in reflecting on the events of the past several days, a verse from Psalms comes to my mind, *When I consider the heavens, the work of Thy fingers, the moon and the stars, which Thou hast ordained; What is man that Thou art mindful of him?*"

The view returned to Armstrong, "The responsibility for this flight lies first with history and the giants of science who have preceded this effort; next with the American people, who have, through their will, indicated their desire; next with four administrations and their Congresses, for implementing that will; and then, with the agency and industry teams that built our spacecraft, the Saturn, the Columbia, the Eagle, and the little EMU, the space suit and backpack that was our small spacecraft out on the lunar surface. We'd like to give a special thanks to all those Americans who built the spacecraft, who did the construction, design, the tests, and put their hearts and all their abilities into those crafts. To those people, tonight, we give a special thank you, and to all the other people that are listening and watching tonight. God bless you. Good night from Apollo 11."

They concluded the telecast with a brief shot out of the window showing Earth from a distance of 91,371 nautical miles.

"I thought that was a mighty fine television presentation," said McCandless. "There's certainly nothing I can add to it from down here."

With the spacecraft still perpendicular to the ecliptic, Collins reinstated the roll for PTC.

Meanwhile, at home

Wednesday was a quiet day for the wives of the crew, the principal event being to attend a luncheon hosted by North American Rockwell. While there, they received a telephone call from President Nixon, who was in San Francisco on his way to the Pacific to welcome their husbands home following splashdown.

ENTRY AND SPLASHDOWN

As the Green Team concluded its last shift of the mission, McCandless wished the astronauts "Godspeed". The White Team was to work a long shift, during the latter part of which the crew would sleep, until the Maroon Team arrived for the re-entry phase. Duke pointed out that owing to thunderstorms in the recovery zone, the aim point would be relocated eastward to where conditions were better. The main risk in the original recovery zone was air turbulence. Just as aircraft avoid such 'air', so too must a returning capsule. The command module's aerodynamic lift was to be used to extend the entry 215 nautical miles farther downrange towards Hawaii. The forecast for the new recovery zone was for scattered cloud, 16-to-24-knot winds, 2-to-4-foot waves and a 5-to-7-foot swell, excellent visibility, and, most importantly, little or no turbulence.

"I'll wish you good night from the White Team for the last time," announced Duke when it was time for the crew to retire for their final sleep period. "It has been a pleasure working with you guys. It was a beautiful show from all three of you. We appreciate it very much, and we'll see you when you get out of the LRL." At that time, Apollo 11 was 74,906 nautical miles from Earth and approaching at 6,954 feet per second. As soon as the telemetry indicated that the astronauts were asleep, the flight controllers teased Duke by projecting onto one of the Eidophor screens a cartoon about his misidentification of the Moon for Earth on the previous day's telecast. It showed the spacecraft midway between the Moon and Earth with the words, 'Neil, I just spotted a continent on the Moon' and 'Charlie, the camera's on Earth now'.

To start flight day 9, Thursday, 24 July, Kranz handed over to Windler, who was to handle the final phase of the historic mission. It was decided to allow the crew an extra hour's sleep, but at 189:30, almost 8 am in Houston, Armstrong called to ask whether they would require to make midcourse correction 7 and was told that their trajectory was satisfactory. Apollo 11 was now 40,961 nautical miles from Earth, had accelerated to 9,671 feet per second, and was less than 6 hours from atmospheric entry.

"While you're eating your breakfast there," Evans called, "I have the Maroon Bugle with the morning news. President Nixon surprised your wives with a phone call from San Francisco just before he boarded a plane to fly out to USS *Hornet* to meet you. They were very touched by your television broadcast yesterday; Jan and Pat watched from Mission Control here. Wally Schirra has been elected to a 5-year term on the board of trustees of the Detroit Institute of Technology. He'll serve on the Institute's development committee. Air Canada says it has taken over 2,300 reservations for flights to the Moon in the last 5 days. It might be noted that more than 100 were by men for their mothers in law! And finally, it appears that rather than killing romantic songs about the Moon, you've inspired hundreds of song writers. Nashville, Tennessee, which probably houses the largest collection of recording companies and song publishers in the country, reports being flooded by Moon songs. Some will make it. The song at the top of the best sellers list this week is, 'In the year 2525'."

"Thank you, very much, Ron," Armstrong acknowledged.

Meanwhile, in the Aldrin home, Joan's father, Michael Archer, was preparing the champagne for the splashdown party – although the best vintage had been put aside for when the astronauts emerged from quarantine. The shooting of fireworks was illegal in Texas other than on the Fourth of July, Christmas and New Year, but the local fire department had issued a waiver to permit a display in the back yard after splashdown.

With about 3 hours to go, Deke Slayton and the backup crew of Lovell, Anders and Haise joined Evans at the CapCom console.

"This is Jim, Mike," called Lovell. "The backup crew is still standing by. I just want to remind you that the most difficult part of your mission is going to be after recovery." He was referring to the quarantine.

"Well, we're looking forward to all parts of it," Collins replied.

"Please don't sneeze," Lovell implored. If, after having been away for so long, they were to show any symptoms at all upon their return, then this would cause the ever-worrying medics to presume immediately that they had contracted 'lunar flu'.

"Keep the mice healthy," Collins said. There were hundreds of white mice in the Lunar Receiving Laboratory, and if even one of them were to show an unusual symptom on being exposed to lunar material, it would surely result in calls for the quarantine to be extended.

As Earth's gravity continued to draw them in, the planet loomed with amazing rapidity. "The Earth is really getting bigger up here and, of course, we see it as a crescent. We've been taking pictures. We have four exposures left and then we'll pack the camera," Collins pointed out.

Finally ready, Collins strapped into the left couch, Armstrong in the centre and Aldrin on the right. They were to fly re-entry in 'shirt sleeves', their pressure suits having been stowed under their couches.

"This is your friendly backup CMP," Anders called. "Have a good trip, and do remember to come in BEF." The acronym BEF stood for 'blunt end forward' – he was reminding them to enter the atmosphere with the heat shield on the base of the capsule facing the direction of travel.

"You better believe it," Collins replied. They were triple-checking every step of the checklist. In preparing to jettison the service module, he mused, "That old service module has taken good care of us; it's been a champ." He made a visual check of the horizon, "The horizon check passes; it's right on the money." As the spacecraft entered Earth's shadow, he noted the time, "The Sun's going down on schedule."

In view of the predicted surface winds and prevailing sea state, each astronaut took an anti-motion-sickness pill. At 3,000 nautical miles from Earth, Apollo 11 was travelling at 26,685 feet per second. Over the ensuing 20 minutes it would be accelerated by almost 10,000 feet per second. As on departing Earth, its trajectory and speed now enabled it to pass through the van Allen belts without significant exposure to radiation. Because the trajectory was west-to-east, the final treat for the Carnarvon station of the Manned Space Flight Network was to monitor the approach. The service module was jettisoned at an altitude of 1,288 nautical miles and a speed of 31,232 feet per second. Its trajectory would cause it to burn up like a meteor, with perhaps just a few small fragments falling into the ocean. Collins used the command module's own thrusters to face its heat shield in the direction of travel. At launch, Columbia had weighed 65,000 pounds, but most of that had been propellant. Alone now, the mass of the command module was just 11,000 pounds.

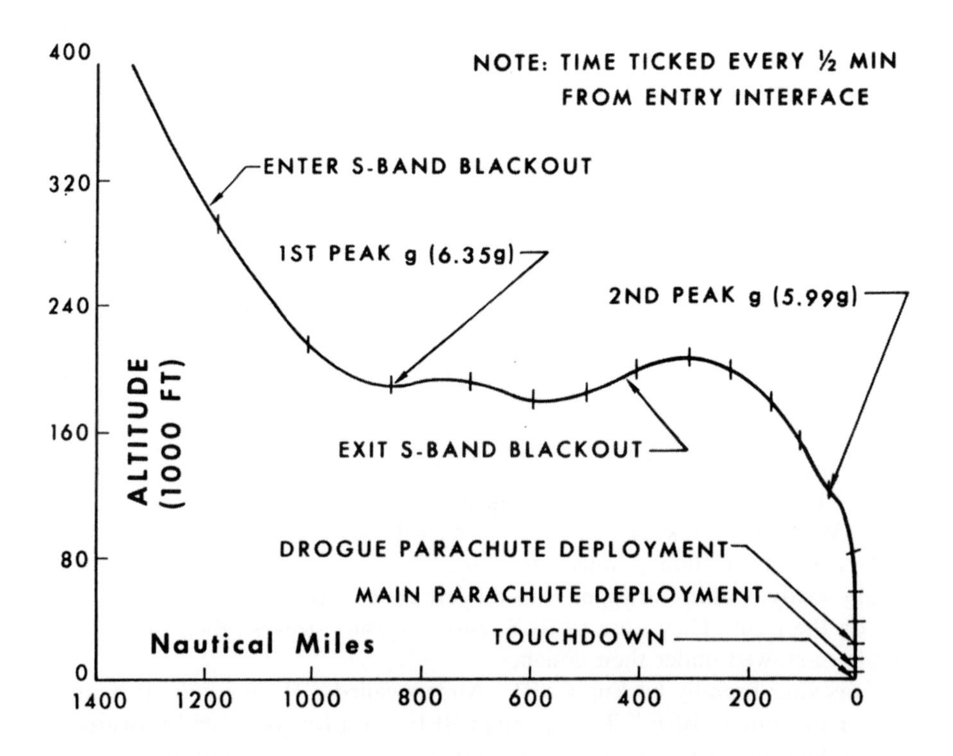

A plot of altitude versus range to splashdown for the nominal entry profile.

"You're looking mighty fine from here. You're cleared for landing," Evans confirmed

"Roger," Collins acknowledged. "Gear's down and locked."

Continuing the aviation theme, Windler advised the Recovery officer, "We're on final for the carrier."

With 7 minutes to go to atmospheric entry, the spacecraft was at an altitude of 800 nautical miles and had accelerated to 33,000 feet per second. Early in the development of the Apollo program, it was decided to use a 'double dip' profile for atmospheric entry on a high-speed return from cislunar space so as to minimise the heating effects. After the entry interface at 400,000 feet, the capsule was to make a shallow penetration of the atmosphere to shed a significant fraction of its energy, then utilise its limited aerodynamic lift to climb back to pursue a 'skip' prior to descent and landing. The nominal entry corridor for the approach was inclined 6.5 degrees below horizontal. If the capsule were to come in significantly shallower, it would bounce off the atmosphere like a stone skimming on a pond, and the astronauts would die when their consumables ran out some hours later. If the entry was centred in this corridor, the peak deceleration would be 6.5 *g*, increasing with the angle. If they were to enter too steeply the capsule would burn up. As a result of the midcourse correction on the transearth coast the angle was –6.48 degrees. The speed at the entry interface was 36,237 feet per second; for a spacecraft returning from low orbit it would have been only 25,000 feet per second. The computer navigated the 'extended' 1,500-nautical-mile profile by adjusting the exit angle from the 'dip' in order to lengthen the 'skip', with Collins monitoring, ready to intervene in the event of a problem.

"You'll be going over the hill shortly. You're looking mighty fine to us," said Evans.

"See you later," Armstrong replied. If things were to go seriously pear-shaped as a result of the failure of part of the entry system, this would be the last thing heard from the crew.

First contact with the atmosphere occurred over the Solomon Islands, east of Australia. The 0.05-*g* light illuminated at about 300,000 feet to indicate that air drag was beginning to decelerate the capsule. Aldrin had placed the 16-millimetre Maurer camera in his window to document their luminescent 'wake', which initially was an inner core of orange/yellow, surrounded by patches of violet, blue and green, beyond which was the black of space.

As the capsule compressed the tenuous upper atmosphere, it formed a shock wave that rapidly raised the temperature of the heat shield to 2,870°C. An Air Force Airborne Lightweight Optical Tracker System KC-135 was on station to photograph the re-entry. Free electrons in the ionised gas prevented radio communications, but a tape was recording telemetry and crew comments. The black-out was expected to last no more than 3 minutes. As the wake intensified, the glow illuminated the cabin like daylight, even though the re-entry was in Earth's shadow. Having been weightless, the deceleration was punishing, but it did not last long. The black out ended on time. Collins was relieved when their velocity fell below orbital, because it meant that they were guaranteed to come down

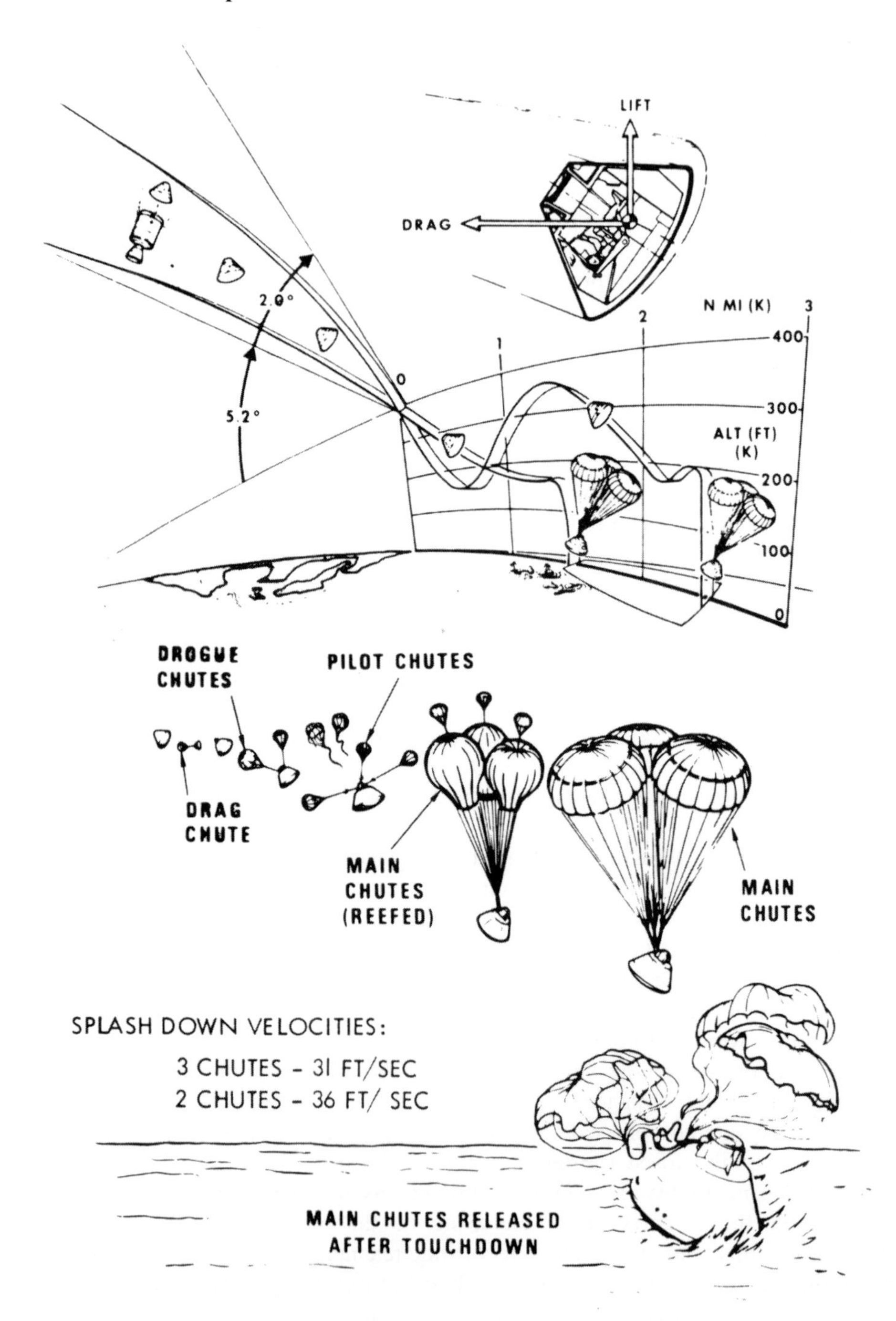

The Apollo spacecraft's Earth Landing System.

somewhere. The recovery force was at the nominal aiming point of 169°W, 13°N, some 850 nautical miles southwest of Honolulu.

At 24,000 feet the forward heat shield was jettisoned to expose the apparatus of the Earth landing system. There was a jolt as mortars deployed the two reefed 16.5-foot-diameter drogue parachutes. Houston made no attempt to communicate, preferring to leave the 'air' clear for the recovery forces. An aircraft reported that it had S-Band contact. Once the drogues had stabilised and decelerated the capsule to a rate of descent at which it was safe to deploy the main parachutes, the drogues were jettisoned and mortars deployed three pilot chutes which, in turn, pulled out the 83-foot-diameter main chutes that were reefed to produce a gradual inflation in three stages, reaching full inflation as the capsule descended through 10,500 feet; only two chutes were necessary for a water landing, the third represented redundancy.

Leading the nine-ship recovery force was the 30,000-ton aircraft carrier USS *Hornet*. It had been commissioned in 1943, served in the Pacific, was modified after the war, and then served off Korea and Vietnam. On Wednesday President Nixon had flown *Air Force One* from San Francisco to Hawaii, and then on to Johnston Island, where he flew by helicopter to the communications ship, USS *Arlington*, which sailed overnight to join the recovery force, finally transferring to *Hornet* an hour before Apollo 11 was due to arrive.

"Apollo 11. This is *Hornet*," called the ship just after hearing a sonic boom.

"Hello, *Hornet*," replied Armstrong. "This is Apollo 11 reading you loud and clear."

One of the Sea King helicopters reported, "Swim One has a visual dead ahead, about a mile."

"300 feet," announced Armstrong, reporting the altimeter reading.

"Roger. You're looking real good," replied Swim One.

The capsule struck the water at 30 feet per second. The couches were mounted together on a frame set on shock absorbers in order to cushion the impact, but the three men grunted at the sudden deceleration.

"Splashdown!" announced Swim One.

The capsule was 15 nautical miles from *Hornet*. The 'extended' entry profile put it 40 seconds behind the flight plan. It was 7.51 am in the local (Hawaiian) time zone. As sunrise was 10 minutes away, the illumination was poor for television.

The capsule was a truncated cone about 10 feet tall and 13 feet in diameter across its base. Its centre of gravity in water meant that it would readily flip over into an apex-down orientation. In view of the 18-knot winds, immediately after splash Aldrin was to insert a circuit breaker alongside his right elbow, and Collins was then to throw a switch to jettison the parachutes to preclude their dragging the capsule over. Armstrong had bet Collins a glass of beer that he would not make it. The shock of impact forced Aldrin's poised hand away from the panel, and by the time he had relocated and pushed in the breaker the wind had caught the parachutes and had flipped the capsule over, leaving the crew hanging in their straps. With a distinct sense of 'up', but one with dark green water in the windows, the cabin suddenly became a very unfamiliar place! The waves were just 3 to 6 feet, but to the astronauts the sea state felt much rougher, and they endeavoured to resist the onset of sea

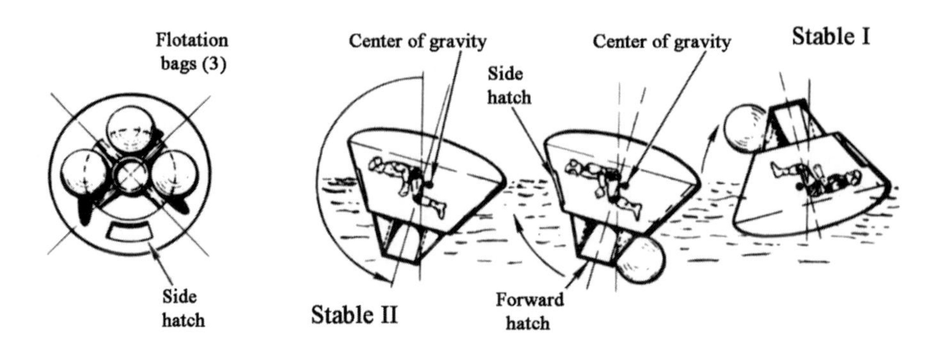

If the CM becomes inverted after splashdown, its offset centre of mass combined with strategic flooding and the inflation of flotation bags enables it to be righted.

sickness while a trio of air bags distributed around the apex inflated to right the capsule – a process that took 8 minutes. As soon as they could, each man took a second anti-motion-sickness pill. As soon as the capsule had flipped upright and exposed its antennas, Armstrong reported to the Air Boss, circling in one of the helicopters, "Everyone inside is okay. Our checklist is complete. Awaiting the swimmers."

QUARANTINE

In July 1964 a conference organised by the Space Science Board of the National Academy of Sciences highlighted the potential hazard of 'back contamination' to Earth from lunar missions, and urged the development of preventative measures. On 8 December NASA asked the Board to submit recommendations. On 23 May 1965 the Life Sciences Committee said that astronauts returning from the Moon must be quarantined for at least three weeks; this period being chosen because it exceeded the incubation of most terrestrial germs. On 14 June Robert R. Gilruth, Director of the Manned Spacecraft Center, set up a committee chaired by Edwin Samfield to oversee the design of the Lunar Sample Receiving Laboratory, and on 1 November formed the Lunar Sample Receiving Laboratory Office as an interim organisational unit. On 13 May 1966 George E. Mueller, Director of the Office of Manned Space Flight, renamed the facility the Lunar Receiving Laboratory (LRL). On 28 July, Gilruth was authorised to issue the contract for its construction. The three-storey building was to have a suite of offices, a vault to store lunar material in vacuum conditions, a laboratory to process samples in isolation chambers, and a human quarantine facility. On 1 August Persa R. Bell was appointed as director. The installation and testing of the internal systems continued through 1968. A review in November 1968 prompted some changes, as did a 30-day simulation in March 1969, but on 5 June the LRL was declared operational. Although Bell saw the likelihood of anything harmful being returned from the Moon as "probably one in one-hundred-billion", the consequences of an unfortunate outcome might be dire. Michael Crichton's novel *The Andromeda Strain*, published in 1969 and a Book Club recommendation for June, served only to heighten public awareness.[6] On the basis that any bugs that could survive on the Moon would *thrive* on Earth, a 21-day quarantine starting the day of the moonwalk was a compromise between those who thought quarantine was unnecessary and those who argued for many months.

However, a chain is only as strong as its weakest link and for quarantine this was the opening of the spacecraft hatch after splashdown. An option would have been for the crew to remain in the capsule until it was hoisted onto a recovery ship which carried the requisite isolation facilities, but as there was no guarantee that the capsule would splash down in proximity to a such a ship, and it was impracticable for the crew to remain in the capsule for a lengthy period awaiting its arrival, it was decided to develop a Biological Isolation Garment, known as a BIG, to be worn during the normal recovery procedure.

After a member of Underwater Demolition Team 11 jumped from Sea King tail number 64, call sign 'Swim Two', and attached a sea anchor to prevent the capsule from drifting, three more members of the team attached a flotation collar to stabilise it against the swell, emplaced two large rafts that were tied to the collar, and then inflated another raft, boarded it and moved off upwind of the capsule. At this point, 25-year-old Lieutenant Clancey Hatleberg from Chippewa Falls, Wisconsin, jumped into the water with four BIGs, one for each of the astronauts and one for himself. Joseph P. Kerwin, a physician recruited as a scientist–astronaut in 1965, had assisted in the design of the grey-green rubberised suit. It had a zipper running diagonally from lower left to upper right, and a hood incorporating a mask with an air filtration system projecting from the chin. Once in the raft located immediately in front of the hatch, Hatleberg put on his suit. Having written his checklist inside his face mask using a grease pencil ('VENTS TAPE INFLATE') he reached up to the apex of the capsule to confirm that the air vent valves were closed – otherwise the biological isolation would be compromised. He then removed the tape from the filters of the garments intended for the crew. Finally, he inflated the attached life preservers. Since the hatch was capable of being opened regardless of pressure differential, the internal pressure had been set lower than ambient in order to ensure that when (some 30 minutes after splash) Hatleberg opened the hatch to pass in the suits the airflow would be inwards to limit the scope for 'bugs' to escape. He immediately closed the hatch and washed it down with a disinfectant. While Aldrin donned his suit on the right couch, first Armstrong and then Collins did so in the lower equipment bay, and despite the capsule being stabilised by the flotation collar they were distinctly wobbly on their feet. Hatleberg communicated with them by way of hand-signals through the hatch window. When they were ready, the crew scrambled out as rapidly as possible, first Armstrong, then Collins and finally Aldrin, and Hatleberg closed the hatch and again disinfected it. As the suit had no facility for communication, no

[6] The widely acclaimed film of Crichton's novel was not released until 1971.

Having decontaminated the astronauts, Clancey Hatleberg transferred everyone to the recovery raft to await helicopter pick up.

words were spoken, but the entire procedure had been rehearsed many times in training. Furthermore, because the suits had no provision for ventilation, the men soon overheated and their face masks misted over, reducing their already limited visibility.[7] All four men wiped each other using cloths saturated with chemicals that were presumed to be lethal to lunar bugs. The disinfecting kit was then weighted and tossed into the sea. Despite the suits having been designed for isolation, in-seeping water made the rubber sticky and uncomfortable. The astronauts moved to the adjacent raft to await retrieval. With the Sun having risen and *Hornet* now less than 1 mile away, the television coverage of the recovery operation was excellent. Sea King tail number 66, call sign 'Recovery One', hoisted the astronauts on board one at a time (the order of their doing so being unclear, since the suits were identical). Earlier in the space program a simple 'horse collar' harness through which an astronaut slipped his arms had been employed, but this device had been superseded by a Billy Pugh net. Named after its inventor, this was a plastic and metal basket, open at one side, into which a man could climb for the hoist to the helicopter. Once the astronauts were clear, the swimmers returned in their raft, disinfected themselves, sank the decontamination raft, and prepared the capsule to be hoisted onto the ship.

The astronauts were greeted in the helicopter by flight surgeon Bill Carpentier, who deflated their life preservers. To reacclimatise to Earth gravity, Collins did a series of deep knee bends. Aldrin joined him. Armstrong did not. The flight to the *Hornet* lasted about 10 minutes. Watching from the island structure were President Nixon, Secretary of State William Rogers, NASA Administrator Thomas O. Paine and Frank Borman. As it landed, the ship's band was playing on deck. Unlike their predecessors, the Apollo 11 crew remained in the helicopter and rode the elevator down to the hangar deck. They detached the mission patches from their BIGs and gave them to the aircraft crew as souvenirs. When the door was opened, Armstrong led the way out, holding onto the rail of the steps for stability. Despite their misted face masks, they were able to see sufficiently well to make their way unaided the short distance to the Mobile Quarantine Facility (MQF), on the way waving to the dimly perceived crowd. By this time, the Mission Operations Control Room was all flags and cigars, many of the VIPs having left the viewing gallery in order to join the celebrations on the main floor, which was standing-room only. While the telecast from *Hornet* was shown on one Eidophor, another screen displayed an Apollo 11 patch and 'Task Accomplished ..... July 1969'. The MQF was a 9-foot-wide 35-foot-long gleaming aluminium commercial travel trailer, minus its wheels, that had been modified for NASA by Airstream and American Standard to provide biological containment. One end was hinged to provide the main door. John K. Hirasaki, the engineer in charge of the facility, welcomed the astronauts and Carpentier on board, then joined them. After the astronauts had doffed their rubber suits, Carpentier

[7] Note that there was a presumption that the astronauts would not get sea sick while wearing their suits, as the mask would have to have been removed in order to vomit, which would have violated the isolation.

Thomas O. Paine (foreground) and Richard Nixon on USS *Hornet* watch the recovery of the Apollo 11 crew.

The astronauts walk from the recovery helicopter to the MQF.

Management Row of the Mission Operations Control Room celebrates the safe recovery of Apollo 11.

'Task Accomplished' is displayed in the Mission Operations Control Room.

checked their blood pressure, temperature, respiration and heart rate. Having taken turns to shower in the utility compartment at the opposite end of the trailer, they donned blue flight suits bearing an Apollo 11 mission patch, a NASA 'meat ball' insignia, and a white pin-on-badge proclaiming 'HORNET PLUS THREE'. They then gathered at the door and drew back the curtain of the full-width window, above which was mounted a sign with 'HORNET + 3'.

As Nixon made his appearance the band struck up *Ruffles and Flourishes*. At the microphone in front of the MQF door, with television cameras running, Nixon, welcomed the crew home – his reputation for cold aloofness yielding to genuine delight. After the initial banter, he said, "I was thinking as you came down and we knew it was a success, it's only been eight days – just a week, a long week – but this is the greatest week in the history of the world since the Creation because, as a result of what has happened this week, the world is bigger infinitely, the world has never been closer together before. And we just thank you for that. And I only hope that all of us in government, all of us in America – that as a result of what you've done – we can do our jobs a little better. We can reach for the stars, just as you have reached so far for the stars."[8] The astronauts, somewhat bewildered, could only say 'Yessir' at the appropriate times.

The welcome over, the astronauts drew their curtain and were given additional medical tests, including riding an ergometer for 18 minutes each while hooked up to monitoring equipment and breathing through a tube to analyse their exhalation, in order to check their physical condition; this program was to be repeated every 24 hours to monitor their recovery. By the time the medicals were over, Nixon's party had left *Hornet* by helicopter. Later, the capsule was retrieved and installed alongside the MQF. When a plastic tunnel had been strung from Columbia's hatch to the MQF, Hirasaki retrieved the rock boxes, film and other essential items and passed them out through a decontamination airlock for air-freight to Houston. The US Navy was 'dry', but when Carpentier pointed out that by their watches (set on Houston time) it was drinking time, they broke out bottles of scotch, bourbon and gin. The trailer had a galley, and a lounge with six comfortable seats and a table. Hirasaki cooked steak with baked potato. Everyone then retired to the two-level bunks on each side of the main compartment, slept for 9 hours, and awoke late. Meanwhile, the rock boxes, the film, and the blood samples drawn immediately upon the astronauts' arrival were divided into two packages and flown to Hawaii by separate helicopters, just in case one was forced to ditch in transit. A Customs official cleared them for immediate entry to US territory. They were then loaded onto separate Air Force C-141 Starlifter transports for passage to the mainland. Paine accompanied the first consignment.

Meanwhile, the astronauts watched recorded Apollo 11 television coverage. On seeing the crowds of people clustered to watch the moonwalk, Aldrin observed to his

[8] In saying the Apollo 11 mission had been "the greatest week since the Creation", Nixon drew criticism in the press for neglecting the significance of Jesus Christ!

commander, "Neil, we missed the whole thing!" To help to 'wind down', Armstrong and Collins undertook a marathon game of gin rummy. On their second evening, Collins visited the capsule and used a pencil to inscribe on the wall above the sextant: "Spacecraft 107 – alias Apollo 11 – alias Columbia. The best ship to come down the line. God Bless Her. Michael Collins, CMP."

On the morning of the third day, Saturday, 26 July, 55 hours after splashdown, *Hornet* steamed into Pearl Harbor, near Honolulu, Hawaii. During a private visit to the MQF, Admiral John Sidney McCain Jr, the Commander in Chief of the Pacific Fleet, told the astronauts they were "lucky sons of bitches". "I would have given anything to go [to the Moon] with you." The ship's crane hoisted the MQF onto a flat-bed truck. With what seemed to be half the population of Honolulu lining the route, the truck took over an hour to drive the 10 miles to Hickham Field, where the MQF was loaded on board a C-141 for a non-stop 6-hour flight to Ellington Air Force Base. On the way, the astronauts signed a stack of pictures that were to be given to people they thought would appreciate a memento, and then they napped. On landing at Ellington just after local midnight there was an hour's delay before the MQF was able to be offloaded – it took three attempts, prompting Carpentier to observe wryly, "They can send men safely to the Moon and back, but they can't get 'em off the damned airplane!" There were welcoming speeches by Louie Welch (the Mayor of Houston) and Robert Gilruth, and an opportunity for their wives to speak over an intercom – all of which was televised, of course. At 2.30 am, after a 1-hour drive, the truck delivered the MQF to the 'garage' of the Lunar Receiving Laboratory, and once this was sealed the astronauts were finally able to leave the utilitarian trailer and take up residence in the living quarters of the quarantine area, which included a kitchen, a dining room, a recreation room, separate bedrooms for each of the astronauts, and a dormitory for everyone else. Ten people were already in quarantine, including two cooks (also described as a chief steward and steward), a doctor serving as a laboratory specialist, a Public Affairs Officer and a NASA photographer. The astronauts' quarantine had actually begun at the moment Eagle landed on the Moon, so they would require to spend only the next 14 days in the LRL. The people who were to be confined with them had already spent 21 days in isolation to enable any bugs *they* might be carrying to be identified, and thereby preclude these spreading and being mistaken for lunar infections.

The film was to be subjected to 48 hours' decontamination in an autoclave. Because a test of this device several days previously had ruined a film, the first Apollo 11 film to be processed was one taken in space, rather than on the lunar surface. If this were to be ruined in processing, the other film would be retained for the full quarantine period and then processed in the normal manner. But all was well. When *Life* magazine got its copy of the Hasselblad film of the moonwalk, its editors set out to find the best photograph of Armstrong to put on the cover of a 'special issue', and were astonished to find that there was not even one.[9] Thus, although

[9] There are several Hasselblad pictures of Armstrong on the lunar surface, but he is in shadow and it was some time before his presence on these frames was noted.

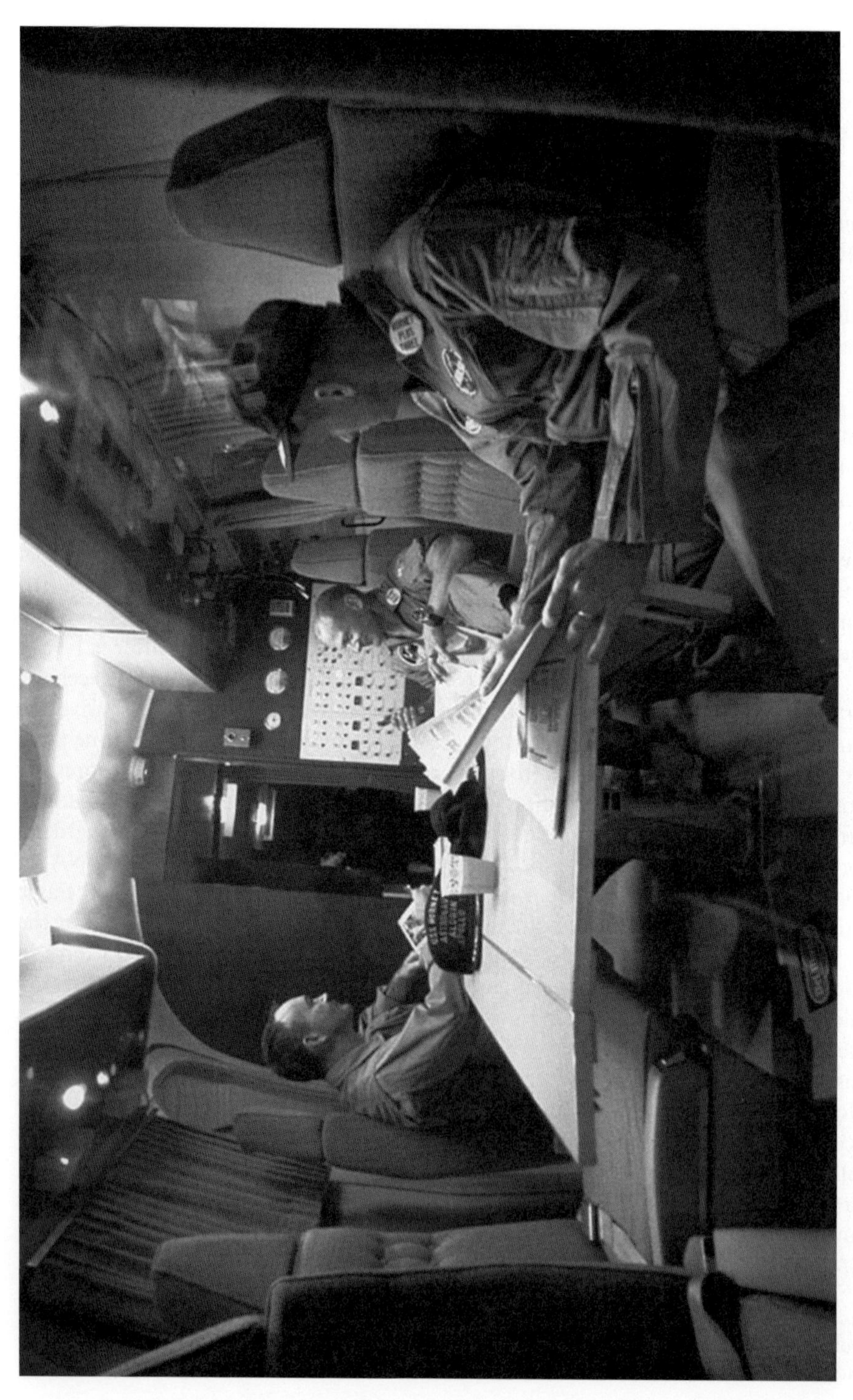

The Apollo 11 crew inside the MQF.

On the arrival of the MQF at Ellington Air Force Base, the Apollo 11 astronauts were greeted by their wives.

Armstrong entered the history books as the first man to set foot on the lunar surface, the iconic image of the mission was of Aldrin.

The main technical task for the astronauts in quarantine was a debriefing, in which they methodically relived the various phases of the mission. In addition to taped and filmed interviews through a window with colleagues in a room beyond the quarantine area, each man wrote detailed reports. When it was transcribed, the oral debriefing ran to some 500 single-spaced pages. To jog their memories, they had their annotated flight plans, and were able to view the pictures and movies. Some of the debriefing was conducted by men assigned to future missions, seeking tips about landing, such as how visibility in the hovering phase was impaired by the dust displaced by the engine plume. On 29 July, Armstrong and Aldrin were told where Eagle had actually landed: the 16-millimetre movie taken by the camera mounted in Aldrin's window enabled Tranquility Base to be located at 0°41′15″N, 23°25′45″E – almost precisely where the geologists had placed it.[10] Although the debriefing was time-consuming, time was the one commodity that the astronauts had in abundance. Armstrong welcomed the seclusion, but his crewmates found it tedious. They were able to receive visits from their families, but only through the glass partition, and on 5 August Armstrong celebrated his birthday in quarantine. It came as no surprise to be informed that, on orders from the White House, they were to spend the next several months undertaking national and international 'goodwill' tours.

Deke Slayton visited one day and suggested that they give some consideration to whether they wished to remain in the 'rotation' for future missions. Although at this point NASA hoped to fly missions through to Apollo 20, to rotate through backup to prime would be tedious, and besides there were other pilots eager for missions.[11] Armstrong, however, was in the same 'trap' as John Glenn, who, on the suggestion of the White House, had been grounded after his Mercury mission on the basis that, as a national icon, he was too valuable to risk on another flight. During a T-38 flight from Houston to the Cape in May, Collins had said to Duke that if Apollo 11 was successful he intended to retire. On telling his wife Pat, she had urged him to refrain from making this official until after the mission, just in case he should have a change of heart. In fact, several days before launch Slayton had asked Collins if he would like to rotate through Apollo 14 backup for the prospect of commanding Apollo 17, but he replied that he would end with Apollo 11. Having walked on the Moon, there was little chance of Aldrin being allowed to go again. If he wished to fly again this would have to be to visit the Skylab space station, which was scheduled for launch no earlier than 1972.[12] Thus, by the end of quarantine all three were ready to stand

[10] The aim point was at 0°42′50″N, 23°42′28″E.

[11] Actually, as Apollo 11 was heading home, NASA decided to withdraw one Saturn V from the lunar program in order to launch the Skylab space station, but this had not yet been announced.

[12] This is what Pete Conrad and Al Bean did after walking on the Moon on Apollo 12. Their CMP, Dick Gordon, remained in the lunar program in the hope of commanding Apollo 18, but this flight was cancelled.

down. They were not alone. Of the Apollo 8 crew, only Jim Lovell was still active, and having backed up Apollo 11 he was in line to command Apollo 14. Of the Apollo 9 crew, Dave Scott was backing up Apollo 12 in the expectation of commanding Apollo 15. Of the Apollo 10 crew, both John Young and Gene Cernan were in the rotation and hoping for their own commands.

As soon as the lunar material arrived in the LRL, samples were pulverised and 'fed' to germ-free mice, birds, fish, insects and plants in an effort to induce some sort of a reaction. As Collins had joked prior to the mission when this procedure was explained to the crew, "Let's hope none of those mice die!" The merest hint of an infection would have caused the quarantine period to be extended, potentially in an open-ended manner. Indeed, on observing that Aldrin's core temperature was now consistently several degrees higher than normal, some of the medics suggested that he should remain in isolation until it could be explained, but Carpentier rejected this, and Berry concurred.[13] As none of the mice developed any symptoms, at 9 pm on Sunday, 10 August, Armstrong, Aldrin and Collins were released and driven away in NASA staff cars, each chased by a television truck, and on arriving home were met by reporters. The solitude of quarantine was over, and they were public property again.

TOURING

On Tuesday, 12 August, Armstrong, Aldrin and Collins presented a 2-hour press conference in the packed Auditorium of the Manned Spacecraft Centre. In addition to answering questions from reporters, they provided spontaneous commentary as photographs were projected and the 16-millimetre movies were played.

At 5 am the next day the three families, together with a flock of Public Affairs people, boarded an aircraft of the Presidential Fleet at Ellington Air Force Base. The first stop was New York, landing at La Guardia, where they were greeted by Mayor John Lindsay and flown by helicopter to the city for a motorcade. The astronauts rode in one open limosine, their wives in a second, and their children in a third. Thousands of people lined the streets. Others waved from windows. Flags were ubiquitous, and ticker tape rained down from the skyscrapers. The cheering in the canyons between the buildings was incredible. At City Hall, Lindsay formally welcomed his guests, and then each astronaut delivered what was to be the first of many speeches to enthralled audiences around the world. After keys to the city had been presented, there was a short drive to the United Nations building and another round of speeches, followed by a helicopter flight back to the airport. When their aircraft landed in Chicago they were received by Mayor Richard Daley, were given an even more rapturous motorcade, accepted more keys and returned to the airport

[13] Aldrin's core temperature would remain elevated for several months, apparently unrelated to the mission.

to fly on to Los Angeles where, after being welcomed to the City of Angels by Mayor Sam Yorty, they took a helicopter to the Century Plaza Hotel, where they were to spend the night. Having freshened up, the astronauts and wives (minus their children) were taken to the Presidential Suite, where they were received by Richard Nixon, his wife Patricia and their daughters Julie and Tricia. After this private welcome, Nixon led them all into a ballroom for a State Banquet arranged in their honour. The tables appeared to stretch as far as the eye could see. There were thousands of guests and, of course, the US television networks. Catching the astronauts unaware, Vice President Spiro T. Agnew presented each man with the Presidential Medal of Freedom, the nation's highest civilian honour. Also present were Gene Kranz and his wife Marta, and Steve Bales of the White Team. Bales accepted a Medal of Freedom on behalf of all the flight controllers. After dinner, the astronauts' families accompanied Nixon's party to a private room for a round of picture taking. Finally, after a very long day, they were shown to their rooms for some well-deserved sleep.

Houston welcomed the Apollo 11 crew home on Saturday, 16 August, with a parade and a star-studded night of entertainment at the Astrodome that featured a performance by Frank Sinatra, who sang *Fly Me To The Moon*.

On Saturday, 6 September, it was time for 'home town' visits. For Armstrong, this was Wapakoneta, Ohio, and for Aldrin it was Montclair, New Jersey. Although Collins's parents had purchased a house in Alexandria on the Potomac while he was a teenager, he did not consider this to be his home. Instead, he opted for New Orleans, Louisiana, the home of the congressman who had nominated him for West Point. On Monday, 15 September, they visited the US Post Office in Washington, DC, where they returned an envelope that bore a stamp drawn up to commemorate the mission, which they had 'cancelled' during the transearth coast. The next day, Tuesday, 16 September, they attended a joint session of Congress, where first Armstrong, Aldrin, and finally Collins read an address, each receiving a rousing ovation.

Planning for the Giant Step Apollo 11 Goodwill Tour began in early September and involved NASA, the White House and the State Department. Armstrong, Aldrin and Collins were advised by Frank Borman, the astronaut with the most experience of touring. An aircraft of the Presidential Fleet left Andrews Air Force Base near Washington on the morning of Monday, 29 September, flew to Ellington Air Force Base to pick up the astronauts, their wives, and their many support staff, and then set off on a hectic schedule, each stop of which involved an official welcome, a motorcade, a press conference, an official dinner or two, and the giving and receiving of gifts.

Italian coverage of the mission had celebrated the fact that Collins was born in Rome. When the tour reached Italy, Collins was summoned to his place of birth, where he unveiled a 3-foot marble plaque that bore an inscription which began: "In this house on 31 October 1930 was born Michael Collins, intrepid astronaut of the Apollo 11 mission", which was all very well, but then, unfortunately, continued with "first man on the moon". The special treat for his wife, a Catholic, was the audience with Pope Paul VI in the Vatican.

The Apollo 11 crew during their New York parade.

Chicago welcomes the Apollo 11 crew.

In Washington, the Apollo 11 astronauts unveil the commemorative stamp.

Neil Armstrong addresses a joint session of Congress.

When visiting Mexico City, the Apollo 11 astronauts donned sombreros.

The Apollo 11 astronauts and their wives receive an audience with Pope Paul VI at the Vatican.

On completing their Giant Step tour, the Apollo 11 crew visited Richard Nixon at the White House.

Table: Giant Step Apollo 11 Goodwill Tour

Date of visit	Place
29–30 September 1969	Mexico City, Mexico
30 September–1 October	Bogota, Columbia
1 October	Brasilia, Brazil
1–2 October	Buenos Aires, Argentina
2–4 October	Rio de Janeiro, Brazil
4–6 October	Las Palmas, Canary Islands
6–8 October	Madrid, Spain
8–9 October	Paris, France
9 October	Amsterdam, Holland
9–10 October	Brussels, Belgium
10–12 October	Oslo, Norway
12–14 October	Cologne, Bonn and Berlin, West Germany
14–15 October	London, United Kingdom
15–18 October	Rome, Italy
18–20 October	Belgrade, Yugoslavia
20–22 October	Ankara, Turkey
22–24 October	Kinshasha, Zaire
24–26 October	Tehran, Iran
26–27 October	Bombay, India
27–28 October	Dacca, East Pakistan
28–31 October	Bangkok, Thailand
31 October	Perth, Australia
31 October–2 November	Sydney, Australia
2–3 November	Agana, Guam
3–4 November	Seoul, South Korea
4–5 November	Tokyo, Japan
5 November	Elmendorf, Alaska
2–3 December 1969	Ottawa and Montreal, Canada

On 20 July 1970, to mark the first anniversary of the lunar landing, Armstrong, Aldrin and Collins flew to Jefferson City, Missouri, where Columbia was on show during its tour of the USA. By now the three men were growing apart, they were no longer a crew, just three amiable strangers who had made a brief, but momentous, journey together.[14]

[14] Columbia is now a permanent exhibit at the National Air and Space Museum in Washington, DC.

10

The Moon revealed

SEA OF TRANQUILITY

The Air Force C-141 Starlifter carrying NASA Administrator Thomas O. Paine and the first rock box landed at Ellington Air Force Base on Friday, 25 July 1969. Awaiting it were Samuel C. Phillips, the Apollo Program Director, Robert R. Gilruth, Director of the Manned Spacecraft Center, and George M. Low, Manager of the Apollo Spacecraft Program Office in Houston. Gilruth and Low posed for photographs holding the box, before taking it to the Lunar Receiving Laboratory on the campus of the Manned Spacecraft Center. The second box arrived later that day. The next day, a member of the 50-strong Preliminary Examination Team used a vacuum chamber with a window and rubberised 'arms' to raise the lid of the first box, and found the interior so coated with black dust as to make it impractical to say anything definitive about the contents! When the boxes were emptied, there was found to be 48 pounds of lunar material in the form of 20 individual rocks and a pile of fragments and grains. One by one, the rocks were cleaned for inspection. At a press conference on 28 July, Persa R. Bell, Director of the Lunar Receiving Laboratory, opined that the rocks had been "beautifully selected". Elbert King, the curator, announced that the first rock to be examined under a microscope appeared to be a granular igneous rock. Gene Shoemaker of the US Geological Survey suggested that it represented a lava flow. But this was only a first impression. Once the material had been catalogued, small samples were issued to 150 principal investigators who had spent years developing the means to subject such material to almost every possible kind of analysis. The investigations proceeded at such a pace that on 15 September NASA was able to announce the preliminary findings and, to follow up, on 4 January 1970 the agency hosted the first of what was to become an annual Lunar Science Conference.[1]

To Harold C. Urey, who favoured the 'cold' Moon theory in which the interior was uniformly composed of 'pristine' material, the dark plains were the result of

[1] These gatherings are now entitled the Lunar and Planetary Sciences Conferences.

George M. Low (left) and Robert R. Gilruth offload the first rock box from an aircraft at Ellington Air Force Base.

impact melting on a vast scale. While the astronauts were out on the surface, Urey had been concerned when Armstrong reported a vesicular rock, encouraged when Armstrong changed his mind, and dismissed Armstrong's later report of a rock he was sure was vesicular. Most of all, Urey was encouraged that they did not report finding the 'frothy vacuum lava' predicted by his leading rival, Gerard P. Kuiper, who favoured the 'hot' Moon theory in which the interior was differentiated and the dark plains were the result of upwellings of lava through fractures in the floors of major impact basins. The rocks proved to be a form of basalt rich in magnesium and iron (and therefore described as being 'mafic') which isotopic dating revealed to have crystallised some 3.84 to 3.57 billion years ago. In terms of texture, it was strikingly similar to terrestrial basalt. It was not impact melt. This meant that the Moon *had* undergone a process of thermal differentiation in which lightweight aluminous minerals had migrated up to the surface and the heavier minerals had sunk into the interior. The fact that some of this denser material had later been erupted indicated

30 January 1970
Vol. 167, No. 3918

SCIENCE

The contents list of the special issue of the journal *Science* devoted to the initial analysis of the Apollo 11 samples.

that the interior had remained 'hot' for a significant period. However, when compared to terrestrial basalt, the lunar variety was enriched in titanium. The titanium-bearing mineral, which was new to mineralogists, was named 'armalcolite', in honour of the astronauts.[2] The lack of oxidised iron meant that the lava was created in a reducing environment (i.e. one devoid of oxygen). The most striking fact was the total absence of hydrous minerals. The lunar basalt was also deficient in volatile metals such as sodium. The low-alkali (i.e. sodium-depleted) lava would have had an extremely low viscosity, which is why it flowed so readily, and why it left so few 'positive-relief' features. The Sea of Tranquility was evidently accumulated by episodic volcanism over a period of several hundred million years. The presence of *two* types of basalt implied either that there were separate reservoirs of magma or that the single source had undergone chemical evolution over time.

As Armstrong later reflected of the lunar surface, "My impression was that we were taking a 'snapshot' of a steady-state process in which rocks are being worn down on the surface of the Moon with time, and other rocks are being thrown out on top as a result of new events somewhere near or far away. In other words, no matter when you had visited this spot before – 1,000 years ago or 100 years ago, or if you come back to it 1,000,000 years from now – you'd see some different things each time but the scene would generally be the same." This was insightful. On the airless Moon there was little chemical erosion. Large impacts simply excavated bedrock, and this was progressively worn down by smaller impacts to produce the regolith, the majority of which was pulverised basalt. There was little meteoritic material. Many of the discrete samples proved to be regolith compacted by shock. When subjected to physical stress these 'regolith breccias' tended to fall apart. The 'glassy material' found in a small fresh-looking crater was regolith that had been heated and fused by a high-energy impact. This impact-driven weathering process was given the name 'gardening'.

There was a small residue of the regolith that was very different in character. On the basis of his analysis of chemical data provided by Surveyor 7, which had landed near the crater Tycho in the southern highlands in 1968, Shoemaker had predicted that 4 per cent of the regolith at the Apollo 11 site would comprise minuscule fragments of light-coloured rock – and this proved to be the case. This light rock was plagioclase feldspar. Terrestrial plagioclase is rich in sodium, but the Moon is depleted in sodium and the lunar variant had calcium, making it calcic-plagioclase. Some of the fragments were sufficiently pure to justify being called anorthosite, this being the name for a rock comprising at least 90 per cent plagioclase, but most were diluted with mafic minerals and therefore were more properly called anorthositic gabbro; like the material Surveyor 7 had analysed. Shoemaker's rationale for there being highland material in the regolith of the Sea of Tranquility was based on the manner in which the most recently formed highland craters splashed out 'rays' of

[2] The name 'armalcolite' was derived from the first letters of the astronauts' surnames. Some years later this mineral was found on Earth, too.

material. Regarding the highlands, it could now be inferred that the primitive crust was composed of anorthositic rock. At the Lunar Science Conference, J.A. Wood noted that if the 'exotic' fragments in the Apollo 11 regolith were indeed highland rock, then their density of 2.9 grams per cubic centimetre (in comparison to the 3.4 average for the Moon) meant that the heat generated by giant impacts during the accretion of the Moon from planetesimals had created a 'magma ocean' which later solidified to form the crust. This was a significant insight into early lunar history.

What a difference one brief field trip had made; its 'ground truth' had scythed through the long-held theories without consideration for the professional standing of their proponents. Previously minor players found themselves in the limelight by virtue of having been proved right. For example, in a paper published a few weeks prior to Apollo 11, Anthony Turkevich reported a study of data from Surveyor 5, which landed in the Sea of Tranquility in 1967, near where Apollo 11 was to try to land, and he predicted the astronauts would return with titanium-enriched basalt.

OCEAN OF STORMS

Apollo 11 had proved the ability of the LM to land on the Moon, but the fact that it came down off target was frustrating. The ability to land within about 1,000 feet of a specific point was a prerequisite to being able to undertake a planned geological traverse. After the flight dynamics team had devised a simple method to correct for the perturbations of the mascons, they were so confident that they reduced the size of the target ellipse. In addition, it was decided to cut the number of backup sites from two to one. There were five prime sites on the short-list for the first landing. The easterly ALS-1 and ALS-2 sites in the Sea of Tranquility had been backed up by ALS-3 in the Meridian Bay, with ALS-4 and ALS-5 in the Ocean of Storms in reserve against a major launch delay. It would have been natural to send Apollo 12 to one of these sites, but the conservative constraints had resulted in the choice of 'open' sites, and the geologists were eager to sample the ejecta of a sizeable crater. In fact, even before Apollo 11, the site selectors had re-examined sites rejected due to the *inconvenient* proximity of a crater, and listed them for a later mission. In the end, however, in order to convincingly demonstrate the ability to address a 'pin-point' target it was decided to land alongside an unmanned probe. The relaxation of the operational constraints allowed the reinstatement of the Surveyor 1 site (ALS-6) in the Ocean of Storms. However, because this was so far west that it did not permit a backup, it was decided instead to visit Surveyor 3 in the eastern Ocean of Storms. Originally designated 3P-9, this site became ALS-7. Pete Conrad and Al Bean landed their LM, 'Intrepid', within 600 feet of Surveyor 3 on 19 November 1969. On their first excursion they deployed the deferred ALSEP, and during a 3-hour traverse the next day they ranged 1,200 feet from home, collected samples, and cut parts off the Surveyor as trophies. Meanwhile, in orbit, Dick Gordon photographed the site being considered for the next mission.

A pre-mission investigation of the morphology of the craters at the Apollo 12 site predicted that the regolith would not exceed 6 feet in thickness, and that the large impacts would have excavated bedrock. Whereas breccias and basalts were represented equally by number in the Apollo 11 samples, just two of the 34 rocks returned by Apollo 12 were breccias. The crystalline rocks were coarser and more texturally diverse. In view of the fact that they contained less titanium, it appeared, on reflection, that the basalt of the Sea of Tranquility was unusually *enriched* in this element. This chemical variation confirmed that the dark plains were not from a single source. Indeed, the fact that *four* kinds of basalt were identified at the Apollo 12 site meant that there had been several distinct flows in this local area.[3] However, the crystallisation dates clustered within a fairly narrow window, which suggested that the extrusions were the result of partial melting of pockets of rock at shallow depth. The initial results were confusing, but it was immediately recognised that something profound had been discovered concerning early lunar history. The first measurement yielded an age of 2.7 ($\pm$0.2) billion years, which meant *a billion years* had elapsed between the extrusions in the Sea of Tranquility and the Ocean of Storms. The next result pushed this up to 3.4 billion years, but as the analyses continued the dates converged on 3.2 billion years. This 500-million-year span in ages for the lavas at the two landing sites indicated that the driving process had been persistent. Geochemist Paul W. Gast made a surprising discovery in the basalts, in the form of an abundance of potassium, phosphorus and some of the 'rare earth' elements. By linking their chemical symbols, Gast coined the label 'KREEP'. On trying to isolate this material, he realised that it was not present as a mineral. The term is an *adjective*, and it is more correct to describe the Ocean of Storms basalts as being KREEPy. By way of an 'instant science' explanation for the media, Gast suggested this chemical additive might have been picked up from the ancient crust that some scientists believed formed the 'basement' of the dark plains, and he even speculated that it might be associated with the putative light-toned basalt believed (by some) to be prevalent in the highlands, but when the material proved to be rich in radioactive elements, in particular thorium and uranium, it was realised that this could not be typical of the crust because the heat of radioactive decay would have prevented the crust from solidifying. This KREEPy additive became a mystery for a subsequent mission to resolve.

After being discarded, the ascent stage of the LM was deliberately de-orbited, and the ALSEP seismometer recorded the crust 'ringing' for nearly an hour with a signature quite unlike a terrestrial signal. At the Lunar Science Conference, Gary V. Latham, the principal investigator for the seismic instruments, noted it had been difficult to tell the difference between a moonquake and an impact until this strike had provided a point of reference, whereupon it was found that surprisingly few of the 150 seismic events on record were internal quakes. It seemed that the crust was brecciated to a depth of about 18 nautical miles, indicating that, after the crust had solidified, further impacts had churned this up to a considerable depth, forming a

[3] These could be characterised in terms of their terrestrial equivalents as olivine basalt, pyroxene basalt, ilmenite basalt and feldspathic basalt.

'megaregolith'. In order to probe to greater depths, it was decided that on future missions the spent S-IVB should be made to impact the Moon.

FRA MAURO FORMATION

The flight dynamics team felt sufficiently confident to further reduce the size of the target ellipse and reject the requirement that the landing site be free of terrain relief, to permit the next mission to tackle a more confined site in rougher terrain. In 1962 Gene Shoemaker and R.J. Hackman had issued a stratigraphic map of part of the Imbrium Basin's rim. In extending this map, R.E. Eggleton classified the peripheral hummocky terrain as ejecta from the Imbrium impact, and called it the Fra Mauro Formation. Although one geological unit, this terrain was distributed in isolated patches around the periphery of the basin. In terms of total area, it was the largest distinct stratigraphic unit on the near side. Contemporary understanding of lunar history was based on how the ejecta from the Imbrium impact had splattered across thousands of miles. Dating this impact was the single most important item on the lunar science agenda, as it would 'lock in' many other structures. It was not just a matter of learning about the Moon. The lunar basins indicated that the early Solar System was an extremely violent place. If the Moon had suffered such an intense bombardment so, too, must Earth. Studying the Moon would provide insight into the early history of our own planet. The terrestrial record of this age is missing, in part because of erosion but mainly because the crust is recycled by plate tectonics. The Moon, however, is so endogenically inert that its face has remained essentially unchanged for billions of years. The task was to find a crater in the hummocky Fra Mauro Formation which had a rocky rim, offered a safe line of approach from the east, and was within a mile of a landing site. A 1,200-foot-diameter pit situated 22 nautical miles north of the large crater Fra Mauro, south of the Imbrium Basin, was chosen. As a result of its shape, the 'drill hole' crater to be sampled was named Cone. The best landing site was on the undulatory plain 1,000 yards further west, but the target was set twice as far out in order to avoid the fringe of Cone's ejecta. So great were the results to be gained from this site that after Apollo 13 had to abort and make an emergency return to Earth, Apollo 14 was reassigned this site and the target moved to the optimal landing place.

On 5 February 1971 Al Shepard and Ed Mitchell landed their LM, 'Antares'. Following the pattern, they deployed their ALSEP on the first day and made the traverse on the second. Since the rocks were consolidations of shattered precursors (i.e. breccias) the analysis was rather more complicated than for previous missions. The primary objective was to date when the fragments had been bound together, in order to date the impact that applied the shock. This was achieved by exploiting the fact that the isotopic 'clocks' used to measure formation date are 'reset' when a rock is melted. This was not an issue for basalts from the dark plains, but the study of a breccia involved dating its individual clasts. The samples tended to cluster in two age ranges, one spanning the interval 3.96 to 3.87 billion years and the other spanning the interval 3.85 to 3.82 billion years. It was therefore inferred that the breccias formed around 3.84 billion years ago as ejecta splashed from the Imbrium impact. The older dates provided the formation ages of the rocks shattered by that impact. It had been hoped that samples taken from right on Cone's rim would characterise the basement on which the Fra Mauro Formation resided, which was expected (by some) to be volcanic. At first, several intriguing samples did look as if they might represent such volcanism, but they proved to be the first instances of another type of breccia. In fact, there proved to be many forms of breccia. The terms 'fragmental breccia' was coined for clasts of shattered rock bound up in a matrix of pulverised rock. As further samples were studied, it was found that fragments of individual minerals could become bound into breccias, showing that not all clasts were lithic. Also, since breccias themselves could be caught in impacts, there were 'breccias of breccias' in which the clasts of one breccia were fragments of earlier breccias, and the term 'one-rock' and 'two-rock' were coined to reflect this history. The samples initially thought to be volcanic were a type of breccia in which clasts were bound in impact-melt.[4] Despite the violence of the shock-melting, the breccias contained very fragile crystals that could only have been formed by diffusion as mineral-rich vapour escaped from the ejecta. This crystallisation process was very similar to sulphur encrustation of volcanic vents on Earth, but in this case the gas was released by the ejecta itself rather than from the ground on which the ejecta sat, indicating that the rubble was hot when it was deposited and then fused as it congealed. Intriguingly, the impact-melt breccias proved to be KREEPy. Analysis revealed that they were originally a gabbro (i.e. a basalt that solidified deep underground rather than on the surface) that derived from the magma ocean. In the process of crystallisation, an element is accepted or rejected according to whether it fits the crystalline structure; elements that do not fit are known as 'incompatibles'. As trace elements tend not to participate in mineralisation, they remain in the melt as the 'compatible' elements are extracted, with the result that their concentration progressively increases. The radioactives at depth helped to maintain this reservoir molten, and were locked in when it finally solidified. The impact that made the Imbrium Basin had penetrated sufficiently deep to excavate and scatter some of this material across the surface; mystery solved.

THE END OF THE BEGINNING

Apollo 14 drew to a conclusion the initial phase of the exploration of the Moon in which astronauts traversed on foot. Even before Apollo 11, NASA had ordered the design of a battery powered Lunar Roving Vehicle to enable the so-called 'J'-class missions to range far and wide across their sites, carry a variety of tools, and return a large amount of material ... but the stories of these missions are for another book.

[4] Impact melt resembles basalt to the extent that it is a solidified rock melt, but endogenic basalt is homogeneous.

Glossary

AGS	Abort Guidance System
ALSCC	Apollo Lunar Surface Close-up Camera
ALSEP	Apollo Lunar Surface Experiments Package
AOS	Acquisition of Signal
AOT	Apollo Optical Telescope
Apogee	The highest point of a spacecraft's orbit around Earth
Apolune	The highest point of a spacecraft's orbit around the Moon
APS	Ascent Propulsion System
ARIA	Apollo Range Instrumentation Aircraft
BEF	Blunt End Forward
BIG	Biological Isolation Garment
CapCom	Capsule Communicator
CDH	Constant Differential Height manoeuvre
CDR	Commander
CM	Command Module
CMP	Command Module Pilot
CMS	Command Module Simulator
Columbia	Apollo 11 Command and Service Modules
CSI	Coelliptic Sequence Initiation manoeuvre
CSM	Command and Service Modules
DAP	Digital Autopilot
delta-H	change in height (altitude)
delta-P	pressure differential
delta-V	change in velocity
DOI	Descent Orbit Insertion
DPS	Descent Propulsion System
DSKY	Display and Keyboard
Eagle	Apollo 11 Lunar Module
EASEP	Early Apollo Surface Experiments Package
EMU	Extravehicular Mobility Unit
F-1	The type of engine on the S-IC
FDAI	Flight Director Attitude Indicator
g	gravity
GET	Ground Elapsed Time

Hasselblad	A still camera
IAU	International Astronomical Union
IMU	Inertial Measurement Unit
J-2	The type of engine on the S-II and S-IVB
KSC	Kennedy Space Center
LEC	Lunar Equipment Conveyor
LLRV	Lunar Landing Research Vehicle
LLTV	Lunar Landing Training Vehicle
LM	Lunar Module
LMP	Lunar Module Pilot
LMS	Lunar Module Simulator
LOI	Lunar Orbit Insertion
LOS	Loss Of Signal
LPD	Landing Point Designator
LRL	Lunar Receiving Laboratory
LRRR	Lunar Ranging Retro Reflector
Maurer	A movie camera
MCC	Mid-Course Correction
MESA	Modular Equipment Stowage Assembly
MILA	Merritt Island Launch Area
MOCR	Mission Operations Control Room
MQF	Mobile Quarantine Facility
MSC	Manned Spacecraft Center
MSFN	Manned Space Flight Network
NASA	National Aeronautics and Space Administration
OPS	Oxygen Purge System
Orbit	An orbit is one revolution plus the time taken to catch up with the axial rotation of the body to which a spacecraft is gravitationally bound, in order to return to the same longitude
PAD	Pre-Advisory Data
Perigee	The lowest point of a spacecraft's orbit around Earth
Perilune	The lowest point of a spacecraft's orbit around the Moon
PDI	Powered Descent Initiation
PGNS	Primary Guidance and Navigation System
PLSS	Portable Life-Support System
PSE	Passive Seismic Experiment
psi	pounds per square inch
PTC	Passive Thermal Control
PTT	Push To Talk
PUGS	Propellant Utilisation Gauging System
RCS	Reaction Control System
REFSMMAT	Reference to the Stable Member Matrix
Revolution	A revolution (measured in inertial coordinates) is the time a spacecraft takes to travel 360 degrees around the body to which it is gravitationally bound

RTCC	Real-Time Computer Complex
S-IC	First stage of the Saturn V
S-II	Second stage of the Saturn V
S-IVB	Third stage of the Saturn V
Saturn V	Apollo launch vehicle
SEQ	Scientific Equipment
SLA	Spacecraft/Lunar Module Adapter
SM	Service Module
SPS	Service Propulsion System
SRC	Sample Return Container
SWC	Solar Wind Collector
TD&E	Transposition, Docking and Extraction
TEI	Transearth Injection
TLI	Translunar Injection
TPF	Terminal Phase Finalisation manoeuvre
TPI	Terminal Phase Initiation manoeuvre
USGS	US Geological Survey
VHF	Very High Frequency
VOX	Voice activated
VTOL	Vertical Take Off and Landing

Mission log

Apollo 11 mission event	GET (h:m:s)	Date (GMT)	Time (h:m:s)
Terminal countdown started.	–028:00:00	14 Jul 1969	21:00:00
Scheduled 11-hr hold at T–9 hr.	–009:00:00	15 Jul 1969	16:00:00
Countdown resumed at T–9 hr.	–009:00:00	16 Jul 1969	03:00:00
Scheduled 1-hr 32-min hold at T–3 hr 30 min.	–003:30:00	16 Jul 1969	08:30:00
Countdown resumed at T–3 hr 30 min.	–003:30:00	16 Jul 1969	10:02:00
Guidance reference release.	–000:00:16.968	16 Jul 1969	13:31:43
S-IC engine start command.	–000:00:08.9	16 Jul 1969	13:31:51
S-IC engine ignition (#5).	–000:00:06.4	16 Jul 1969	13:31:53
All S-IC engines thrust OK.	–000:00:01.6	16 Jul 1969	13:31:58
Range zero.	000:00:00.00	16 Jul 1969	13:32:00
All holddown arms released (1st motion).	000:00:00.3	16 Jul 1969	13:32:00
Liftoff (umbilical disconnected) (1.07 g).	000:00:00.63	16 Jul 1969	13:32:00
Tower clearance yaw maneuver started.	000:00:01.7	16 Jul 1969	13:32:01
Yaw maneuver ended.	000:00:09.7	16 Jul 1969	13:32:09
Pitch and roll maneuver started.	000:00:13.2	16 Jul 1969	13:32:13
Roll maneuver ended.	000:00:31.1	16 Jul 1969	13:32:31
Mach 1 achieved.	000:01:06.3	16 Jul 1969	13:33:06
Maximum dynamic pressure (735.17 lb/ft^2).	000:01:23.0	16 Jul 1969	13:33:23
Maximum bending moment (33,200,000 lbf-in).	000:01:31.5	16 Jul 1969	13:33:31
S-IC center engine cutoff command.	000:02:15.2	16 Jul 1969	13:34:15
Pitch maneuver ended.	000:02:40.0	16 Jul 1969	13:34:40
S-IC outboard engine cutoff.	000:02:41.63	16 Jul 1969	13:34:41
S-IC maximum total inertial acceleration (3.94 g).	000:02:41.71	16 Jul 1969	13:34:41
S-IC maximum Earth-fixed velocity. S-IC/S-II separation command.	000:02:42.30	16 Jul 1969	13:34:42
S-II engine start command.	000:02:43.04	16 Jul 1969	13:34:43
S-II ignition.	000:02:44.0	16 Jul 1969	13:34:44
S-II aft interstage jettisoned.	000:03:12.3	16 Jul 1969	13:35:12
Launch escape tower jettisoned.	000:03:17.9	16 Jul 1969	13:35:17
Iterative guidance mode initiated.	000:03:24.1	16 Jul 1969	13:35:24
S-IC apex.	000:04:29.1	16 Jul 1969	13:36:29
S-II center engine cutoff.	000:07:40.62	16 Jul 1969	13:39:40
S-II maximum total inertial acceleration (1.82 g).	000:07:40.70	16 Jul 1969	13:39:40
S-IC impact (theoretical).	000:09:03.7	16 Jul 1969	13:41:03
S-II outboard engine cutoff.	000:09:08.22	16 Jul 1969	13:41:08

Apollo 11 mission event – *continued*	GET (h:m:s)	Date (GMT)	Time (h:m:s)
S-II maximum Earth-fixed velocity. S-II/S-IVB separation command.	000:09:09.00	16 Jul 1969	13:41:09
S-IVB 1st burn start command.	000:09:09.20	16 Jul 1969	13:41:09
S-IVB 1st burn ignition.	000:09:12.20	16 Jul 1969	13:41:12
S-IVB ullage case jettisoned.	000:09:21.0	16 Jul 1969	13:41:21
S-II apex.	000:09:47.0	16 Jul 1969	13:41:47
S-IVB 1st burn cutoff.	000:11:39.33	16 Jul 1969	13:43:39
S-IVB 1st burn maximum total inertial acceleration (0.69 g).	000:11:39.41	16 Jul 1969	13:43:39
Earth orbit insertion. S-IVB 1st burn maximum Earth-fixed velocity.	000:11:49.33	16 Jul 1969	13:43:49
Maneuver to local horizontal attitude started.	000:11:59.3	16 Jul 1969	13:43:59
Orbital navigation started.	000:13:21.1	16 Jul 1969	13:45:21
S-II impact (theoretical).	000:20:13.7	16 Jul 1969	13:52:13
S-IVB 2nd burn restart preparation.	002:34:38.2	16 Jul 1969	16:06:38
S-IVB 2nd burn restart command.	002:44:08.2	16 Jul 1969	16:16:08
S-IVB 2nd burn ignition (STDV open).	002:44:16.2	16 Jul 1969	16:16:16
S-IVB 2nd burn cutoff.	002:50:03.03	16 Jul 1969	16:22:03
S-IVB 2nd burn maximum total inertial acceleration (1.45 g).	002:50:03.11	16 Jul 1969	16:22:03
S-IVB 2nd burn maximum Earth-fixed velocity.	002:50:03.5	16 Jul 1969	16:22:03
S-IVB safing procedures started.	002:50:03.8	16 Jul 1969	16:22:03
Translunar injection.	002:50:13.03	16 Jul 1969	16:22:13
Maneuver to local horizontal attitude started.	002:50:23.0	16 Jul 1969	16:22:23
Orbital navigation started.	002:50:23.9	16 Jul 1969	16:22:23
Maneuver to transposition and docking attitude started.	003:05:03.9	16 Jul 1969	16:37:03
CSM separated from S-IVB.	003:15:23.0	16 Jul 1969	16:47:23
CSM separation maneuver ignition.	003:17:04.6	16 Jul 1969	16:49:04
CSM separation maneuver cutoff.	003:17:11.7	16 Jul 1969	16:49:11
CSM docked with LM/S-IVB.	003:24:03.7	16 Jul 1969	16:56:03
CSM/LM ejected from S-IVB.	004:17:03.0	16 Jul 1969	17:49:03
CSM/LM evasive maneuver from S-IVB ignition.	004:40:01.72	16 Jul 1969	18:12:01
CSM/LM evasive maneuver from S-IVB cutoff.	004:40:04.65	16 Jul 1969	18:12:04
S-IVB maneuver to lunar slingshot attitude initiated.	004:41:07.6	16 Jul 1969	18:13:07
S-IVB lunar slingshot maneuver – LH_2 tank CVS opened.	004:51:07.7	16 Jul 1969	18:23:07
S-IVB lunar slingshot maneuver – LOX dump started.	005:03:07.6	16 Jul 1969	18:35:07
S-IVB lunar slingshot maneuver – LOX dump ended.	005:04:55.8	16 Jul 1969	18:36:55
S-IVB lunar slingshot maneuver – APS ignition.	005:37:47.6	16 Jul 1969	19:09:47
S-IVB lunar slingshot maneuver – APS cutoff.	005:42:27.8	16 Jul 1969	19:14:27
S-IVB maneuver to communications attitude initiated.	005:42:48.8	16 Jul 1969	19:14:48
TV transmission started (recorded at Goldstone and transmitted to Houston at 011:26).	010:32	17 Jul 1969	00:04
TV transmission ended.	010:48	17 Jul 1969	00:20
Midcourse correction ignition.	026:44:58.64	17 Jul 1969	16:16:58
Midcourse correction cutoff.	026:45:01.77	17 Jul 1969	16:17:01
TV transmission started.	030:28	17 Jul 1969	20:00
TV transmission ended.	031:18	17 Jul 1969	20:50
TV transmission started.	033:59	17 Jul 1969	23:31

Apollo 11 mission event – *continued*	GET (h:m:s)	Date (GMT)	Time (h:m:s)
TV transmission ended.	034:35	18 Jul 1969	00:07
TV transmission started.	055:08	18 Jul 1969	20:40
CDR and LMP entered LM for initial inspection.	055:30	18 Jul 1969	21:02
TV transmission ended.	056:44	18 Jul 1969	22:16
CDR and LMP entered CM.	057:55	18 Jul 1969	23:27
Equigravisphere.	061:39:55	19 Jul 1969	03:11:55
Lunar orbit insertion ignition.	075:49:50.37	19 Jul 1969	17:21:50
Lunar orbit insertion cutoff.	075:55:47.90	19 Jul 1969	17:27:47
Sighting of an illumination in the Aristarchus region. 1st time, a lunar transient event sighted by an observer in space.	077:13	19 Jul 1969	18:45
TV transmission started.	078:20	19 Jul 1969	19:52
S-IVB closest approach to lunar surface.	078:42	19 Jul 1969	20:14
TV transmission ended.	079:00	19 Jul 1969	20:32
Lunar orbit circularization ignition.	080:11:36.75	19 Jul 1969	21:43:36
Lunar orbit circularization cutoff.	080:11:53.63	19 Jul 1969	21:43:53
LMP entered LM for initial power-up and system checks.	081:10	19 Jul 1969	22:42
LMP entered CM.	083:35	20 Jul 1969	01:07
CDR and LMP entered LM for final preparations for descent.	095:20	20 Jul 1969	12:52
LMP entered CM.	097:00	20 Jul 1969	14:32
LMP entered LM.	097:30	20 Jul 1969	15:02
LM system checks started.	097:45	20 Jul 1969	15:17
LM system checks ended.	100:00	20 Jul 1969	17:32
CSM/LM undocked.	100:12:00.0	20 Jul 1969	17:44:00
CSM/LM separation maneuver ignition.	100:39:52.9	20 Jul 1969	18:11:52
CSM/LM separation maneuver cutoff.	100:40:01.9	20 Jul 1969	18:12:01
LM descent orbit insertion ignition (DPS).	101:36:14	20 Jul 1969	19:08:14
LM descent orbit insertion cutoff.	101:36:44	20 Jul 1969	19:08:44
LM acquisition of data.	102:17:17	20 Jul 1969	19:49:17
LM landing radar on.	102:20:53	20 Jul 1969	19:52:53
LM abort guidance aligned to primary guidance.	102:24:40	20 Jul 1969	19:56:40
LM yaw maneuver to obtain improved communications.	102:27:32	20 Jul 1969	19:59:32
LM altitude 50,000 ft.	102:32:55	20 Jul 1969	20:04:55
LM propellant settling firing started.	102:32:58	20 Jul 1969	20:04:58
LM powered descent engine ignition.	102:33:05.01	20 Jul 1969	20:05:05
LM fixed throttle position.	102:33:31	20 Jul 1969	20:05:31
LM face-up maneuver completed.	102:37:59	20 Jul 1969	20:09:59
LM 1202 alarm.	102:38:22	20 Jul 1969	20:10:22
LM radar updates enabled.	102:38:45	20 Jul 1969	20:10:45
LM altitude less than 30,000 ft and velocity less than 2,000 ft/sec (landing radar velocity update started).	102:38:50	20 Jul 1969	20:10:50
LM 1202 alarm.	102:39:02	20 Jul 1969	20:11:02
LM throttle recovery.	102:39:31	20 Jul 1969	20:11:31
LM approach phase entered.	102:41:32	20 Jul 1969	20:13:32
LM landing radar antenna to position 2.	102:41:37	20 Jul 1969	20:13:37
LM attitude hold mode selected (check of LM handling qualities).	102:41:53	20 Jul 1969	20:13:53
LM automatic guidance enabled.	102:42:03	20 Jul 1969	20:14:03

Apollo 11 mission event – *continued*	GET (h:m:s)	Date (GMT)	Time (h:m:s)
LM 1201 alarm.	102:42:18	20 Jul 1969	20:14:18
LM landing radar switched to low scale.	102:42:19	20 Jul 1969	20:14:19
LM 1202 alarm.	102:42:43	20 Jul 1969	20:14:43
LM 1202 alarm.	102:42:58	20 Jul 1969	20:14:58
LM landing point redesignation.	102:43:09	20 Jul 1969	20:15:09
LM altitude hold.	102:43:13	20 Jul 1969	20:15:13
LM abort guidance attitude updated.	102:43:20	20 Jul 1969	20:15:20
LM rate of descent landing phase entered.	102:43:22	20 Jul 1969	20:15:22
LM landing radar data not good.	102:44:11	20 Jul 1969	20:16:11
LM landing data good.	102:44:21	20 Jul 1969	20:16:21
LM fuel low-level quantity light.	102:44:28	20 Jul 1969	20:16:28
LM landing radar data not good.	102:44:59	20 Jul 1969	20:16:59
LM landing radar data good.	102:45:03	20 Jul 1969	20:17:03
1st evidence of surface dust disturbed by descent engine.	102:44:35	20 Jul 1969	20:16:35
LM lunar landing.	102:45:39.9	20 Jul 1969	20:17:39
LM powered descent engine cutoff.	102:45:41.40	20 Jul 1969	20:17:41
Decision made to proceed with EVA prior to first rest period.	104:40:00	20 Jul 1969	22:12:00
Preparation for EVA started.	106:11:00	20 Jul 1969	23:43:00
EVA started (hatch open).	109:07:33	21 Jul 1969	02:39:33
CDR completely outside LM on porch.	109:19:16	21 Jul 1969	02:51:16
Modular equipment stowage assembly deployed (CDR).	109:21:18	21 Jul 1969	02:53:18
First clear TV picture received.	109:22:00	21 Jul 1969	02:54:00
CDR at foot of ladder (starts to report, then pauses to listen).	109:23:28	21 Jul 1969	02:55:28
CDR at foot of ladder and described surface as "almost like a powder".	109:23:38	21 Jul 1969	02:55:38
1st step taken lunar surface (CDR). "That's one small step for a man...one giant leap for mankind".	109:24:15	21 Jul 1969	02:56:15
CDR started surface examination and description, assessed mobility and described effects of LM descent engine.	109:24:48	21 Jul 1969	02:56:48
CDR ended surface examination. LMP started to send down camera.	109:26:54	21 Jul 1969	02:58:54
Camera installed on RCU bracket, LEC stored on secondary strut of LM landing gear.	109:30:23	21 Jul 1969	03:02:23
Surface photography (CDR).	109:30:53	21 Jul 1969	03:02:53
Contingency sample collection started (CDR).	109:33:58	21 Jul 1969	03:05:58
Contingency sample collection ended (CDR).	109:37:08	21 Jul 1969	03:09:08
LMP started egress from LM.	109:39:57	21 Jul 1969	03:11:57
LMP at top of ladder. Descent photographed by CDR.	109:41:56	21 Jul 1969	03:13:56
LMP on lunar surface.	109:43:16	21 Jul 1969	03:15:16
Surface examination and examination of landing effects on surface and on LM started (CDR, LMP).	109:43:47	21 Jul 1969	03:15:47
Insulation removed from modular equipment stowage assembly (CDR).	109:49:06	21 Jul 1969	03:21:06
TV camera focal distance adjusted (CDR).	109:51:35	21 Jul 1969	03:23:35
Plaque unveiled (CDR).	109:52:19	21 Jul 1969	03:24:19
Plaque read (CDR).	109:52:40	21 Jul 1969	03:24:40

Apollo 11 mission event – *continued*	GET (h:m:s)	Date (GMT)	Time (h:m:s)
TV camera redeployed. Panoramic TV view started (CDR).	109:59:28	21 Jul 1969	03:31:28
TV camera placed in final deployment position (CDR).	110:02:53	21 Jul 1969	03:34:53
Solar wind composition experiment deployed (LMP).	110:03:20	21 Jul 1969	03:35:20
US flag deployed (CDR, LMP).	110:09:43	21 Jul 1969	03:41:43
Evaluation of surface mobility started (LMP).	110:13:15	21 Jul 1969	03:45:15
Evaluation of surface mobility end (LMP).	110:16:02	21 Jul 1969	03:48:02
Presidential message from White House and response from CDR.	110:16:30	21 Jul 1969	03:48:30
Presidential message and CDR response ended.	110:18:21	21 Jul 1969	03:50:21
Evaluation of trajectory of lunar soil when kicked (LMP) and bulk sample collection started (CDR).	110:20:06	21 Jul 1969	03:52:06
Evaluation of visibility in lunar sunlight (LMP).	110:10:24	21 Jul 1969	03:42:24
Evaluation of thermal effects of Sun and shadow inside the suit (LMP).	110:25:09	21 Jul 1969	03:57:09
Evaluation of surface shadows and colors (LMP).	110:28:22	21 Jul 1969	04:00:22
LM landing gear inspection and photography (LMP).	110:34:13	21 Jul 1969	04:06:13
Bulk sample completed (CDR).	110:35:36	21 Jul 1969	04:07:36
LM landing gear inspection and photography (CDR, LMP).	110:46:36	21 Jul 1969	04:18:36
Scientific equipment bay doors opened.	110:53:38	21 Jul 1969	04:25:38
Passive seismometer deployed.	110:55:42	21 Jul 1969	04:27:42
Lunar ranging retroreflector deployed (CDR).	111:03:57	21 Jul 1969	04:35:57
1st passive seismic experiment data received on Earth.	111:08:39	21 Jul 1969	04:40:39
Collection of documented samples started (CDR/LMP).	111:11	21 Jul 1969	04:43
Solar wind composition experiment retrieved (LMP).	111:20	21 Jul 1969	04:52
LMP inside LM.	111:29:39	21 Jul 1969	05:01:39
Transfer of sample containers reported complete.	111:35:51	21 Jul 1969	05:07:51
CDR inside LM, assisted and monitored by LMP.	111:37:32	21 Jul 1969	05:09:32
EVA ended (hatch closed).	111:39:13	21 Jul 1969	05:11:13
LM equipment jettisoned.	114:05	21 Jul 1969	07:37
LM lunar liftoff ignition (LM APS).	124:22:00.79	21 Jul 1969	17:54:00
LM orbit insertion cutoff.	124:29:15.67	21 Jul 1969	18:01:15
Coelliptic sequence initiation ignition.	125:19:35	21 Jul 1969	18:51:35
Coelliptic sequence initiation cutoff.	125:20:22	21 Jul 1969	18:52:22
Constant differential height maneuver ignition.	126:17:49.6	21 Jul 1969	19:49:49
Constant differential height maneuver cutoff.	126:18:29.2	21 Jul 1969	19:50:29
Terminal phase initiation ignition.	127:03:51.8	21 Jul 1969	20:35:51
Terminal phase initiation cutoff.	127:04:14.5	21 Jul 1969	20:36:14
LM 1st midcourse correction.	127:18:30.8	21 Jul 1969	20:50:30
LM 2nd midcourse correction.	127:33:30.8	21 Jul 1969	21:05:30
Braking started.	127:36:57.3	21 Jul 1969	21:08:57
Terminal phase finalize ignition.	127:46:09.8	21 Jul 1969	21:18:09
Terminal phase finalize cutoff.	127:46:38.2	21 Jul 1969	21:18:38
Stationkeeping started.	127:52:05.3	21 Jul 1969	21:24:05
CSM/LM docked.	128:03:00	21 Jul 1969	21:35:00
CDR entered CM.	129:20	21 Jul 1969	22:52
LMP entered CM.	129:45	21 Jul 1969	23:17

Apollo 11 mission event – *continued*	GET (h:m:s)	Date (GMT)	Time (h:m:s)
LM ascent stage jettisoned.	130:09:31.2	21 Jul 1969	23:41:31
CSM/LM final separation ignition.	130:30:01.0	22 Jul 1969	00:02:01
CSM/LM final separation cutoff.	130:30:08.2	22 Jul 1969	00:02:08
Transearth injection ignition (SPS).	135:23:42.28	22 Jul 1969	04:55:42
Transearth injection cutoff.	135:26:13.69	22 Jul 1969	04:58:13
Midcourse correction ignition.	150:29:57.4	22 Jul 1969	20:01:57
Midcourse correction cutoff.	150:30:07.4	22 Jul 1969	20:02:07
TV transmission started.	155:36	23 Jul 1969	01:08
TV transmission ended.	155:54	23 Jul 1969	01:26
TV transmission started.	177:10	23 Jul 1969	22:42
TV transmission ended.	177:13	23 Jul 1969	22:45
TV transmission started.	177:32	23 Jul 1969	23:04
TV transmission ended.	177:44	23 Jul 1969	23:16
CM/SM separation.	194:49:12.7	24 Jul 1969	16:21:12
Entry.	195:03:05.7	24 Jul 1969	16:35:05
Drogue parachute deployed	195:12:06.9	24 Jul 1969	16:44:06
Visual contact with CM established by aircraft.	195:07	24 Jul 1969	16:39
Radar contact with CM established by recovery ship.	195:08	24 Jul 1969	16:40
VHF voice contact and recovery beacon contact established.	195:14	24 Jul 1969	16:46
Splashdown (went to apex-down).	195:18:35	24 Jul 1969	16:50:35
CM returned to apex-up position.	195:26:15	24 Jul 1969	16:58:15
Flotation collar inflated.	195:32	24 Jul 1969	17:04
Hatch opened for crew egress.	195:49	24 Jul 1969	17:21
Crew egress.	195:57	24 Jul 1969	17:29
Crew on board recovery ship.	196:21	24 Jul 1969	17:53
Crew entered mobile quarantine facility.	196:26	24 Jul 1969	17:58
CM lifted from water.	198:18	24 Jul 1969	19:50
CM secured to quarantine facility.	198:26	24 Jul 1969	19:58
CM hatch reopened.	198:33	24 Jul 1969	20:05
Sample return containers 1 and 2 removed from CM.	200:28	24 Jul 1969	22:00
Container 1 removed from mobile quarantine facility.	202:00	24 Jul 1969	23:32
Container 2 removed from mobile quarantine facility.	202:33	25 Jul 1969	00:05
Container 2 and film flown to Johnston Island.	207:43	25 Jul 1969	05:15
Container 1 flown to Hickam Air Force Base, Hawaii.	214:13	25 Jul 1969	11:45
Container 2 and film arrived in Houston, TX.	218:43	25 Jul 1969	16:15
Container 1, film, and biological samples arrived in Houston.	225:41	25 Jul 1969	23:13
CM decontaminated and hatch secured.	229:28	26 Jul 1969	03:00
Mobile quarantine facility secured.	231:03	26 Jul 1969	04:35
Mobile quarantine facility and CM offloaded.	250:43	27 Jul 1969	00:15
Safing of CM pyrotechnics completed.	252:33	27 Jul 1969	02:05
Mobile quarantine facility arrived at Ellington AFB, Houston.	280:28	28 Jul 1969	06:00
Crew in Lunar Receiving Laboratory, Houston.	284:28	28 Jul 1969	10:00
CM delivered to Lunar Receiving Laboratory.	345:45	30 Jul 1969	23:17
Passive seismic experiment turned off.	430:26:46	03 Aug 1969	11:58:46
Crew released from quarantine.	–	10 Aug 1969	–
CM delivered to contractor's facility in Downey, CA.	–	14 Aug 1969	–
EASEP turned off by ground command.	–	27 Aug 1969	–

Further reading

A Man on the Moon: The Voyages of the Apollo Astronauts, Andrew Chaikin. Michael Joseph, 1994.

'All we did was fly to the Moon': Astronaut Insignias and Call Signs, Richard L. Lattimer. The Whispering Eagle Press, Florida, 1985.

Apollo: The Definitive Sourcebook, Richard W. Orloff and David M. Harland. Springer–Praxis, 2006.

Apollo: The Race to the Moon, Charles Murray and Catherine Bly Cox. Simon & Schuster, 1989.

Apollo 11: Preliminary Science Report. NASA SP-214, 1969.

Apollo 11: The NASA Mission Reports, vol. 1, Robert Godwin (Ed.). Apogee Books, 1999.

Apollo 11: The NASA Mission Reports, vol. 2, Robert Godwin (Ed.). Apogee Books, 1999.

Apollo 11: The NASA Mission Reports, vol. 3, Robert Godwin (Ed.). Apogee Books, 2002.

Assault on the Moon, Eric Burgess. Hodder and Stoughton, 1966.

Carrying the Fire: An Astronaut's Journeys, Michael Collins. W.H. Allen, 1975.

Decision to Go to the Moon, John Logsdon. MIT Press, 1970.

DEKE!, Donald K. Slayton with Michael Cassutt. Forge/Tom Doherty, 1995.

Exploring the Moon: The Apollo Expeditions, David M. Harland. Springer–Praxis, 1999.

Failure is Not and Option, Gene Kranz. Simon & Schuster, 2000.

First Man: The Life of Neil A. Armstrong, James R. Hansen. Simon & Schuster, 2005.

First on the Moon: A Voyage with Neil Armstrong, Michael Collins and Edwin E. Aldrin Jr, with Gene Farmer and Dora Jane Hamblin. Michael Joseph, 1970.

Flight: My Life in Mission Control, Chris Kraft. Plume/Penguin–Putnam, 2002.

Flying to the Moon: An Astronaut's Story (2nd edn), Michael Collins. Sunburst Books, 1994.

How NASA Learned to Fly in Space: An Exciting Account of the Gemini Missions, David M. Harland. Apogee Books, 2004.

Journey to Tranquillity: The History of Man's Assault on the Moon, Hugo Young, Brian Silcock and Peter Dunn. Jonathan Cape, 1969.

Lunar Sourcebook: A User's Guide to the Moon, Grant H. Heiken, David T. Vaniman and Bevan M. French (Eds). Cambridge University Press, 1991.

Men from Earth, Buzz Aldrin and Michael McConnell. Bantam Press, 1989.
Proceedings of the First Lunar Science Conference (3 vols), A.A. Levinson (Ed.). Pergamon Press, 1970.
Return to Earth, Buzz Aldrin with Wayne Warga. Bantam Press, 1974.
The Big Splat, Or How Our Moon Came to Be, Dana Mackenzie. Wiley, 2003.
The Invasion of the Moon 1969: The Story of Apollo 11, Peter Ryan. Penguin, 1969.
The Moonlandings: An Eyewitness Account, Reginald Turnill. Cambridge University Press, 2003.
The Race: The Definitive Story of America's Battle to Beat Russia to the Moon, James Schefter. Century, 1999.
The Unbroken Chain, Guenter Wendt and Russell Still. Apogee Books, 2001.
To a Rocky Moon: A Geologist's History of Lunar Exploration, Don E. Wilhelms. The University of Arizona Press, 1993.
We Reach the Moon, John Noble Wilford. Bantam Books, 1999.
Where No Man Has Gone Before: A History of Apollo Lunar Exploration Missions, William David Compton. SP-4214, NASA, 1989.

Websites

For the full transcript of Eagle's powered descent and the lunar surface activities, see the Apollo Lunar Surface Journal:
http://www.hq.nasa.gov/alsj/
For high resolution imagery see:
http://www.apolloarchive.com/apollo_gallery.html

Index

Printing: Mercedes-Druck, Berlin
Binding: Stein+Lehmann, Berlin